D1767755

INTERNATIONAL SERIES OF MONOGRAPHS IN
NATURAL PHILOSOPHY
GENERAL EDITOR: D. TER HAAR

VOLUME 29

SUPERCONDUCTIVITY AND QUANTUM FLUIDS

SUPERCONDUCTIVITY AND QUANTUM FLUIDS

BY

ZYGMUNT M. GALASIEWICZ

Professor of Physics, Wrocław

PERGAMON PRESS

OXFORD · LONDON · EDINBURGH · NEW YORK
TORONTO · SYDNEY · PARIS · BRAUNSCHWEIG

PWN—POLISH SCIENTIFIC PUBLISHERS

WARSZAWA

Pergamon Press Ltd., Headington Hill Hall, Oxford
4 & 5 Fitzroy Square, London W.1
Pergamon Press (Scotland) Ltd., 2 & 3 Teviot Place, Edinburgh 1
Pergamon Press Inc., Maxwell House, Fairview Park, Elmsford, New York 10523
Pergamon of Canada Ltd., 207 Queen's Quay West, Toronto 1
Pergamon Press (Aust.) Pty. Ltd., 19a Boundary Street, Rushcutters Bay, N.S.W. 2011, Australia
Pergamon Press S.A.R.L., 24 rue des Écoles, Paris 5e
Vieweg & Sohn GmbH, Burgplatz 1, Braunschweig

First edition 1970

Library of Congress Catalog Card No. 73-103002

PRINTED IN POLAND
08 013089 5

Contents

PREFACE vii

PART I. SUPERCONDUCTIVITY

INTRODUCTION 3

CHAPTER I. THEORY OF THE GROUND STATE 33

§ 1. The generalized Bogoliubov transformations and the principle of compensation of "dangerous" diagrams 33
§ 2. The generalized variational principle in many-body theory 38
§ 3. Solution of the general variational principle describing the superconducting state 42
§ 4. Stability of the solution describing the superconducting state 55

CHAPTER II. THERMODYNAMICS 62

§ 1. The mean value of the Hamiltonian 62
§ 2. Temperature dependence of the energy gap 66
§ 3. Thermodynamic potential 70
§ 4. Entropy and specific heat 78
§ 5. Critical magnetic field 81

CHAPTER III. COLLECTIVE OSCILLATIONS 85

§ 1. The approximate second quantization Hamiltonian 85
§ 2. Diagonalization of the asq Hamiltonian and collective oscillations 89

CHAPTER IV. ELECTRODYNAMICS 103

§ 1. The asq Hamiltonian in the case of weak external fields 103
§ 2. Sum rule 107
§ 3. Gauge invariance of the current and the Meissner effect 108

Contents

PART II. QUANTUM FLUIDS

INTRODUCTION 117

CHAPTER V. BASIC IDENTITIES AND RELATIONS 132

§ 1. Time derivatives of some "local quantities" for Bose systems 132
§ 2. Time derivatives of some "local quantities" for Fermi systems 138
§ 3. The relations between the retarded Green functions and the variations of average values 142
§ 4. Gauge transformations and Green functions 147

CHAPTER VI. ORDINARY BOSE AND FERMI FLUIDS 153

§ 1. Hydrodynamic equations for ordinary Bose and Fermi fluids 153
§ 2. Linearized hydrodynamic equations and the retarded Green functions 163

CHAPTER VII. THE BOSE SUPERFLUID 171

§ 1. Hydrodynamic equations without viscous terms 171
§ 2. Hydrodynamic equations with viscous terms 184
§ 3. Hydrodynamic equations in the acoustic approximation 188
§ 4. The solution of the acoustic equations and the calculation of the Green functions 192

CHAPTER VIII. THE FERMI SUPERFLUID 209

§ 1. Hydrodynamic equations 209
§ 2. The linearized hydrodynamic equations and the Green functions 220

REFERENCES 227

INDEX 235

Other Titles in the Series 239

Preface

THE author spent two years, 1958 and 1959, at the Joint Institute for Nuclear Research in Dubna, U.S.S.R., where he had the opportunity of working on some problems in the theory of superconductivity suggested to him by Professor N. Bogoliubov. During this period the author became familiar with the research of Bogoliubov and his students in these topics, and later, from 1963 to 1967, studied their work in the field of superfluidity. Consequently, the trend represented by Bogoliubov is strongly reflected in the author's choice of problems and methods, as well as in the literature quoted by the author in his lectures at the University of Wrocław, at the Institute of Low Temperatures of the Polish Academy of Sciences and at the Winter Schools of Theoretical Physics organized in Karpacz by the University of Wrocław. The present book grew from these lectures, and is designed to present the microscopic theory of superconductivity and superfluidity. Special care is taken to give an explicit derivation of the formulae. The author makes no pretention of presenting a complete list of references and gives preference to the work due to Bogoliubov's school. The book does not include applications; however, monographs and review articles where those applications can be found are quoted in the references at the close of the book.

The author is deeply indebted to Professor Bogoliubov for suggesting to him many problems in the theory of superconductivity and superfluidity as well as for many fruitful discussions. The author would also like to thank cordially Dr. J. Czerwonko for several inspiring discussions during the preparation of the manuscript, as well as for his critical remarks while reading it. Finally, the author would like to thank his wife for her considerable help in preparing this book.

ZYGMUNT GALASIEWICZ

PART I

SUPERCONDUCTIVITY

Introduction

THROUGHOUT the nineteenth century physicists were interested in the liquefaction of all known gases, because this achievement would eventually permit the study of low temperature phenomena. Faraday's achievements in this field in 1845 (the liquefaction of all known gases with the exception of He, H_2, O_2, N_2, CH_4, NO and CO) were not supplemented until 1883 when Wróblewski and Olszewski condensed oxygen and nitrogen (boiling temperatures of 90.1°K and 77.3°K, respectively). But thereafter progress was more rapid and in 1898 Dewar condensed hydrogen (boiling temperature 20.38°K). In 1908 Kamerlingh Onnes liquefied helium (boiling temperature 4.22°K), the last "permanent" gas whose condensation furnished essential difficulties. The study of the properties of metals at very low temperatures was finally possible.

New as a fact, the concept of the liquefaction of gases dated back to the second century A.D. The famous ancient satirist Lucian of Samosata first mentioned liquefaction in describing an imaginary journey. In his account a traveler, having arrived on the moon, perceives that the moon's inhabitants drink liquid air: "Their drink is air compressed into a vessel, and out of which they squeeze a kind of moisture like dew". However, the satirist's prevision is to be doubted since he must have created this incident as yet another fabulous event in an unbelievable narrative. The difficulties which provoked physicists to name oxygen and nitrogen "permanent" gases may at times have driven them as well to consider liquefied gas a fiction.

In the experimentation which followed Onnes' achievement, it was observed with particular interest that, as lower and lower metal temperatures are created, the electric resistance decreases. This decrease was later found to terminate at a constant value, the "residual" resistance, which is independent of the temperature. The value of the "residual" resistance depends on impurities and diminishes when the latter are reduced.

An early theory which helped to explain this phenomenon was offered by Bloch in 1928. In his quantum-mechanical theory of normal metals (Bloch, 1928) the resistance is interpreted as a consequence of the scattering of electrons by lattice ions if there are deviations from the perfect lattice configuration. The perfect lattice is the space-periodic motionless lattice which gives a periodic potential constant in time. The problem of the motion of electrons in such a potential is a one-electron problem. The solution gives the one-electron wave functions called Bloch waves (free travelling waves in the metal), with energy eigenvalues giving the band structure of the single-electron spectrum. Thus in the case of an ideal lattice we have no scattering.

Deviations from the perfect lattice are caused by oscillations of the lattice ions about the equilibrium positions (dynamical deviations), or by chemical impurities and other lattice imperfections (statistical deviations). The oscillatory motion of the ions is thermal and vanishes when the temperature goes to absolute zero. So the dynamical deviations are temperature dependent, and the resistance connected with the scattering by these deviations is a decreasing function of the temperature. The statistical deviations do not depend on the temperature. The resistance connected with the scattering caused by these deviations is therefore called the residual resistance.

The energy of interacting ions can be represented as the energy of independent oscillators, vibrating at their normal frequencies. The corpuscular analogues of these oscillations are the sound quanta, phonons. The interaction of the electrons with free phonons (scattering) causes transitions of the electrons between states with definite momentum. If an electric field is introduced, one direction is distinguished and an ordered motion of the electrons is induced. Then, if the field is afterwards removed, the transitions between the states will lead to the vanishing of the ordering, and consequently, the electric current will vanish.

Therefore, if the dissipative processes (appearing as electric resistance or friction) are forbidden, an ordered persistent current can remain. It is evident that the weaker the interaction of electrons with the ion lattice, the smaller the electric resistance. In good conductors such as Au, Cu, Ag, Pt, the electron–phonon interaction must be weak.

Profiting directly from his achievement, Kamerling Onnes had meanwhile measured the resistance of a sample of nearly pure platinum at liquid helium temperatures. He concluded that the resistance would go to zero if the temperature were further decreased, but found instead that the resistance went to a constant value. Aware that the metal's purity might be essential, and having attempted in vain to procure a purer sample of platinum, he chose then to experiment upon a sample of purer mercury. Down to the temperature 4.12°K he found that for mercury, similar to other metals, the resistance depends upon the temperature. At the temperature 4.12°K the resistance dropped suddenly to zero. Thus, in 1911, Kamerlingh Onnes discovered the phenomenon called superconductivity (Kamerling Onnes, 1911a).

Soon afterwards it was confirmed that many other metals are superconductors. At present twenty-three are known. It was surprising that the best conductors such as Au, Cu, Ag, Pt are not superconductors. Superconductivity is observed not only in pure metals but also in many alloys. Usually, at least one of the components must be a superconductor but there are some alloys, for instance Au_2Bi, where neither component is superconducting.

The temperature Θ_c at which a metal reaches the superconducting state is called the "transition temperature". The following are the transition (or critical) temperatures for some metals:

Metal	Nb	Pb	Hg	Al	Zn	Hf
Θ_c(°K)	9.22	7.26	4.12	1.14	0.79	0.3

In 1913 Kamerlingh Onnes made the further discovery that the superconducting state is destroyed by a sufficiently strong magnetic field. The field H_c necessary to destroy superconductivity is called the critical field. This field is a function of temperature. The critical temperature $\Theta = \Theta_c$ corresponds to $H_c = 0$. If we wish to reach the superconducting state in the presence of a magnetic field we must cool the sample to a temperature $\Theta < \Theta_c$. (For the present, we are considering the case of a cylindrical sample with an external magnetic field parallel to the axis of cylinder.) The critical fields for pure metals are relatively small, for example:

Metal	Cd	Sn	Ta	V	Nb
$H_c(0)$ (gauss)	28.8	304.5	975	1200	2600

We see that niobium is distinguished by its relatively large critical field. This increased range is characteristic of niobium as a so-called type II superconductor, a class of superconductors that will be discussed below.

The existence of a critical field corresponds to the existence of a critical current. The field connected with the current destroys the supercurrent. The practical use of pure superconductors as electric wires or solenoids is thus seriously restricted.

The temperature dependence of the critical field $H_c = H_c(\Theta)$ is given in Fig. 1.

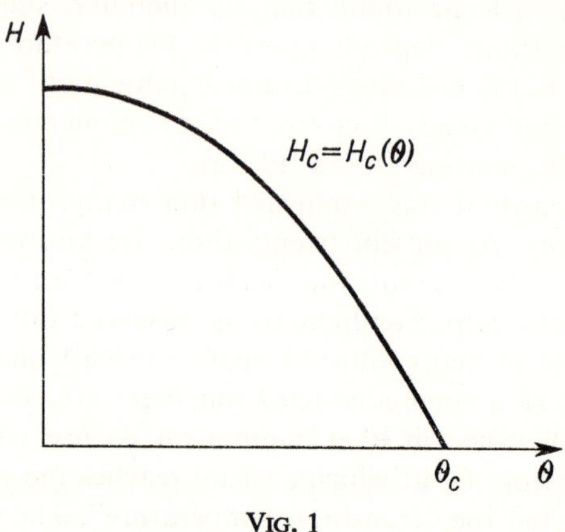

FIG. 1

We see that the superconducting state does not exist for fields $H > H_c(0)$ and, independently, for temperatures $\Theta > \Theta_c$.

Attempts to explain the nature of superconductivity based on the observed electric property, i.e. the disappearance of the electric resistance, did not give satisfactory results as can be seen by the following argument. Consider the application of Ohm's law $\boldsymbol{E} = \boldsymbol{j}/\sigma$ to a sample in the superconducting state, i.e. $\Theta < \Theta_c$. The assumption of perfect conductivity requires that σ is infinite and hence that $\boldsymbol{E}$ vanish. If this is the case, Maxwell's equation

$$\nabla \times \boldsymbol{E} = -\frac{1}{c}\frac{\partial \boldsymbol{B}}{\partial t} \tag{1}$$

yields

$$\boldsymbol{B} = \text{const.} \tag{2}$$

The value of $\boldsymbol{B}$ in the sample after the transition to the superconducting state must therefore be equal to the value of $\boldsymbol{B}$ in the sample just at the time of the transition since, by assumption, at that time σ became infinite. Consider now how, according to this result, one could describe the state of a metal by variation of two external parameters, temperature Θ and magnetic field H (the third parameter, pressure, is of less importance). We may wish to pass in the (Θ, H)-plane from the point A_1 (coordinates $H = H_1 = 0$, $\Theta = \Theta_1 > \Theta_c$) to the point A_3 (coordinates $H = H_3 < H_c$, $\Theta = \Theta_3 < \Theta_c$). Point A_1 lies in the normal phase and A_3 in the superconducting phase. We then pass from the point A_1 to A_3 along the curve $A_1 A_2 A_3$ (see Fig. 2), by first lowering the tem-

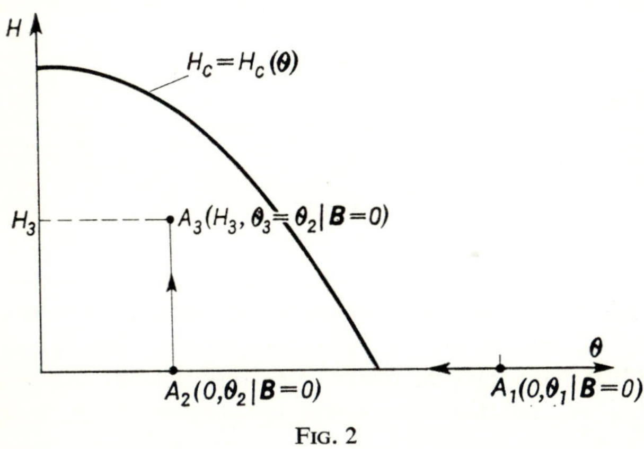

FIG. 2

perature from $\Theta_1 > \Theta_c$ to $\Theta_2 < \Theta_c$, with the external field, and $\boldsymbol{B}$, equal to zero. Next, at constant temperature Θ_2, we increase the magnetic field from zero to H_3 at A_3. From (2) it follows that at the point A_3 characterized by the values of the external parameters H_3, Θ_2, the magnetic induction $\boldsymbol{B}$ in the sample is equal to zero. Let us now pass from A_1 to A_3 along the path $A_1 A_4 A_3$ (see Fig. 3). At the temperature Θ_1 we increase the magnetic field from zero to H_4 at the point A_4. Now we pass to the point A_3 at a constant magnetic field by lowering the temperature from Θ_1 to Θ_3. From eqn. (2) we see that the magnetic induction inside the sample is now $\boldsymbol{B} = H_3 \neq 0$.

Now it is clear that assuming infinite conductivity as the basic property of the superconducting state prohibits the interpretation of the curve $H_c = H_c(\Theta)$ in Fig. 1 as a phase diagram since the same external parameters do not give the same internal states. It follows that the superconducting state cannot be a state of thermodynamic equilibrium, and cannot be treated by means of the thermodynamics of reversible processes.

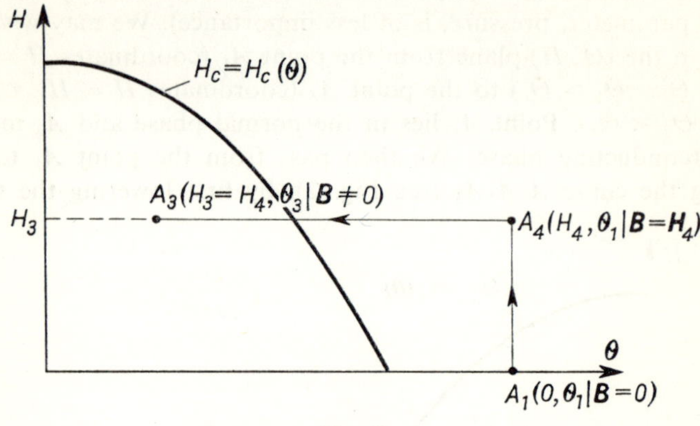

FIG. 3

Under these circumstances, the experiment of Meissner and Ochsenfeld (1933) was of fundamental importance for an understanding of the nature of superconductivity. This experiment showed that if at a temperature $\Theta_1 > \Theta_c$ one switches on a magnetic field $H_3 < H_c$ and then lowers the temperature to $\Theta_3 < \Theta_c$, the magnetic field is expelled from the superconductor. This means that if in Fig. 3 we pass from the point A_4 to A_3 then, upon crossing the curve $H_c = H_c(\Theta)$, the induction inside the metal will become equal to zero.

Due to the Meissner–Ochsenfeld (M–O) effect, the external parameters (H_3, Θ_3) determine in a unique, i.e. path-independent, way the magnetic induction inside the metal. Moreover, the superconducting state is stable and can be described by means of the thermodynamics of reversible processes. The discovery of such an important magnetic property of superconductors as the M–O effect allows us to use the "diamagnetic" picture of the magnetic properties of a superconductor (for a cylindric sample in a parallel magnetic field). In this picture we have:

(i) Inside the superconductor

$$\boldsymbol{B}_i = 0, \quad \boldsymbol{H}_i \neq 0, \quad \text{and} \quad \boldsymbol{I}_i \neq 0, \tag{3}$$

where $\boldsymbol{B}$ is the magnetic induction, $\boldsymbol{H}$ is the magnetic field, and $\boldsymbol{I}$, the magnetization. Since

$$\boldsymbol{B}_i = \boldsymbol{H}_i + 4\pi \boldsymbol{I}_i, \qquad \boldsymbol{I}_i = \chi \boldsymbol{H}_i, \tag{4}$$

it follows from (3) and (4) that

$$\chi = -\frac{1}{4\pi} < 0. \tag{5}$$

(ii) At the surface of the superconductor

$$\boldsymbol{j}_s = 0.$$

(iii) Outside the superconductor

$$\boldsymbol{B}_0 = \boldsymbol{H}_0 + \boldsymbol{H}_I, \qquad \mu = 1,$$

where $\boldsymbol{H}_I$ is a magnetic field connected with the existence of the magnetization of the sample.

The fact that $\chi < 0$ implies that $|\boldsymbol{B}| < |\boldsymbol{H}|$ and hence, that we are dealing with diamagnetic phenomena. The extreme case is $|\boldsymbol{B}| = 0$. We call this case perfect diamagnetism.

The failure to explain superconductivity on the basis of electric properties (perfect conductivity) suggested that magnetic properties such as demonstrated in the M–O effect are of a more fundamental character.

Before discussing this problem more fully, we shall consider, on the basis of the diamagnetic picture, the connexion between the internal and external magnetic field in non-cylindrical superconducting samples. The strength of the internal magnetic field H_i is generally different from the external strength. From the theoretical as well as the experimental point of view, an interesting example is that of an ellipsoid of revolution. (Let the length of the semi-major axis be a, and of the semi-minor one, b.) We assume that this ellipsoid is located in an applied external magnetic field H_0 parallel to the semi-major axis.

It can be shown that inside a sample the strength of the magnetic field is given by

$$\boldsymbol{H}_i = \boldsymbol{H}_0 - 4\pi D \boldsymbol{I}_i \tag{6}$$

where D is called the demagnetizing factor and is given by the formula

$$D = \left(\frac{1}{e^2} - 1\right)\left(\frac{1}{2e}\ln\frac{1+e}{1-e} - 1\right), \qquad e = \sqrt{1 - \left(\frac{b}{a}\right)^2}. \tag{7}$$

From this formula we can calculate D for several special cases.

(i) For an infinite cylinder with radius b, which corresponds to $a = \infty$, $a/b = \infty$, and $e = 1$, we have $D = 0$. We use the relation $\lim_{e\to 1}(1-e)\ln(1-e) = 0$.

(ii) For a sphere of radius a corresponding to $a = b$, $b/a = 1$, and $e = 0$, we obtain $D = 1/3$. We use the approximation $\ln \frac{1+e}{1-e} \approx 2e\left(1+\frac{1}{3}e^2\right)$ for small e.

(iii) For an infinite sheet of thickness a, we set $b = \infty$, $a/b = 0$, then $e = i\infty$, and we obtain $D = 1$.

As the quotient a/b varies from ∞ to 0, D varies from 0 to 1. We see that the cylindrical sample is distinguished, since for it $D = 0$ and $H_i = H_0$.

For the superconducting state it follows from (4) and (5) that

$$\boldsymbol{I}_i = -\frac{1}{4\pi}\boldsymbol{H}_i. \tag{8}$$

From (6) and (8) we have

$$\boldsymbol{H}_i = \boldsymbol{H}_0\frac{1}{1-D}. \tag{9}$$

We wish to examine more fully the total external field at the surface of the ellipsoid. It follows from (9) that the internal field H_i is parallel to H_0 which, we recall, is along the semi-major axis. We treat the interior of the ellipsoid as a medium of permeability $\mu_2 = 0$. In this medium

$$\boldsymbol{B}_i = 0, \quad \boldsymbol{H}_i = \boldsymbol{H}_0\frac{1}{1-D}.$$

In the exterior we have $\mu_1 = 1$ or $B_0 = H_0' = H_0+H_I$. We know from electrodynamics that, when we pass from one medium to another with $\mu_1 \neq \mu_2$, the vector $\boldsymbol{B}$ has a continuous normal component, while $\boldsymbol{H}$ has a continuous tangential component. Hence

$$B_{0n} = H'_{0n} = B_{in} = 0,$$

$$H'_{0t} = H_{it} = H_{0t}\frac{1}{1-D} = \boldsymbol{H}_0\boldsymbol{t}\,\frac{1}{1-D}, \tag{10}$$

$$H'_{0t} = \frac{H_0\sin\beta}{1-D}, \quad |\boldsymbol{t}| = 1,$$

where β is the angle between the external field and the normal to the ellipsoid. Therefore, at the equator (where the field is perpendicular to the normal) the internal field attains its maximum value for the surface of the ellipsoid

$$H'_{eq} = \frac{H_0}{1-D}, \tag{11}$$

and at the poles it achieves its minimum value

$$H'_p = 0.$$

Since $0 \leqslant D \leqslant 1$, we have

$$\frac{1}{1-D} \geqslant 1 \tag{12}$$

where the equality applies to the cylindrical sample. For a sphere we have

$$H'_{eq} = \tfrac{3}{2} H_0.$$

If we now take $H_0 = \frac{2}{3} H_c < H_c$ then

$$H'_{eq} = H_c > H_0. \tag{13}$$

In general, if

$$H_0 = (1-D) H_c,$$

the relation (13) follows from (11).

Therefore, even though the applied external field is weaker than the critical field, the field reaches the critical value at the equator. Such a field destroys the superconducting state and consequently, there should be a transition to the normal state, in which the field can penetrate into the sample. The question arises, in what way does the ellipsoid make the transition into the normal state when we increase the applied field from $H_0 = (1-D)H_c$ to $H_0 = H_c$.

Landau (1937) showed that from the point of view of thermodynamic stability, the transition from a superconducting to a normal state proceeds as follows. If the field reaches the critical value at at least one point on the surface of an ellipsoid, this solid decomposes at once into alternating normal and superconducting layers. Such a state of a superconductor is called the intermediate state. Upon increasing the strength of the applied field, the total volume of the normal layers increases and for $H_0 = H_c$ the entire ellipsoid reaches the normal state. The existence of the layer structure was observed experimentally by Meshkovsky and Shalnikov (1947).

In a "diamagnetic" picture we treat the magnetization $\boldsymbol{I}$ as the magnetic moment per unit volume. If the external field (parallel to the axis of the cylinder) is constant then the magnetization is also constant. If the magnetic moment $\boldsymbol{I}$ is located in the field $\boldsymbol{H}$, the energy is given by

$$E = -\int \frac{\boldsymbol{HI}}{2} \mathrm{d}V = -\frac{\boldsymbol{HI}}{2} V = -\chi \frac{H^2}{2} V. \tag{14}$$

Since for metals in the normal state $\mu \approx 1$, hence $\chi \approx 0$ ($|\chi| = 10^{-6}$–10^{-7}) and $E_{In} \approx 0$. However, in the superconducting state $\chi = -\frac{1}{4\pi}$ and

$$E_{Is} = \frac{H^2}{8\pi} V > 0. \tag{15}$$

We see that in the normal state in the presence of a magnetic field the energy of the cylinder does not change while in the superconducting state it increases. One can therefore understand that in a sufficiently strong magnetic field the superconducting state is unstable and will, in fact, be destroyed by this field. In this way, the experimental results Kamerlingh Onnes obtained in 1913 are clarified by the diamagnetic picture.

As we emphasized, the M–O effect leads to the conclusion that the superconducting state is a state of thermodynamic equilibrium. Along the curve $H_c = H_c(\Theta)$ the superconducting and normal phases are in equilibrium. This is analogous to the fact that the liquid and the gas phases are in equilibrium along the curve $p = p(\Theta)$ (p = pressure).

Keesom and van den Ende (1932) showed that in the absence of a magnetic field, the specific heat of electrons suffers a jump on passing from the superconducting to the normal phase. This is characteristic of second-order phase transitions.

In general, at a second-order phase transition the energy and specific volume are unchanged and no heat is absorbed or given off. However, the specific heat and the coefficients of heat expansion and compressibility change discontinuously at the transition point. Another example of a second-order phase transition is that from the liquid He I into superfluid He II.

First-order phase transitions are characterized by a discontinuous change of energy and of specific volume. Therefore, such transitions are accompanied by a giving off or absorption of heat, the so-called latent heat.

If the transition from the superconducting phase to the normal one is the result of increasing the external magnetic field then heat has to be supplied and the transition is a first-order one.

After the discovery of the M–O effect, Gorter and Casimir (1934a), when developing the ideas contained in the earlier works of Keesom, Rutgers and Gorter, obtained the basic equations for the thermodynamics of superconductors. Their derivation was based on the existence of an equilibrium between the superconducting and the normal phase for $H_c = H_c(\Theta)$. We shall give the corresponding derivations by Landau and Lifshitz (1957).

We see from (15) that the free energy of a cylinder in the superconducting state increases by $\frac{H^2}{8\pi}V$ after the application of a magnetic field H. Hence

$$F_s = F_{s0}(V, \Theta) + \frac{H^2}{8\pi}V \tag{16}$$

where F is the free energy. By differentiating with respect to volume and by using the fact that $-\left(\frac{\partial F}{\partial V}\right)_\Theta = p$ we obtain for the pressure

$$p = p_0(V, \Theta) - \frac{H^2}{8\pi}. \tag{17}$$

The subscript "0" indicates the corresponding values for $H = 0$. The thermodynamic potential is equal to

$$\begin{aligned} \psi_s &= F_s + pV = F_{s0}(V, \Theta) + p_0 V = \psi_{s0}\left(p + \frac{H^2}{8\pi}, \Theta\right), \\ V &= V(p, \Theta). \end{aligned} \tag{18}$$

For the critical field we must have

$$\psi_s = \psi_n = \psi_{s0}\left(p + \frac{H_c^2}{8\pi}, \Theta\right). \tag{19}$$

We can see from (14) that the external field does not change the free energy in the normal state, and therefore

$$\psi_n = \psi_{n0}.$$

Observing that $\left(\frac{\partial \psi}{\partial p}\right)_\Theta = V$ and expanding the right-hand side of (19) in powers of H_c^2 we obtain

$$\psi_{n0}(p, \Theta) - \psi_{s0}(p, \Theta) = \frac{H_c^2}{8\pi} V(p, \psi). \tag{20}$$

From this equation we can obtain further relations by differentiating once or twice with respect to the temperature along the curve $H_c = H_c(\Theta)$ and making use of

$$S = -\left(\frac{\partial \psi}{\partial \Theta}\right)_{p,H}, \quad c_p = \Theta\left(\frac{\partial S}{\partial \Theta}\right)_{p,H} \tag{21}$$

where S is the entropy and c_p the specific heat.

Consequently,

$$S_n - S_s = -\frac{VH_c}{4\pi}\frac{dH_a}{d\Theta} > 0, \quad \frac{dH_c}{d\Theta} < 0, \tag{22}$$

and

$$c_s - c_n = \frac{V\Theta}{4\pi}\left[H_c \frac{d^2 H_c}{d\Theta^2} + \left(\frac{dH_c}{d\Theta}\right)^2\right]. \tag{23}$$

We must stress that these equations are valid along the curve $H_c = H_c(\Theta)$. Since $H_c(\Theta_c) = 0$, it follows that at the critical temperature we have $S_n = S_s$. From the curve in Fig. 1 we see that $\frac{dH_c}{d\Theta}$ is always negative. From (22) it follows that $S_s < S_n$ and hence that the superconducting state is more ordered than the normal one. Moreover, for $\Theta = \Theta_c$ (i.e. $H_c = 0$) we obtain from (23)

$$c_s - c_n = \frac{V\Theta_c}{4\pi}\left(\frac{dH_c}{d\Theta}\right)^2 > 0. \tag{24}$$

Therefore, the theory gives a jump in the specific heat, as has been observed experimentally.

From (22) we obtain an expression for the latent heat Q

$$Q = \Theta(S_n - S_s) = -V\Theta \frac{H_c}{4\pi}\frac{dH_c}{d\Theta} > 0. \tag{25}$$

Let us consider this equation for the case $\Theta = \Theta_c$, $H_c = 0$. Then $Q = 0$, and there is no latent heat. Both $c_s - c_n \neq 0$ and $Q = 0$ are characteristic of second-order phase transitions.

If we are at the point of the curve $H_c = H_c(\Theta)$ other than

$\Theta = \Theta_c$, $H_c = 0$, we see from (25) that $Q > 0$. In the course of isothermic transitions from the superconducting to the normal state heat is absorbed.

The discovery that superconductors have the property of ideal diamagnetism was extremely important as it allowed us to apply the thermodynamics of reversible processes to the superconducting state. This suggested that the magnetic as opposed to electric properties of superconduction are of fundamental importance and that the Maxwell equations must be completed by other equations in a manner which would describe the M–O effect. The full system of equations describing the electrodynamics of the superconducting state was given by H. and F. London (1935). In these equations there appears an important parameter λ, called the penetration depth. This parameter, of the order of 10^{-6} cm, gives a theoretical value for the depth to which the magnetic field is effectively restricted inside the body. We see that the property of perfect diamagnetism follows from London's equations only for samples of dimensions much greater than λ. The penetration of external magnetic fields into the surface layer of a superconductor was observed experimentally, but the experimental penetration depth was approximately five times greater than λ. Since in superconductors we have two currents, a supercurrent with density $\boldsymbol{j}_s$ and a normal current with density $\boldsymbol{j}_n$, the number of equations describing the electrodynamics of the superconducting state must be greater than the number of equations describing the electrodynamics of the normal state. The total current is the sum of the two currents. Ohm's law $\boldsymbol{j} = \sigma \boldsymbol{E}$ is not valid for superconductors. It is replaced by the set of equations

$$\boldsymbol{j} = \boldsymbol{j}_n + \boldsymbol{j}_s, \tag{26}$$

$$\boldsymbol{j}_n = \sigma \boldsymbol{E}, \tag{27}$$

$$\operatorname{curl}(\Lambda \boldsymbol{j}_s) = -\frac{1}{c}\boldsymbol{H}, \tag{28}$$

$$\frac{\partial(\Lambda \boldsymbol{j}_s)}{\partial t} = \boldsymbol{E}, \quad \Lambda = \frac{m}{ne^2} \tag{29}$$

where n is the density of electrons.

Equations (26)–(29) together with Maxwell equations and the equations connecting $\boldsymbol{B}$, $\boldsymbol{H}$ and $\boldsymbol{D}$, $\boldsymbol{E}$ give the complete set of equations for the electromagnetic field in superconductors.

Equation (28) and Maxwell equation $c\,\mathrm{curl}\,\boldsymbol{H} = 4\pi \boldsymbol{j}_s$ give

$$\nabla^2 \boldsymbol{H} - (1/\lambda^2)\boldsymbol{H} = 0, \quad \lambda = \sqrt{\frac{\Lambda c^2}{4\pi}} \tag{30}$$

where λ is the penetration depth.

Consider a superconducting half-space $z \geqslant 0$: if the magnetic field is parallel to the surface of the superconductor, the solution of (30) has the form

$$H(z) = H(0)e^{-z/\lambda}. \tag{31}$$

We see from (31) that the magnetic field penetrates into the superconductor practically to a depth λ. The solution (31) illustrates how London's equations describe the M–O effect.

Equations (28) and (29) connect the supercurrent with the magnetic and the electric field, respectively. Equation (27) gives the connexion between the normal current and the electric field while (29) gives the connexion between the supercurrent and the electric field. From (27) and (29) we see that for $\boldsymbol{E} = 0$ the normal current is equal to zero and the supercurrent is constant in time. The supercurrent is connected with the magnetic field by means of (28). Equation (28) is one of the equations added to the Maxwell equations in order to describe the perfect diamagnetism of large samples. The supercurrent is also called the diamagnetic current. The diamagnetic current can be expressed in terms of the vector potential $\boldsymbol{A}$. From $\boldsymbol{H} = \mathrm{curl}\,\boldsymbol{A}$, equation (28), the condition $\mathrm{div}\,\boldsymbol{A} = 0$ and the condition that at the surface the normal component of $\boldsymbol{A}$ is equal to zero we have

$$\boldsymbol{j}_s(\boldsymbol{r}) = -\frac{1}{\Lambda c}\boldsymbol{A}(\boldsymbol{r}). \tag{32}$$

A more general, non-local connexion between $\boldsymbol{j}_s$ and $\boldsymbol{A}$ of the form

$$\boldsymbol{j}_s(\boldsymbol{r}) = -\frac{1}{\Lambda c}\hat{L}\boldsymbol{A}(\boldsymbol{r}) \tag{33}$$

$$= -\frac{3}{\Lambda c 4\pi \xi_0}\int \frac{\boldsymbol{R}[\boldsymbol{R}, \boldsymbol{A}(\boldsymbol{r}')]\exp(-R/\xi_0)\exp(-R/l)\,\mathrm{d}^3\boldsymbol{r}'}{R^4},$$

$$R = |\boldsymbol{r} - \boldsymbol{r}'|$$

was given by Pippard (1953). $\hat{L}$ is the integral operator with the exponential factor $\exp\left[-\frac{|\boldsymbol{r}-\boldsymbol{r}'|}{\xi_0}\right]$. On account of the exponential factor the integration in (33) is effectively restricted to a sphere of radius ξ_0 around the point $\boldsymbol{r}$. We call $\xi_0 \sim 10^{-4}$ cm the coherence length, l is the mean free path. If the potential $A(r)$ changes substantially over the length ξ_0, (32) differs essentially from (33). If we can assume that A does not change over the length ξ_0 then (32) is equivalent to (33).

As we noticed earlier, the superconducting state is more ordered than the normal state. The parameter ξ_0 alone characterizes the radius of this ordering. The coherence length finds a direct interpretation in the microscopic theory of superconductivity as the radius of a so-called Cooper pair, which will be discussed later.

Pippard's non-local theory proved superior to London's theory in explaining a number of experiments. For example Pippard's theory gives a greater value for λ than the London theory and the former agrees with experimental data. In the theory of superconductivity the penetration depth was therefore joined by a second important parameter, the coherence length. Experimental evidence for the existence of a surface energy between the superconducting and normal phase prompted the introduction of ξ_0. From the value of the surface energy we can calculate the thickness of the layer between the normal and the superconducting phase. It is of the order of 10^{-4} cm. During the discussion of the microscopic theory we will return to the problem of the surface energy and its connexion with the radius of Cooper pairs.

After Landau theoretically predicted, in 1937, the structure of the intermediate state, the problem of the surface energy between the superconducting and the normal phase got special interest. The London theory yielded a negative surface energy and thus implied that the state with negative surface energy would be energetically privileged and the destruction of the superconducting phase in the bulk sample would not occur for $H = H_c$. The sample may, instead, decompose into regions, and therefore the magnetic field should penetrate inside the sample. These difficulties led to a generalization of London's theory, which was given by Ginzburg and Landau (1950). Their object was to take into account quantum effects. The contribution of quantum effects can be estimated on the basis of an uncertainty relation, and should lead to a positive surface energy as can be seen by the following simple argument. The supercurrent flows in a thin surface layer whose thickness is equal to the penetration depth λ.

Hence the uncertainty in the momentum of superconducting electrons of the density ϱ_s is $\sim \hbar/\lambda$ and therefore the uncertainty in the energy density is

$$\Delta E \sim (\hbar/\lambda)^2 \varrho_s/m^* = 10^4 \text{ erg/cm}^3$$

where $m^* \sim 10^{-27}$ g is the effective mass of the superconducting electrons, $\varrho_s \sim 10^{21}$. As the magnetic energy is given by $H^2/8\pi$, we see that for fields $H \sim 5\times10^2$ gauss, $\Delta E > H^2/8\pi$. Following the general theory of second-order phase transitions, these authors took ϱ_s as a positive parameter in their theory. (Compare the case of ferromagnetism where the square of the spontaneous magnetization I^2 is such a parameter.) The density of superelectrons, constant in the London theory, is expressed by means of the so-called effective wave function

$$\varrho_s(r) = |\psi_s(r)|^2. \tag{34}$$

The equations for ψ_s, ψ_s^*, and $\boldsymbol{A}(r)$ were obtained by minimizing the free energy with respect to ψ_s, ψ_s^* and $\boldsymbol{A}$. Converted to dimensionless quantities, the equations have the form

$$\left(\frac{i}{\varkappa}\nabla'+\boldsymbol{A}'\right)^2 \psi_s' = \psi_s' - \psi_s'|\psi_s'|^2, \tag{35}$$

$$\nabla'^2\boldsymbol{A}' = |\psi_s'|^2\boldsymbol{A}' + \frac{i}{2\varkappa}(\psi_s'^*\nabla'\psi_s' - \psi_s'\nabla'\psi_s'^*), \tag{36}$$

and the condition at the surface of the sample is

$$\boldsymbol{n}(-i\nabla'\psi_s' - \varkappa\boldsymbol{A}'\psi_s') = 0, \tag{37}$$

$$\varkappa = \frac{\sqrt{2}e}{hc} H_{cm}\lambda_0^2. \tag{38}$$

Here, ∇' means differentiation with respect to the dimensionless quantity $\boldsymbol{r}' = \boldsymbol{r}/\lambda_0$, where λ_0 is London's penetration depth given by (30). In the G–L theory, the penetration depth is not constant, but rather depends upon the magnetic field, a prediction confirmed by experiments. From (35), (36) we see that ψ' depends on $\boldsymbol{A}'$ or on $\boldsymbol{H}'$. Hence ϱ_s depends on the magnetic field, and the penetration depth depends on ϱ_s.

The G–L equation for the effective wave function of a superelectron is a non-linear quantum-mechanical equation. It is very interesting that all superconducting electrons, which obey Fermi statistics ($\varrho_s \sim 10^{21}$), are described by the same wave function.

Such a situation is characteristic of particles obeying Bose statistics, in the case of Bose–Einstein condensation. We have then a state occupied by a macroscopically large number of particles, the condensate. The success of the G–L theory suggested that in the transition to the superconducting state some kind of Bose–Einstein condensation might have taken place. The G–L equations have later been derived by Gor'kov (1959) from the microscopic theory of superconductivity.

The possibility of obtaining an expression for the diamagnetic current (see (32)) by means of the quantum-mechanical formula for the current was discussed by F. London in 1935. The discovery of the M–O effect and the work of Bohr (1911) and van Leeuwen (1931) showing that a classical system cannot be diamagnetic, made clear the need for a quantum-mechanical description of superconductivity.

One sees from (35)–(37) that in the theory there appears a new dimensionless parameter $\varkappa$, expressed in terms of two quantities characteristic of the superconductors: H_{cm}, the critical field for large specimens, and λ_0. The parameter $\varkappa$ is of great importance since, as Ginzburg and Landau showed, for $\varkappa < 1/\sqrt{2}$ the surface tension between the normal and the superconducting phase is positive while, for $\varkappa > 1/\sqrt{2}$, it is negative. As we emphasized earlier, when the surface energy is negative, the penetration of a magnetic field into some regions of the superconductor is energetically favoured.

Ginzburg and Landau have elaborated their theory for $\varkappa < 1/\sqrt{2}$; however, this theory does not explain the experimental values obtained by Zavaritsky (1952) for the critical field of thin layers. This fact led Abrikosov and Zavaritsky to hypothesize that for thin layers $\varkappa > 1/\sqrt{2}$. In 1952 they conjectured the existence of two different groups of superconductors: superconductors of type I, for which $\varkappa < 1/\sqrt{2}$, and superconductors of type II, for which $\varkappa > 1/\sqrt{2}$. All known pure metallic superconductors (with the exception of niobium) are of type I. Niobium and all superconducting alloys are of type II. For Nb we have $\varkappa = 1.2$ and for the alloy Ti+25% V we have $\varkappa = 96 \gg 1$. The supposition that $\varkappa > 1/\sqrt{2}$ for alloys was first made by Landau.

The division of superconductors into two different groups

proved to be well founded. Superconductors of type II differ from superconductors of type I with respect to magnetic and thermodynamic properties. The differences in their magnetic properties are especially pronounced. The theory of superconductors of type II was given by Abrikosov (1952). In these superconductors the transition from the superconducting state to the normal state does not take place suddenly when $H = H_c$, as may be observed for superconductors of type I. The transition takes place in some sufficiently large interval of the magnetic field. This follows from the fact that for type II superconductors there exist two critical fields: a lower one H_{c1} and an upper one H_{c2}. An external magnetic field $H_e < H_{c1}$ does not penetrate into the superconductor. For fields $H_{c1} < H_e < H_{c2}$, the magnetic field begins gradually to penetrate into the superconductor, which for $H_e = H_{c2}$ goes to the normal state. It is a fact of considerable importance that for fields near H_{c2} and relatively weak currents no electrical resistance is observed. This is true even if H_{c2} is quite strong. The value of H_{c2} increases with the concentration of defects and can reach hundreds of kOe. For example, for the alloys Nb_3Sn, V_3Ga we have $H_{c2} \sim 100$—300 kOe. Very large critical fields for alloys were first observed by de Haas and Voogd in 1930, while they investigated the alloys of Bi with Pb. They found one alloy with $H_{c2} \sim 15$ kOe (de Haas and Voogd, 1931). It was not until 1960 that an attempt was made to apply superconductors of the second type to superconducting solenoids. This was done by Autler (1960). A solenoid from niobium wire gave a magnetic field of 4.3 kOe. In 1961 others obtained a wire from the alloy Nb_3Sn which remained in the superconducting state until 88 kOe, with a current density 10^5 A/cm^2 (Kunzler *et al.*, 1961). In 1963 a superconducting state in a field of 101 kOe was obtained (Martin *et al.*, 1963).

Since superconductors of type II in large fields exhibit no electric resistance, it is of considerable importance that an arbitrary superconductor of type I can be transformed into a superconductor of the type II either by adding impurities or by causing defects in the crystal lattice.

As the magnetic field is increased from H_{c1} to H_{c2} the picture of the penetration of magnetic field into the specimen is the following. When $H_e = H_{c1}$ there is a first- or second-order phase transition (for $\varkappa \gg 1$ of second order). For $H_{c1} < H_e < H_{c2}$

separate magnetic flux filaments penetrate into the superconductor. The magnetic field has its maximum value in the centre of the filament where

$$H_{c1} = \frac{H_{cm}}{\sqrt{2}\,\varkappa}\,[\ln(\varkappa+1.8)+0.08] \tag{39}$$

and vanishes in a distance equal to the penetration depth.

Later it was seen that the parameter $\varkappa$ is a function of temperature, and moreover, that $\varkappa$ can be expressed in terms of two important parameters: λ_0 and $\xi_0 = \frac{0.18hv}{\Theta_c}$ where v is the mean velocity of electrons. In particular, we have (see, for example, Abrikosov (1965))

$$\varkappa(\Theta) = \varkappa(\Theta_c)A(\Theta) \tag{40}$$

where $A(\Theta)$ is a function of temperature, decreasing from 1.25 to 1 for $0 \leqslant \Theta \leqslant \Theta_c$. For pure superconductors (for which the mean free path of electrons is $l \gg \lambda$)

$$\varkappa(\Theta_c) = \varkappa_0(\Theta_c) = \frac{\lambda_0}{\xi_0}. \tag{41}$$

For alloys

$$\varkappa(\Theta_c) = \varkappa_0(\Theta_c)+7.5\times 10^3 R_n\gamma^{1/2} \tag{42}$$

where R_n is the residual resistance in the normal state, γ is the coefficient in the formula $c_n = \gamma\Theta$ giving the temperature dependence of the electronic specific heat.

In the microscopic theory of superconductivity the binding energy of Cooper pairs is equal to zero in the centre of the filament (i.e. there is no superconducting state). If we increase the magnetic field, the filaments become more numerous, tend to approach one another, and form (in a cross-section perpendicular to the filaments) a periodic structure: a triangular lattice. By further increasing the magnetic field a first-order phase transition takes place and the triangular structure is changed to a quadratic structure. If the field is further increased the filaments come closer together until their distances are equal to the coherence length (or to the radius of the Cooper pair). Then a further increase in the external field results in an increase of the value of the magnetic field in the space between the filaments. Finally, for $H_e = H_{c2}$ a transition to the normal state takes place. It is a second-order transition.

The state which exists for $H_{c1} < H_n < H_{c2}$ is called the mixed state.

The critical field H_{c2} is given by the formula

$$H_{c2} = H_{cm}\sqrt{2}\varkappa \tag{43}$$

where H_{cm} is found from

$$\frac{H_{cm}^2}{8\pi} = F_n - F_s = -\int_0^{H_{c2}} I(H)\,\mathrm{d}H. \tag{44}$$

In practice, experiments give the right-hand side of (44), i.e. we find H_{c2} and the dependence $I = I(H)$. A graphical integration gives us H_{cm}. From (43) we can find $\varkappa$.

When describing particular phenomena, various generalizations of the phenomenological theory have been necessary to achieve greater agreement with experimental results.

Yet attempts to create a microscopic theory of superconductivity met with no success. An especially significant step toward understanding the essence of superconductivity was the theoretical prediction and experimental confirmation in 1950 of the so-called isotope effect. The theoretical prediction of this effect was made by Fröhlich (1950). He improved upon the Bloch theory by constructing a Hamiltonian which explicitly described the electron–ion interaction. This Hamiltonian was expressed in the second quantization framework and contained terms corresponding to free electrons, to free phonons, to the electron–phonon interaction and to the Coulomb interaction of electrons. In Bloch's theory of the conductivity of metals the essential matrix elements were those describing electron transitions together with free phonon emission and absorption; the full Hamiltonian was not considered. In Fröhlich's formulation the problem was treated in the fashion of quantum field theory and hence virtual transitions were present. It became evident that the virtual exchange of phonon results in an effective attraction between those electrons whose momenta are near the momentum at the Fermi surface. The basic result of Fröhlich's theory is that the transition temperature into the superconducting state is inversely proportional to the square root of the mass M of the metallic ion in the crystal lattice, i.e.

$$\Theta_c \sim 1/\sqrt{M}.$$

Independently of these theoretical considerations, Maxwell (1950), as well as Reynolds *et al.* (1950), investigated experimentally the transition temperatures for samples made of different isotopes of the same metal. They found, for three samples made from different isotopes of Sn, that beyond all doubt each sample has a different transition temperature. Experiments also confirmed the dependence of Θ_c on M as found by Fröhlich.

If the transition temperature into the superconducting state depends on the mass of the atoms of which the crystal lattice is built, then the interaction of electrons with the lattice must play a fundamental role in the transition into the superconducting state. For this reason the discovery of the isotope effect was decisive for a microscopic theory. However, in spite of this success, Fröhlich did not explain the phenomenon of superconductivity.

Ginzburg (1953) suggested a model of superconductor with a gap in the energy spectrum. In the same year Goodman (1953) put forth the hypothesis that the dependence of the specific heat on the temperature in the form $\exp\left[-\frac{\alpha}{k\Theta}\right]$ is characteristic of a model of superconductors with an energy gap. Goodman's hypothesis resulted from investigations of the thermal conduction of superconductors and consisted in a simple interpretation of Koppe's theory (Koppe, 1950) which gave an exponential dependence of the specific heat on temperature. The two-fluid theory of Gorter and Casimir (1934b) indicated that the dependence of the specific heat on the temperature is of the form $\sim \Theta^3$. The first clear discrepancy from this dependence was observed for niobium by Brown, Zemansky and Boorse (1953). The first papers which showed unambiguously the exponential dependence of the specific heat were the experiments performed in 1954 with vanadium (Corak *et al.*, 1956) and Sn (Corak and Satterthwaite, 1954). Subsequently, such a dependence was established for several elements. As a result, the models of superconductors with an energy gap became especially interesting. In particular, Bardeen (1955) showed that models with an energy gap led to a non-local theory of the type created by Pippard. Migdal (1958) noted that a model with an energy gap cannot be solved by perturbation theory.

Thus, forty-four years after discovery of the phenomenon of superconductivity, the theoretical situation was as follows. It was clear that the microscopic theory has to be non-local and to give the relation (33) and its particular example (32). It must imply an energy gap in the spectrum of elementary excitations; in the neighbourhood of absolute zero the dependence of the specific heat on temperature must be exponential, and at the critical temperature the specific heat must have a jump. It has to describe a phenomenon analogous to the Bose–Einstein condensation.

The theoretical work of Cooper (1956) was an essential contribution to a further understanding of the phenomenon of superconductivity.

In this paper, he took under consideration a single pair composed of two mutually interacting electrons, above the quiescent Fermi sea. The electrons filling the Fermi sea do not interact with the pair. On the other hand, they block the levels below the Fermi surface. Cooper proved that in the case of an arbitrary weak attraction the pair is in a bound state. As was shown previously by Fröhlich, such an attraction can be caused by the exchange of virtual phonons. There are singlet pairs with total spin zero and triplet pairs with total spin one. The diameter of a pair is very large, the order of magnitude being 10^{-4} cm. This implies, in the more general case of many pairs, an overlap of the pairs' wave functions and does not allow a system with pairing to be treated as an ideal Bose gas.

On the basis of the Cooper's results, Bardeen, Cooper and Schrieffer (1957a) and Bogoliubov (1958a, b) constructed a microscopic theory of superconductivity which gave a satisfactory explanation of this phenomenon.

In the model considered by Cooper, the artificially singled-out pair is in a bound state. In the general case, however, in the volume associated with the diameter of a pair there are about one million other pairs. Hence, instead of bound states, we have a state with strongly correlated electron pairs, the so-called state with coupled or correlated pairs. The analysis of the singularities of the *S*-matrix for the problem of superconductivity shows that the Cooper and the BCS–Bogoliubov models of a superconductor lead to different types of singularities (Mathur *et al.*, 1966).

The state of correlated pairs is a condensed state, and the pairs form the condensate. However, the one-particle elementary excitations do not have an energy spectrum characteristic of bosons. It follows from this and from the previous remarks that we are not dealing with an ordinary Bose–Einstein condensation. Therefore, in his monograph Blatt (1964) suggested calling the condensation of electron pairs "Schafroth condensation" in connexion with some interesting works of Schafroth devoted to this problem.

It can be shown that the energy of a system in the superconducting state is smaller than the energy in the normal state. The difference of the energies, called the energy of condensation, is very small and of the order 4×10^3 erg/cm^3. For comparison, we note that the kinetic energy is 10^{11} erg/cm^3, larger by eight orders of magnitude. When passing into the superconducting state the system loses part of its energy for the creation of bound pairs. The existence of the binding energy of a pair implies the existence of an energy gap in the spectrum of one particle excitations. The width of the energy gap is equal to twice the binding energy. If we multiply the binding energy of a pair by half the number of electrons in the neighbourhood of the Fermi surface in the superconducting state, then we obtain just the energy of condensation. As was pointed out above, an effective attraction occurs for electrons whose momenta are near the momentum at the Fermi surface p_F. This is clear in view of the Pauli exclusion principle. For not very strong interactions only the electrons in the neighbourhood of the Fermi surface can change their momenta. Under these circumstances the pairs with net momentum equal to zero will be singled out. This is also apparent from Fig. 4 (see Cooper

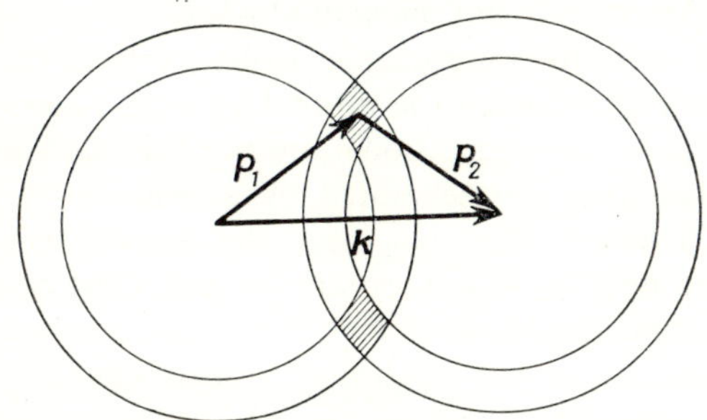

FIG. 4

(1960)). In the spherical shell (we show a cross-section) near the Fermi surface one has electrons which can form bound pairs. The shaded region indicates those electrons which can form pairs with a given momentum $\boldsymbol{k}(|\boldsymbol{p}_1| \sim |\boldsymbol{p}_2| \sim p_F)$. The larger $\boldsymbol{k}$, the smaller is the shaded region. The maximum, when the entire shell is shaded, occurs for $\boldsymbol{k} = 0$. This corresponds to pairing of particles with opposite momenta.

The existence of the binding energy is a collective effect, the energy being larger if a greater number of pairs are bound. The energy is maximum at absolute zero. As the temperature increases, thermal motion leads to the disintegration of pairs and so to a decrease in their number. This, in turn, causes the binding energy of the remaining pairs to decrease.

If an external electric field causing an ordered motion of the system is present, all correlated pairs have the same (non-vanishing) momentum. The state of correlated pairs, either the "resting" or current-carrying variety, is very stable. The pairs are then neither scattered nor excited. Dissipative processes, which are essentially one-particle processes, can thus occur only if a pair is destroyed, an action which requires a finite energy loss.

The greater the drift velocity of pairs, the smaller the loss of energy. Thus, if the current is not too strong, dissipative processes are forbidden and we have a persistent current of correlated pairs, the supercurrent.

Good electric conductivity is the result of a weak interaction of the electrons with the crystal lattice. However, a weak interaction with the lattice cannot lead to the condensation of pairs. Thus the microscopic theory explains why the best conductors such as Au, Ag, Pt, are not superconductors.

As we have seen, the microscopic theory explains the existence of a surface energy between the normal and the superconducting phases. Consider a one-dimensional model. The electrons become correlated when they are separated by a distance of the order of a pair's diameter. Electrons near the surface can correlate with their left neighbour if they are on the right-hand surface or with their right neighbour if they are on the left-hand surface, having no right, or no left neighbour, respectively.

Inside the metal electrons can correlate with neighbours on both sides. Thus the binding energy is an increasing function of the distance from the surface. This function can be approximated by a step function which is equal to zero within the distance of the pair diameter, and equal to the binding energy at greater distances from the surface. It is clear that the strength of a magnetic field penetrating into the metal can also be expressed by a step function vanishing at distances from the surface larger than the penetration depth. If these two step functions do not overlap, there is a region in which both the binding energy of the pairs and the magnetic field strength are equal to zero. Because of the vanishing of the binding energy, this region should be actually a normal, that is, a non-superconducting region. However, since the magnetic field also vanishes in this region, it must be a superconducting region. Therefore, this region is to be identified as a layer separating the normal and the superconducting phases. As there is no magnetic energy in this layer, we must assume the presence of another type of energy, namely the positive surface energy. This energy is negative if the two step functions overlap.

The theoretical foundations of the microscopic theory of superconductivity will be given in the subsequent chapters. Before proceeding, we would like to mention briefly a number of interesting experimental and theoretical results obtained after the microscopic theory of superconductivity had been formulated.

On the basis of the BCS theory, Yosida (1958) calculated the magnetic susceptibility χ in the superconducting state and obtained $\chi_s = 0$. The experimental data obtained a year earlier by Reif (1957) contradicted this result. The change of the spin susceptibility due to the superconducting transition can be obtained experimentally from the Knight shift measurements. In 1949 Knight noticed that the frequency ν_m of the magnetic nuclear resonance in a metal is greater than the resonance frequency ν_f for the same nuclei in free ions. The Knight shift is defined as $K = \frac{\nu_m - \nu_f}{\nu_m}$ and is proportional to the paramagnetic spin susceptibility.

Reif had examined the Knight shift in colloids of mercury and obtained $K_s(0)/K_n = 2/3 = \chi_s(0)/\chi_n$, where $K_s(0)$ and K_n are Knight shifts in the superconducting and normal states respectively. Subsequent measurements on Sn gave $K_s(0)/K_n = 3/4 \approx 2/3$ (Androes and Knight, 1959) and on V gave $K_s(0)/K_n = 1$ (Noer and Knight, 1961). The investigations of the superconducting intermetallic compounds of V_3X (where X = Si, Ga, Ge, As, ...) (Clogston *et al.*, 1962) led to non-vanishing Knight shifts.

Several attempts were made to explain the discrepancy between the microscopic theory of superconductivity and experiment. We assume that spin-orbit coupling is important for heavier elements such as Hg and Sn (Ferrell, 1959; Anderson, 1959). For small "particles" of metals this coupling leads to the spin-reversing collisions with the specimen surfaces. In consequence it leads to a finite spin susceptibility.

Balian and Werthamer (1963) considered the superconducting state with "p"-pairing and an isotropic energy gap. They obtained the result $\chi_s = {}^2/_3\chi_n$. But, as emphasized in Schrieffer's monograph, even a small amount of disorder destroys such a state. On the other hand, from Pines' paper (Pines, 1958) we learn that the criterion of superconductivity, which is associated with a certain parameter $\tilde{V}$ measuring the average strength of the net interaction between electrons, is evidently not fulfilled in the case of Hg. One can perhaps assume that interactions in Hg lead to "p"-pairing which is not considered in the parameter $\tilde{V}$.

The spin-orbit coupling depends very strongly on the atomic number, thus it is not important in lighter elements such as vanadium and aluminium. However, it was argued (Clogston *et al.*, 1962) that in vanadium the entire Knight shift is due to the temperature-independent Kubo–Obata d-band orbital paramagnetism (Kubo and Obata, 1956). Since this paramagnetism is temperature independent it should be identical in both the normal and the superconducting state.

One explained the results of the measurements on the compounds V_3X on the basis of the temperature-independent orbital paramagnetism and temperature-dependent d-electron spin susceptibility. The last contribution is reduced by at least 75 per cent in the superconducting state. For this reason the ratio $K_s(0)/K_n$ for V_3X is smaller than for vanadium.

The measurements of the Knight shift in aluminium were very important from the point of view of the spin-orbit coupling and orbital paramagnetism theories. In Al, the spin-orbit coupling should be small. On the other hand, aluminium has no d-electrons, which are responsible for orbital paramagnetism. Therefore one can assume that the only contribution to the paramagnetism arises from the spin of the conduction s-electrons.

The first measurements performed on Al gave a finite Knight shift $K_s(0)/K_n = 3/4$ (Hammond and Kelly, 1964). These results complicated the theoretical description of paramagnetic susceptibility in superconductors for several years. Recently, however, the same authors repeated their measurements on a new sample and found a vanishing Knight shift in aluminium (Hammond and

Kelly, 1967). The temperature dependence of the shift differs from the dependence calculated on the basis of the microscopic theory. The authors were able to account for this deviation in terms of the depairing effect of the magnetic field considered by Fulde and Maki (1965a, b). The depairing effect causes a field-dependent increase in the susceptibility as compared to the one calculated by Yoshida.

The last results for Al, a metal particularly suited for these purposes, seem to show that there is no discrepancy between the theoretical calculations of the paramagnetic susceptibility of *s*-electrons, based on the microscopic theory of superconductivity, and the experimental data. The additional effects seem to be connected with orbital paramagnetism due to the *d*-band electrons.

The theory of superconductivity exerted an influence on nuclear theory. In particular, the difference between the single-particle spectrum of even–even and neighbouring odd nuclei has long been observed. The energy of the first excited level of a nucleus with an odd nucleon appears to be several times smaller than the energy of the corresponding level for the neighbouring even–even nucleus. Moreover, there was a difference between the experimental and theoretical values for the moments of inertia, the latter values being obtained from calculations based on the "cranking" model. This model gives the nucleus the moment of inertia of a rigid body.

Bohr, Mottelson and Pines (1958) assumed the existence of "residual forces" in nuclei, which lead to the proton–proton and neutron–neutron pairing. In this case, the excitation of an even–even nucleus is possible only after an energy loss equal to the energy necessary to break a pair. In the odd nuclei the unpaired nucleon can be excited directly. Moreover, the theory states that the moment of inertia of a system with a condensate of pairs is three times smaller than the moment of inertia of a "rigid" body. This is in agreement with experimental data.

Abrikosov and Gor'kov (1960) gave a theory of superconductors with magnetic impurities, i.e. atoms with a magnetic moment. According to this theory, an increase of impurities lowers the transition temperature. If the concentration N of impurities amounts to 1 per cent, the superconducting state vanishes. For $N = 0.9$ per cent the gap in the energy spectrum vanishes. In the latter case we are dealing with a so-called "gapless" superconductivity. This effect was observed experimentally by Reif and Woolf (1962). The specific heat in this case is a linear function of temperature.

Giaever (1960) gave a simple demonstration of the existence of the energy gap. He had investigated electron tunnelling first between a normal and a superconducting metal separated by a thin oxide layer and, subsequently, between two superconducting metals. In the first case, the tunnelling current flowed only if the quantity eU (where U is the applied voltage) exceeded the energy gap. In the second case, eU had to exceed the sum of the energy gaps. Single electrons as well as pairs participate in the tunnelling effect.

Josephson (1962) considered the process of bound pairs tunnelling (see also Feynman, 1963). He concluded that the tunnelling current can flow even when no voltage ($U = 0$) is applied because no energy is needed to break a pair. On the other hand, for $U \neq 0$ the superfluous energy of a pair $2eU$ is emitted with the frequency $\omega = 2eU/\hbar$ or there is an alternating current of pairs with the same frequency. For example, for $U = 10^{-4}$ V we have $\omega \sim 10^{11}$ sec^{-1}. The amplitude of the oscillations is very small (see the experiments of Anderson and Rovell (1963), and Shapiro *et al.* (1964)).

In 1961 it was demonstrated experimentally that the quantization of the magnetic flux is connected with the pairing of electrons (Deaver and Fairbank, 1961; Doll and Näbauer, 1961). Quantization of magnetic flux was suggested by London (1950). If we consider a multiply-connected superconductor, the surface supercurrent produces a magnetic flux Φ passing through the holes in the superconductor. According to F. London's quantum-mechanical considerations of the superconducting state, the flux Φ must be quantized in the units ch/e. Experiments (Deaver and Fairbank, 1961; Doll and Näbauer, 1961) give quantization of flux in units $ch/2e$. They agree with Onsager's predictions of 1959, that due to the pairings of electrons the charge must be taken to be $2e$. Theoretical considerations based on the microscopic theory of superconductivity (Byers and Yang, 1961) led to flux quantization in units $\Phi = ch/2e$, where the charge $2e$ could be associated with a Cooper pair.

Geballe *et al.* (1961) observed that there is no isotopic effect for rhutenium or osmium. One hypothesis which explains this phenomenon states that the effective attraction which leads to the condensation is not a consequence of an interaction with phonons. Earlier, other mechanisms leading to an effective attraction were proposed. Akhiezer and Pomerantsuk (1959) proved that in ferromagnets there exists an additional attraction between the conduction electrons, due to the virtual emission and absorbtion of spin waves. The attraction takes place only in the triplet state, provided the sum of the components of the spins along the axis of easy magnetization is equal to zero. In the singlet state, this mechanism leads to repulsion. It was shown (Akhiezer, A. I. and I. A., 1962; Matthias, Suhl and Corenzwit, 1958; Matthias and Suhl, 1960) that superconductivity and ferromagnetism can coexist within the same region of space (see also Vonsovsky and Svirsky (1964a)).

Privorotsky (1962) considered a similar problem for antiferromagnets (see also Baltensperger and Sträsler (1963)). Vonsovsky and Svirsky (1964b) showed that for triplet pairs the Coulomb interaction leads to an effective attraction.

Another possible mechanism for the origin of superconductivity was proposed by Little (1965). He considered long organic molecules (for example, DNA). The effective attraction between the electrons is caused by Coulomb forces in the following manner. An electron moving along the main chain of the molecule polarizes the side chains and induces positive charges at the branching points of these chains. The positive charges, in turn, attract electrons, leading to an effective attraction among the electrons. The attractive energy is large, of the order of 1.5–2 eV, which results in a large critical temperature, $\Theta_c \sim 2000°K$. From the practical point of view, the possibility of superconductors existing at such high temperatures is very interesting, since at present the highest critical temperature for a superconducting alloy, Nb_3Sn, is $\Theta_c = 18.2°K$, and for a composition $4Nb_3Al:1Nb_3Ge$, it is $\Theta_c = 20.05°K$.

Ginzburg and Kirzhnits (1964) considered the possibility of the transition of electrons occupying the Tamm surface levels to the superconducting state. The transition may take place due to the exchange of surface phonons. Also, the authors considered possible the existence of surface (two-dimensional) superconductivity for dielectrics. Experiments with graphite lamellar compounds show that as yet no firm conclusion can be reached on the existence of two-dimensional superconductivity (Salzano and Strongin, 1967).

Superconductivity

The most interesting theoretical investigations exploring possible one- and two-dimensional superconductivity are based on an examination of Yang's criterion for long-range order (Yang, 1962). Yang's criterion for the existence of long-range order in condensed Fermi systems is that the off-diagonal elements of the two-particle correlation function are finite in the limit $|\mathbf{r}_1 - \mathbf{r}_2| \rightarrow \infty$.

Rice (1965), using the Ginzburg–Landau theory, has indicated that this criterion is not satisfied in one and two dimensions. This result (Rice, 1965) is strictly true in the neighbourhood of the critical temperature. Recently Kadanoff and Kane (1967) examined the long-range order in superfluid helium by means of the single-particle Green functions in the hydrodynamic approximation. The result indicates that there is no long-range order in the one-dimensional case. For two dimensions either the phase transition will not exist or, if it does exist, it will have a rather different character from that in three dimensions. If an ordered state appeared, its hydrodynamic behaviour would not be described by the two-fluid equations in their usual form. The two-particle Green functions for condensed Fermi systems have, in the hydrodynamic approximation, the same form as the single-particle Green functions for Bose systems (Galasiewicz, 1966a). The conclusions (Rice, 1965) therefore seem to have quite a general character.

Gurevich, Larkin and Firsov (1962) considered, from the theoretical point of view, possible superconductivity in polar semiconductors. The most favourable conditions for the existence of superconductivity are: a sufficiently large carrier concentration in the conduction band and a sufficiently strong electron–phonon interaction. For these reasons Cohen (1964) considered a many-valley band structure model of superconductors. In this model the largest contribution to the attractive electron–electron interaction arises from the exchange of "inter-valley" phonons. This interaction can be larger than the repulsive Coulomb interaction and hence can induce a superconducting transition. As expected, most materials discussed by Cohen (1964) will be type II superconductors if they are superconducting. The transition temperature will be $\sim 0.1°K$. Superconductivity was observed in germanium telluride (Hein *et al.*, 1964) and in reduced strontium titanate ($SrTiO_3$) (Schooley, Hosler and Cohen, 1964). The transition temperature (0.1–0.28°K) was found to be a function of the carrier concentration. An extreme example of a type II superconductor, $SrTiO_3$, behaves in agreement with theoretical predictions.

CHAPTER I

Theory of the ground state

§ 1 The generalized Bogoliubov transformation and the principle of compensation of "dangerous" diagrams

In the early stage of the development of many-body theory, we used to reduce it to a single-particle theory. The potential representing the interaction with other particles was averaged and treated as the potential of an external field. A real progress was achieved by introducing the self-consistent potential obtained from the variational Hartree–Fock (H–F) method. The H–F method partially takes into account only parallel spin correlations (through an attractive Coulomb interaction). Some theoretical attempts were also made to consider opposite spin correlations (Wigner, 1934), but they did not really go beyond the H–F theory. The new theory of superconductivity shows that for this phenomenon correlations of electrons with opposite spins and momenta are dominant. The mathematical formulation of this fact is, in the variant proposed by Bogoliubov, a special canonical transformation. By means of this transformation we switch from originally considered Fermi particles (electrons) to Fermi quasi-particles. A special Bogoliubov transformation connects fermions with opposite spins and momenta. Correspondingly, instead of the particle vacuum we obtain the quasi-particle vacuum. From the "point of view" of particles, the quasi-particle vacuum is a ground state for correlated pairs considered in the BCS theory. The transformation corresponding to the H–F method does not change the vacuum of particles.

In subsequent papers (Bogoliubov, Tolmachev and Shirkov, 1958; Bogoliubov, 1959) Bogoliubov proposed a general canonical transformation and formulated a new variational principle.

This variational principle is an essential generalization of the H–F principle. It gives a fundamental treatment of the correlations with parallel and antiparallel spins and different momenta. The generalized canonical transformation changes the vacuum of particles to the ground state of correlated pairs (Bloch and Messiah, 1962). Among the solutions of the generalized variational principle one finds also the solution corresponding to the H–F method (describing the ground state) and an essentially different solution describing the superconducting state.

Consider a system of identical Fermi particles. The Hamiltonian of the system in the second quantization representation has the form

$$\hat{H} = \sum_{f,f'} T(f,f')a_f^+ a_{f'} + \tfrac{1}{2} \sum_{f_1,f_2,f_1',f_2'} U(f_1,f_2,f_2',f_1')a_{f_1}^+ a_{f_2}^+ a_{f_2'} a_{f_1'}, \tag{I.1}$$

$$T(f,f') = E(f,f') - \lambda, \qquad \hbar = 1,$$

where λ is the chemical potential, f is a set of indices characterizing one-particle states, the $E(f,f')$ are matrix elements of one-particle states (free particles or particles in an external field), the U are matrix elements describing two-particle interactions, and a_f^+, a_f are fermion creation and annihilation operators. These operators obey the following anticommutation relations

$$\begin{aligned} a_f a_{f'}^+ + a_{f'}^+ a_f &= \delta(f-f'), \\ a_f a_{f'} + a_{f'} a_f &= 0, \\ a_f^+ a_{f'}^+ + a_{f'}^+ a_f &= 0. \end{aligned} \tag{I.2}$$

$\delta(f-f')$ is a Kronecker symbol, e.g. for $f = (\boldsymbol{p}, \sigma)$ where $\boldsymbol{p}$ is the momentum and σ the spin index we have $\delta(f-f') = \delta_{p,p'}\delta_{\sigma,\sigma'}$.

As in the papers mentioned above (Bogoliubov, Tolmachev and Shirkov, 1958; Bogoliubov, 1959), we change to the new Fermi amplitudes α^+, α (creation and annihilation of quasi-particles) by means of a generalized Bogoliubov transformation

$$a_f = \sum_{\nu} (u_{f\nu}\alpha_\nu + v_{f\nu}\alpha_\nu^+). \tag{I.3}$$

The transformation (I.3) is canonical when the functions $\{u, v\}$ are connected by the conditions

$$\sum_{\nu} \{u_{f\nu}u_{f'\nu}^* + v_{f\nu}v_{f'\nu}^*\} = \delta(f-f'), \tag{I.4a}$$

$$\sum_{\nu}\{u_{f\nu}v_{f'\nu}+u_{f'\nu}v_{f\nu}\} = 0, \tag{I.4b}$$

or by the equivalent conditions

$$\sum_{f}\{u^*_{f\nu_1}u_{f\nu_2}+v^*_{f\nu_2}v_{f\nu_1}\} = \delta(\nu_1-\nu_2), \tag{I.5a}$$

$$\sum_{f}\{v^*_{f\nu_1}u_{f\nu_2}+v^*_{f\nu_2}u_{f\nu_1}\} = 0. \tag{I.5b}$$

Now we define the new vacuum state $|C\rangle$ in the α-representation by

$$\alpha_\nu|C\rangle = 0, \quad \langle C|\alpha^+_\nu \equiv 0. \tag{I.6}$$

We transform the Hamiltonian (I,1) by means of the transformation (I.3) and decompose the products $a^+_f a_{f'}$, $a^+_{f_1}a^+_{f_2}a_{f_3}a_{f_4}$ into normal products. The ordering is with respect to the operators α^+, α. (the operators α^+ stand to the left of the operators α, with the minus sign for an odd number of permutations). From Wick's theorem we have

$$a^+_f a_{f'} = N(a^+_f a_{f'})+\underline{a^+_f a_{f'}}, \tag{I.7a}$$

$$\begin{aligned} a^+_{f_1}a^+_{f_2}a_{f'_2}a_{f'_1} &= N(a^+_{f_1}a^+_{f_2}a_{f'_2}a_{f'_1})+\underline{a^+_{f_1}a^+_{f_2}}N(a_{f'_2}a_{f'_1}) \\ &+\underline{a_{f'_2}a_{f'_1}}N(a^+_{f_1}a^+_{f_2})+\underline{a^+_{f_1}a_{f'_1}}N(a^+_{f_2}a_{f'_2})+\underline{a^+_{f_2}a_{f'_2}}N(a^+_{f_1}a_{f'_1}) \\ &-\underline{a^+_{f_1}a_{f'_2}}N(a^+_{f_2}a_{f'_1})-a^+_{f_2}a_{f'_1}N(a^+_{f_1}a_{f'_2})+\underline{a^+_{f_1}a^+_{f_2}}\,\underline{a_{f'_2}a_{f'_1}} \\ &-\underline{a^+_{f_1}a_{f'_2}}\,\underline{a^+_{f_2}a_{f'_1}}+\underline{a^+_{f_1}a_{f'_1}}\,\underline{a^+_{f_2}a_{f'_2}} \end{aligned} \tag{I.7b}$$

where

$$\underline{\hat{A}\hat{B}} \overset{\text{df}}{=} \hat{A}\hat{B}-N(\hat{A}\hat{B}). \tag{I.8}$$

It can be shown that

$$\begin{aligned} \underline{a^+_f a_{f'}} &= \langle C|a^+_f a_{f'}|C\rangle = \sum_{\nu} v^*_{f\nu}v_{f'\nu} \equiv F(f,f') = F^*(f',f), \\ \underline{a_f a_{f'}} &= \langle C|a_f a_{f'}|C\rangle = \sum_{\nu} u_{f\nu}v_{f'\nu} = \Phi(f,f') = -\Phi(f',f). \end{aligned} \tag{I.9}$$

Finally, the Hamiltonian of our system, expressed in terms of the operators α, α^+, has the form

$$\hat{H} = \langle \hat{H} \rangle_0 + \sum_{\nu,\mu} \Omega(\nu, \mu) \alpha_\nu^+ \alpha_\mu + \sum_{\nu,\mu} R_{\nu\mu}^* \alpha_\mu \alpha_\nu$$

$$+ \sum_{\nu,\mu} R_{\nu\mu} \alpha_\nu^+ \alpha_\mu^+ + \tfrac{1}{2} \sum_{(f)} U(f_1, f_2, f_2' f_1') N(a_{f_1}^+ a_{f_2}^+ a_{f_2'} a_{f_1'}) \quad \text{(I.10)}$$

where $\langle \hat{H} \rangle_0 = \frac{1}{2} \sum_{f,f'} [T(f,f') + \zeta(f,f')] F(f,f')$

$$+ \tfrac{1}{2} \sum_{(f)} U(f_1, f_2, f_2', f_1') \Phi^*(f_2 f_1) \Phi(f_2' f_1') \quad \text{(I.11a)}$$

(we have now put $\langle C| \; |C\rangle \equiv \langle \rangle_0$),

$$\Omega(\nu, \mu) = \sum_{f,f'} \zeta(f,f') (u_{f\nu}^* u_{f'\mu} - v_{f\mu}^* v_{f'\nu}) + \sum_{(f)} U(f_1, f_2; f_2', f_1')$$

$$\times [\Phi^*(f_2, f_1) u_{f_1'\mu} v_{f_2'\nu} + \Phi(f_2', f_1') u_{f_1\nu}^* v_{f_2\nu}^*], \quad \text{(I.11b)}$$

$$R_{\nu,\mu} = \sum_{f,f'} \zeta(f,f') u_{f\nu}^* v_{f'\mu} + \tfrac{1}{2} \sum_{(f)} U(f_1, f_2; f_2', f_1')$$

$$\times [\Phi^*(f_2, f_1) v_{f_1'\mu} v_{f_2'\nu} + \Phi(f_2', f_1') u_{f_2\mu}^* u_{f_1\nu}^*], \quad \text{(I.11c)}$$

$$\zeta(f,f') = T(f,f') + \sum_{f_1 f_1'} [U(f_1, f; f', f_1') - U(f_1, f; f_1', f')] F(f_1, f_1'). \quad \text{(I.11d)}$$

From (I.10) we see that only the antisymmetric part of $R_{\mu\nu}$ is relevant since the symmetric part automatically vanishes from the Hamiltonian. The functions $\{u, v\}$ are connected by (I.4a, b) or (I.5a,b). However, these equations do not determine $\{u, v\}$ in a unique way. If the indices f and ν can take on n values, finding the solution $\{u_{f\nu}, v_{f'\mu}\}$ is equivalent to determining $4n^2$ real functions. Formula (I.4a) gives n^2 and formula (I.5a) $n(n+1)$ relations between real functions, a total of $2n^2+n$ relations. Other relations give the so-called equations for the compensation of "dangerous diagrams" (Bogoliubov, Tolmachev and Shirkov, 1958) (the arguments for this expression will be given in § 3). These equations follow from the requirement that in the Hamiltonian (I.10) terms leading to the creation of pairs of fermions, i.e. terms proportional to the products $\alpha_\mu \alpha_\nu$, do not appear. The compensation equations have the form

$$R_{\mu\nu} \equiv \langle \alpha_\mu \alpha_\nu \hat{H} \rangle_0 = 0. \quad \text{(I.12)}$$

The vanishing of the antisymmetric part of $R_{\mu\nu}$ additionally gives $n(n-1)$ relations between real functions. Therefore, eqns. (I.4a,b and the compensation equations give $3n^2$ equations for $4n^2$ real functions (n^2 functions or additional relations between these functions can be chosen quite arbitrarily). We now prove that the functions $\{u, v\}$ are defined up to an arbitrary canonical transformation $\{\varphi\}$

$$\alpha_\nu = \sum_g \varphi_{\nu g} \beta_g, \qquad \alpha_\nu^+ = \sum_g \varphi_{\nu g}^* \beta_g^+$$

where φ, φ^* are connected by means of the conditions

$$\sum_g \varphi_{\nu_1 g}^* \varphi_{\nu_2 g} = \delta(\nu_1 - \nu_2), \qquad \sum_\nu \varphi_{\nu g_1}^* \varphi_{\nu g_2} = \delta(g_1 - g_2)$$

(n^2 independent conditions for $2n^2$ real functions, i.e. n^2 functions can be chosen arbitrarily). After performing the canonical transformation $\{\varphi\}$, eqn. (I.3) can be written in the form

$$a_f = \sum_g \{\bar{u}_{fg} \beta_g + \bar{v}_{fg} \beta_g^+\} \tag{I.13a}$$

where

$$\bar{u}_{fg} = \sum_\nu u_{f\nu} \varphi_{\nu g}, \qquad \bar{v}_{fg} = \sum_\nu v_{f\nu} \varphi_{\nu g}^*. \tag{I.13b}$$

Comparing (I.13a) and (I.3) we see that the functions $\{\bar{u}_{fg}, \bar{v}_{fg}\}$, as well as the functions $\{u_{f\nu}, v_{f\nu}\}$, can be determined from eqns. (I.4a,b) and (I.12) with an arbitrary choice of n^2 real functions (or n^2 connexions among them). Hence we have shown that the eqns. (I.4a,b), (I.12) define the functions $\{u, v\}$ of the canonical transformation $\{\varphi\}$ which changes the choice of the n^2 arbitrary functions.

From the definition (I.9) and the condition that $\{\varphi\}$ must be a canonical transformation we can see that the functions $\{F, \Phi\}$ do not depend on the choice of $\{\varphi\}$. From eqns. (I.12) for $\{u, v\}$ we can obtain the corresponding equations for $\{F, \Phi\}$ defined by (I.9).

We introduce the quantities R_1, R_2:

$$\begin{aligned} R_1(f_3, f_4) &\equiv \sum_{\nu, \mu} R_{\nu\mu} u_{f_3 \nu} u_{f_4 \mu} = 0, \\ R_2(f_3, f_4) &\equiv \sum_{\nu, \mu} R_{\nu\mu}^* v_{f_3 \nu} v_{f_4 \mu} = 0. \end{aligned} \tag{I.14}$$

It is easy to show that

$$
\begin{aligned}
&R_1(f_4,f_3)-R_1(f_3,f_4)+R_2(f_4,f_3)-R_2(f_3,f_4)\\
&\quad=\sum_{f_1,f_2} U(f_3,f_4;f_2,f_1)\Phi(f_1,f_2)+\sum_{f'}[\zeta(f_3,f')\Phi(f',f_4)\\
&\quad+\zeta(f_4,f')\Phi(f_3,f')]-\sum_{f'f_1,f_2}\{U(f',f_4;f_2,f_1)F(f',f_3)\\
&\quad+U(f_3,f';f_2,f_1)F(f',f_4)\}\Phi(f_1,f_2)\equiv A(f_3,f_4)=0.
\end{aligned}
\tag{I.15}
$$

The state $|C\rangle$ was defined as the vacuum state for the quasi-particles corresponding to the operators α, α^+. For the particles (a, a^+) the state $|C\rangle$ is a trial state, depending on $\{u, v\}$, which we obtain from the variational principle stating that there must be a minimum for the mean value of the Hamiltonian (I.1). Thus $|C\rangle$ is not an eigenstate of (I.1) and of the total particle number operator $\hat{N}=\sum_f a_f^+ a_f$. As we wish to consider a ground state with a given number N of particles, we must add to (I.1) the term $-\lambda\hat{N}$ where λ is a Lagrange multiplier. We find it from the condition

$$
\langle\hat{N}\rangle_0 = N. \tag{I.16}
$$

It was for this reason that we wrote the Hamiltonian in the form (I.1) where the chemical potential is the Lagrange multiplier.

In the next section we will show that the principle of compensation of "dangerous diagrams" is equivalent to the variational principle which minimizes the mean value of the Hamiltonian.

§ 2 The generalized variational principle in many-body theory

Let us write the expression for the mean value (I.11a) in a form which displays its explicit dependence on F and Φ (where $\{F, \Phi\}$ are defined by (I.9)):

$$
\begin{aligned}
\langle\hat{H}\rangle_0 &= \tfrac{1}{2}\sum_{f,f'} T(f,f')\,[F(f,f')+F^*(f',f)]\\
&\quad+\tfrac{1}{2}\sum_{f_1,f_2,f_1',f_2'}[U(f_1',f_2;f_2',f_1)\\
&\quad-U(f_1',f_2;f_1,f_2')]\,F^*(f_1,f_1')\,F(f_2,f_2')
\end{aligned}
\tag{I.17}
$$

$$+\frac{1}{2}\sum_{f_1,f_2,f'_1,f'_2} U(f_1,f_2;f'_2,f'_1)\Phi(f'_2,f'_1)\Phi^*(f_2,f_1).$$

In various papers (Bogoliubov, 1959; Valatin, 1958; Bogoliubov and Soloviev, 1959) the functions $\{F,\Phi\}$ are introduced only for simplifying the notation [$\{u,v\}$ are the unknown functions, determined by the vanishing of the first variation of the mean value (I.17) considered as a function of $\{u,v\}$, while taking into account the subsidiary conditions (I.4a,b)]. In order that the solution obtained really corresponds to a minimum, the second variation of $\langle\hat{H}\rangle_0$, with the additional terms from the conditions (I.4a,b) must be positive. The second variation is a quadratic form in the variations $(\delta u,\ \delta v)$, with coefficients depending on the functions $\{u,v\}$.

We will now give the formulation of the generalized variational principle (Galasiewicz, 1963a). The basic idea of this formulation stems from the fact that the mean value $\langle\hat{H}\rangle_0$ is a quadratic form of the functions $\{F,\ \Phi\}$. These functions have to be determined from the condition of the vanishing of the first variation of (I.17) and with the help of additional equations which correspond to the equations (I.4a,b) in the method given by Bogoliubov (1959) or Valatin (1958). We give now a short derivation of these equations. We multiply (I.5a) by $v_{f_1\nu_2}v^*_{f_2\nu_1}$ and (I.5b) by $v_{f_2\nu_1}v_{f_1\nu_2}$ and sum over ν_1,ν_2. After using (I.9) we have

$$\xi(f_1,f_2)\equiv\sum_f F^*(f,f_1)F(f,f_2)+\sum_f \Phi^*(f,f_1)\Phi(f,f_2)$$
$$-\frac{1}{2}[F(f_1,f_2)+F^*(f_2,f_1)]=0, \tag{I.18a}$$

$$\eta(f_1,f_2)\equiv\sum_f [F^*(f_1,f)\Phi(f_2,f)+F^*(f_2,f)\Phi(f_1,f)]=0. \tag{I.18b}$$

We find the functions $\{F,\ \Phi\}$ from the variational principle

$$\delta W[F,\ \Phi]=0 \tag{I.19}$$

where

$$W[F,\ \Phi]\equiv\langle\hat{H}\rangle_0+\sum_{f_1,f_2}\lambda(f_1,f_2)\xi(f_1,f_2)$$
$$+\sum_{f,f_1,f_2}\mu^*(f_1,f_2)F^*(f_1,f)\Phi(f_2,f)$$
$$+\sum_{f,f_1,f_2}\mu(f_1,f_2)\Phi^*(f_1,f)F(f_2,f) \tag{I.20}$$

$$+\sum_{f_1,f_2}\gamma^*(f_1,f_2)\Phi(f_1,f_2)+\sum_{f_1,f_2}\gamma(f_1,f_2)\Phi^*(f_1,f_2)$$
$$+\sum_{f_1,f_2}\alpha(f_1,f_2)\,[F(f_1,f_2)-F^*(f_2,f_1)].$$

For the Lagrange multipliers λ, μ, α, γ we have the following relations:

$$\lambda(f_1,f_2)=\lambda^*(f_2,f_1),\quad \alpha(f_1,f_2)=-\alpha^*(f_2,f_1),$$
$$\mu(f_1,f_2)=\mu(f_2,f_1),\quad \gamma(f_1,f_2)=\gamma(f_2,f_1),$$

which follow from the properties of F, Φ. We now treat the variations δF, $\delta\Phi$, δF^*, $\delta\Phi^*$ as independent. In (I.19) the coefficients of these variations must vanish. After elimination of α and γ we have

$$\zeta(f_1,f_2)-\lambda(f_1,f_2)+\sum_f\lambda(f_1,f)F(f_2,f)+\sum_f\lambda(f,f_2)F(f,f_1)$$
$$+\sum_f\mu^*(f,f_2)\Phi(f,f_1)+\sum_f\mu(f,f_1)\Phi^*(f,f_2)=0, \tag{I.21a}$$

$$\sum_{f_1',f_2'}U(f_1',f_2';f_1,f_2)\Phi^*(f_2',f_1')+\sum_f\lambda(f,f_2)\Phi(f_1f,)$$
$$-\sum_f\lambda(f,f_1)\Phi(f_2,f)+\sum_f\mu^*(f,f_1)F^*(f,f_2)$$
$$-\sum_f\mu^*(f,f_2)F^*(f,f_1)=0. \tag{I.21b}$$

Moreover, we obtain the equations which are the complex conjugate of (I.21a,b). Now we wish to eliminate the Lagrange multipliers λ and μ. We proceed as follows:

(i) in (I.21a) we replace f_2 by f', then we multiply the equation by $\Phi(f',f_2)$ and sum over f';

(ii) in (I.21a) we replace f by f', multiply the equation by $\Phi(f_1,f')$ and sum over f';

(iii) in the equation complex conjugate to (I.21b) we replace f_2 by f' and f_1 by f_2, multiply the equation by $F(f',f_1)$ and sum over f';

(iv) we add all these equations to the equation which is the complex conjugate of (I.13b). After using (I.9) we obtain

$$A(f_1,f_2)=0, \tag{I.22}$$

i.e. formula (I.15).

So the principle of the compensation of "dangerous diagrams" is equivalent to the variational principle in many-body theory.

The multipliers λ, μ can be eliminated in another way:

(i) in (I.21a) we replace f_2 by f', multiply the equation by $F(f_2, f')$ and sum over f';

(ii) in (I.21a) we replace f_1 by f', multiply the equation by $-F(f', f_1)$ and sum over f';

(iii) in the equation complex conjugate to (I.21b) we replace f_2 by f', multiply the equation by $\Phi^*(f_2, f')$ and sum over f';

(iv) in (I.21b) we replace f_2 by f' and f_1 by f_2, multiply the equation by $-\Phi(f_1, f')$ and sum over f';

(v) we sum all these equations and after using (I.9) we have

$$B(f_1, f_2) \equiv \sum_f [\zeta(f_1, f) F(f_2, f) - \zeta(f, f_2) F(f, f_2)] + \sum_{f, f_1', f_2'} U(f, f_1; f_2', f_1') \Phi(f_2', f_1') \Phi^*(f_2, f) - \sum_{f, f_1', f_2'} U(f_1', f_2'; f_2, f) \Phi^*(f_2', f_1') \Phi(f_1, f) = 0. \tag{I.23}$$

Equations (I.22, 23) are not independent. We can express their dependence in the following form (Bogoliubov, 1959):

$$\sum_{f_1, f_2} \{\Phi(f_1, f_2') \Phi^*(f_2, f_1') B(f_1, f_2) + \Phi(f_1, f_2') F(f_1', f_2) A^*(f_1, f_2) + \Phi^*(f_2, f_1') F(f_1, f_2') A(f_1, f_2) + F(f_1, f_2') F(f_1', f_2) B^*(f_1, f_2)\}. \tag{I.24}$$

We would like to show now that among the solutions of the variational principle discussed here there are solutions given by the H–F method. Consider a system of orthogonal and normalized functions $\varphi_{f\nu}$, i.e.

$$\sum_\nu \varphi^*_{f\nu} \varphi_{f'\nu} = \delta(f - f'). \tag{I.25}$$

We do not specify ν, but divide the set of these indices into two separate parts F and G. Let n indices belong to F (corresponding to the Fermi sphere), the remaining indices belonging to G. We take

$$\begin{aligned} &u_{f\nu} = 0, \quad v_{f\nu} = v_{f\omega} = \varphi_{f\omega} \quad \text{for} \quad \nu = \omega \in F, \\ &u_{f\nu} = \varphi_{f\nu}, \quad v_{f\nu} = 0 \quad \text{for} \quad \nu \in G. \end{aligned} \tag{I.26}$$

We see that for the functions (I.26) the condition (I.25) is identical to (I.4a) and (I.4b) is fulfilled by (I.26) as an identity. Moreover, for (I.26) we have

$$\Phi(f_1,f_2)\equiv 0, \qquad F(f_1,f_2)=\sum_{\omega\varepsilon F}\varphi^*_{f_1\omega}\varphi_{f_2\omega}. \tag{I.27}$$

The variational principle for the functions (I.26) with the subsidiary condition (I.25) is equivalent to the H–F principle. We see that the new variational principle should describe more correlations among particles and its solutions should contain more physical content. Taking into account (I.27) we obtain for the H–F case the mean value (I.16) in the form

$$\begin{aligned}\langle\hat{H}\rangle_0 = \tfrac{1}{2}\sum_{f,f'}T(f,f')\,[F(f,f')+F^*(f,f')]\\ -\sum_{f_1,f_2,f_1',f_2'}[U(f_1'f_2;f_2',f_1)\\ -U(f_1',f_2;f_1,f_2')]F^*(f_1,f_1')F(f_2,f_2').\end{aligned} \tag{I.28}$$

Equations (I.23) now take the form

$$\sum_f[\zeta(f_1,f)F(f_2,f)-\zeta(f,f_2)F(f,f_1)]=0 \tag{I.29}$$

and the subsidiary conditions (I.18a) the form

$$\sum_f F(f_1,f)F(f,f_2)-F(f_1,f_2)=0. \tag{I.30}$$

Now we will consider a special solution of (I.18a,b) and eqns. (I.15) (or (I.29)), different from the H–F solution, for which δW is equal to zero. Whether this solution actually corresponds to a minimum of $\langle\hat{H}\rangle_0$ is a question to be solved after we examine the sign of $\delta^2 W$. This problem will be considered in § 4.

§ 3 Solution of the general variational principle describing the superconducting state

We wish to prove that among the solutions of eqns. (I.18a, b), (I.22) is a solution describing the ground state ($\Theta = 0$°K) of the Fermi system.

As quantum indices characterizing one-particle states we take $f=(\boldsymbol{p},\sigma)$ where $\boldsymbol{p}$ is the momentum and σ is the spin index. Moreover, we put in the Hamiltonian (I.1):

$$E(f,f') = E(p)\,\delta(f-f'),$$
$$T(f,f') = [E(p)-\lambda]\,\delta(f-f') = \varepsilon(p)\,\delta(f-f'), \qquad \text{(I.31a)}$$
$$U(f_1,f_2;f_2',f_1')$$
$$= \frac{1}{V}I(\boldsymbol{p}_1,\boldsymbol{p}_2;\boldsymbol{p}_2',\boldsymbol{p}_1')\,\delta(\boldsymbol{p}_1+\boldsymbol{p}_2-\boldsymbol{p}_1'-\boldsymbol{p}_2')\,\delta(\sigma_1-\sigma_1')\,\delta(\sigma_2-\sigma_2')$$

where V is the volume, I is real and invariant with respect to the transformation $\boldsymbol{p}\to-\boldsymbol{p}$. The Hamiltonian (I.1) now takes the form

$$\hat{H} = \sum_{p,\sigma}(E(p)-\lambda)a^+_{p\sigma}a_{p\sigma}+\frac{1}{2V}\sum_{p_1+p_2=p_1'+p_2'} I(\boldsymbol{p}_1,\boldsymbol{p}_2;\boldsymbol{p}_2',\boldsymbol{p}_1')$$
$$\times a^+_{p_1\sigma_1}a^+_{p_2\sigma_2}a_{p_2'\sigma_2'}a_{p_1'\sigma_1'}. \qquad \text{(I.31b)}$$

We look for a solution of the form

$$F(f,f') = F(p)\,\delta(f-f'), \quad \Phi(f,f') = \Phi(f)\,\delta(f+f'),$$
$$\Phi(p,-) = \Phi(p), \quad \Phi(p,+) = -\Phi(p) \qquad \text{(I.32)}$$

where $F(p)$ and $\Phi(p)$ are invariant with respect to $\boldsymbol{p}\to-\boldsymbol{p}$. The functions (I.32) satisfy (I.18b) identically and (I.18a) takes the form

$$F(p) = F(p)^2+\Phi(p)^2. \qquad \text{(I.33)}$$

Equation (I.22) now has the form

$$2\zeta(p)\Phi(p)+[1-2F(p)]\frac{1}{V}\sum_{p'}I(\boldsymbol{p},-\boldsymbol{p};-\boldsymbol{p}',\boldsymbol{p}')\Phi(p') = 0 \qquad \text{(I.34)}$$

where

$$\zeta(p) = E(p)-\lambda+\frac{1}{V}\sum_{p'}\{2I(\boldsymbol{p},\boldsymbol{p}';\boldsymbol{p}',\boldsymbol{p})-I(\boldsymbol{p},\boldsymbol{p}';\boldsymbol{p},\boldsymbol{p}')\}F(p')$$
$$= \tilde{E}(p)-\lambda. \qquad \text{(I.35)}$$

Hence the functions $F(p)$, $\Phi(p)$ are found from eqns. (I.33, 34)

Defining

$$C(p) = -\frac{1}{V}\sum_{p'}I(\boldsymbol{p},-\boldsymbol{p};-\boldsymbol{p}',\boldsymbol{p}')\Phi(p'), \qquad \text{(I.36)}$$

from (I.33, 34) we obtain the solution

$$F(p) = \frac{1}{2}\left\{1-\frac{\zeta(p)}{\Omega(p)}\right\}, \quad \Phi(p) = \frac{C(p)}{2\Omega(p)},$$
$$\Omega(p) = \sqrt{\zeta^2(p)+C^2(p)}. \qquad \text{(I.37a)}$$

Another "trivial" solution has the form

$$\Phi(p) = 0 \to C(p) = 0, \quad F(p) \equiv \Theta_F(p) = \begin{cases} 1 \text{ for } p \leqslant p_F, \\ 0 \text{ for } p > p_F. \end{cases} \tag{I.37b}$$

The function $C(p)$ is obtained from the equation

$$C(p) = -\frac{1}{V}\sum_{p'} I(\boldsymbol{p}, -\boldsymbol{p}; -\boldsymbol{p}', \boldsymbol{p}')\frac{C(p')}{2\Omega(p')}. \tag{I.38}$$

It can be shown that looking for the solution F, Φ in the form (I.32) is equivalent to taking the transformation (I.3) with coefficients

$$\begin{aligned} u_{f\nu} &= u(p)\delta(f-\nu), \quad v_{f\nu} = v(f)\delta(f-\nu), \\ u(p)^2 + v(p)^2 &= 1, \quad v(p, +) = v(p) = -v(p, -). \end{aligned} \tag{I.39}$$

In this case the general Bogoliubov transformation gives a special transformation of the form

$$a_{p\sigma} = u(p)\alpha_{p\sigma} + v(p, \sigma)\alpha^{+}_{-p, -\sigma}, \tag{I.40}$$

(see also Valatin, 1958) and the compensation equation now has the form

$$\langle \alpha_{-p, \sigma}\alpha_{p\sigma}\hat{H}\rangle_0 \equiv R_{-p, -\sigma; p, \sigma} = 0. \tag{I.41}$$

If an external inhomogeneous field acts on the system, the first term in the Hamiltonian (I.1) is not diagonal. For example, for a weak electromagnetic field (we can omit terms proportional to A_2, where A is the vector potential) a term of the form

$$-\frac{e}{2m}\{(\boldsymbol{p}A) + (A\boldsymbol{p})\} \tag{I.42}$$

appears in the Hamiltonian. In the second quantization representation this term has the form

$$\sum_{p, q, \sigma} (\boldsymbol{q} - 2\boldsymbol{p}) A(q) a^{+}_{-p\sigma} a_{q-p, \sigma} \tag{I.43}$$

where $A(q)$ is a Fourier component of $A(r)$. After applying the special Bogoliubov transformation (I.40) we obtain the additional term

$$\sum_{p, q, \sigma} (\boldsymbol{q} - 2\boldsymbol{p}) A(q) v(p, \sigma) u(q-p) \alpha_{p, -\sigma}\alpha_{q-p, \sigma}. \tag{I.44}$$

The principle of the compensation of dangerous diagrams leads to the vanishing of terms contributing to the creation of pairs with momenta $q \neq 0$. Hence

$$v(p, \sigma)u(q-p) = 0. \tag{I.45}$$

(I.45) and (I.39) give the trivial solution for the functions $\{u, v\}$. Therefore, a general canonical transformation has been proposed which allows for compensated terms leading to the creation of pairs with $\boldsymbol{q} \neq 0$ in a non-trivial way.

For the special case defined by (I.39), the quantity $\Omega(\mu, \nu)$ described by (I.11b) takes the form

$$\Omega(\nu, \mu) = \Omega(p)\,\delta(\nu-\mu)$$

where

$$\Omega(p) = \sqrt{\zeta^2(p)+C^2(p)}. \tag{I.46}$$

Thus (I.46) is identical with Ω appearing in formula (I.37). In this case, by considering the principle of compensation and by omitting the last term describing the interaction of quasi-particles (important in the problem of the collective oscillations), the Hamiltonian (I.10) describes free quasi-particles with an energy $\Omega(p)$ given by (I.46). If $\tilde{E}(p) \approx E(p)$, then $\zeta(p)$ is the energy of the particles on a scale where the Fermi energy is zero. As we will show later, $\lambda = \tilde{E}(p_F)$, i.e. λ is the energy at the Fermi surface. Consider now the energy spectrum of the quasi-particles, corresponding to the operators α, α^+, defined by (I.46). It is a spectrum with an energy gap characteristic of superconductors. (Considering, in (I.10), the last term by means of a perturbation theory method, does not change the one-particle spectrum in an essential way.) In order to know the spectrum (I.46) one must calculate the quantity $C(p)$ from (I.38). One can try to solve this equation after specifying the matrix elements $I(p_1, p_2, p_3, p_4)$, i.e. after stating which interaction is the most important for superconductivity. In their paper, BCS (Bardeen, Cooper and Schrieffer, 1957b) take for I the expression describing the interaction between electrons, caused by the exchange of virtual phonons. For momenta near p_F it is an attractive interaction of the form

$$I(\boldsymbol{p}_1, \boldsymbol{p}_2; \boldsymbol{p}_2, \boldsymbol{p}_1') = -g^2 \frac{\omega^2(\boldsymbol{p}_1'-\boldsymbol{p}_1)}{\omega^2(\boldsymbol{p}_1'-\boldsymbol{p}_1)-[E(p_1)-E(p_1')]^2}\,\delta(\boldsymbol{p}_1+\boldsymbol{p}_2-\boldsymbol{p}_1'-\boldsymbol{p}_2'). \tag{I.47}$$

Bogoliubov (1958a,b) starts from Fröhlich's Hamiltonian

$$\hat{H} = \sum_{p,\sigma} (E(p)-\lambda) a^+_{p\sigma} a_{p\sigma} + \sum_q \omega(q) b^+_q b_q$$
$$+g \sum_{p,q,\sigma} \left[\frac{\omega(q)}{2V}\right]^{1/2} (a^+_{p,\sigma} a_{p+q,\sigma} b^+_q + a^+_{p+q,\sigma} a_{p,\sigma} b_q) \quad \text{(I.48)}$$

where g is the coupling constant for the phonon field and b^+, b are the phonon creation and anihilation operators. After performing the transformation (I.40), the Hamiltonian (I.48) takes the form

$$\hat{H} = U + \hat{H}_0 + \hat{H}_{\text{Int}},$$
$$\hat{H}_{\text{Int}} = \hat{H}_3 + \hat{H}_1 + \hat{H}_2, \quad \text{(I.49)}$$

where

$$U = 2\sum_k (E(k)-\lambda) v_k^2,$$

$$\hat{H}_0 = \sum_{k,\sigma} (E(k)-\lambda)(u_k^2 - v_k^2) \alpha^+_{k\sigma} \alpha_{k\sigma} + \sum_q \omega(q) b^+_q b_q,$$

$$\hat{H}_3 = 2\sum_k (E(k)-\lambda) u_k v_k (\alpha^+_{k+} \alpha_{-k-} + \alpha_{-k-} \alpha_{k+}),$$

$$\hat{H}_1 = \sum_{\substack{k,k',q \\ k-k'=q}} g \left\{\frac{\omega(q)}{2V}\right\}^{1/2} [u_k v_{k'} (\alpha^+_{k+} \alpha^+_{-k'-} + \alpha^+_{-k'+} \alpha^+_{k-}) \quad \text{(I.50)}$$
$$+ u_{k'} v_k (\alpha_{-k-} \alpha_{k'+} + \alpha_{k'-} \alpha_{-k+})] b^+_q + \text{h.c.},$$

$$\hat{H}_2 = \sum_{\substack{k,k',q \\ k-k'=q}} g \left\{\frac{\omega(q)}{2V}\right\}^{1/2} [u_k u_{k'} (\alpha^+_{k+} \alpha_{k'+} + \alpha^+_{k-} \alpha_{k'-})$$
$$- v_k v_{k'} (\alpha^+_{-k'-} \alpha_{-k-} + \alpha^+_{-k'+} \alpha_{-k+})] b^+_q + \text{h.c.},$$

$$\omega(q) = \omega(|k-k'|).$$

The vacuum state of the system is defined by $|C\rangle|\text{ph}\rangle$, where $|\text{ph}\rangle$ is the phonon vacuum. We see that the term $\hat{H}_3$ in the Hamiltonian (I.49) acting on the vacuum leads to the creation of a pair of particles with opposite spins and momenta without, however, creating a phonon. If we wish to calculate the contribution of $\hat{H}_3$ to the energy of the vacuum state we must consider intermediate states in which there are no phonons. Then the energy denominators in the perturbation expansion are of the form

$$2(E(k)-\lambda)(u_k^2-v_k^2) = 2\varepsilon_1(k). \tag{I.51}$$

If $\lambda \approx E(k_F)$ (as will be demonstrated later), the denominators are singular when $k \approx k_F$. Hence the name "dangerous" applied to the diagrams leading to the creation of a pair of quasi-particles with opposite spins and momenta. The process of the creation of a pair can be described also by means of the common action on $|C\rangle$ of the operators $\hat{H}_1$ and $\hat{H}_2$. For a quasi-particle of given spin and momentum, the second quasi-particle, with opposite spin, can, by absorbing or emitting one phonon, undergo a transition to the state with momentum opposite to the momentum of the first quasi-particle. This process is of second order in the coupling constant. One can also consider the higher order processes in which, by successive creation and absorption of phonons, one arrives at the same final pair state. Calculating the energy of the ground state one sees that the process of virtual creation of a pair $(\boldsymbol{p}, +; -\boldsymbol{p}, -)$ is dangerous for the same reasons as the process caused by $\hat{H}_3$. However, the terms describing the contribution of $\hat{H}_1$, $\hat{H}_2$ to this process contain unknown functions u_k, v_k.

We can impose on these functions new conditions following from the requirement that the coefficient in the expression $R_{k,\sigma,-k,-\sigma}\,\alpha_{k\sigma}^{+}\,\alpha_{-k-\sigma}^{+}|C\rangle$ describing the contribution of $\hat{H}_3$ and $\hat{H}_1$, $\hat{H}_2$ to the process of the creation of a pair must vanish. The equation

$$R_{k,\sigma;-k,-\sigma} = 0 \tag{I.52}$$

so obtained is called by Bogoliubov the equation of compensation of dangerous diagrams.

Let us consider the diagram arising from $\hat{H}_3$ (Fig. 5*a*) and the compensating diagrams which are second order in the coupling constant (Fig. 5*b*,*c*)

Those diagrams were introduced by Hugenholtz (quoted here after Bogoliubov (1958a, b)). Straight lines denote particles and wavy lines phonons. At point 1 (Fig. 5*b*) $\hat{H}_1$ contributes $\sim u_{k'}v_k(\alpha_{k'+}^{+}\alpha_{-k-}^{+} + \alpha_{-k+}^{+}\alpha_{k'-}^{+})b_q$ and at point 2, $\hat{H}_2$ contributes $\sim b_q^{+}(u_k u_{k'}\alpha_{k+}^{+}\alpha_{k'+} - v_k v_{k'}\alpha_{-k'+}^{+}\alpha_{-k+})$. On the other hand, at point 1 (Fig. 5*c*) $\hat{H}_2$

contributes $\sim b_q^+(u_k u_{k'} \alpha_{k-}^+ \alpha_{k'-} - v_k v_{k'} \alpha_{-k'-}^+ \alpha_{-k-})$ and at point 2, $\hat{H}_1$ contributes $\sim u_k v_{k'}(\alpha_{k'+}^+ \alpha_{-k-}^+ + \alpha_{-k+}^+ \alpha_{k'-}^+) b_q$. The same terms come from the processes in Fig. 5*b* and 5*c*. The compensation equation (I.52) then takes the form

$$2(\tilde{E}_1(k)-\lambda)u_k v_k = \frac{u_k^2-v_k^2}{V}\sum_{k'} g\,\frac{\omega(k-k')}{\omega(k-k')+\varepsilon_1(k_1)+\varepsilon_1(k')}\,u_{k'}v_{k'}, \tag{I.53}$$

$$\tilde{E}_1(k) = E(k) - \frac{1}{2V}\sum_{k'} g^2\,\frac{\omega(k-k')}{\omega(k-k')+\varepsilon_1(k)+\varepsilon_1(k')}\,(u_{k'}^2-v_{k'}^2),$$

$$\varepsilon_1(k) = (E(k)-\lambda)\,(u_k^2-v_k^2).$$

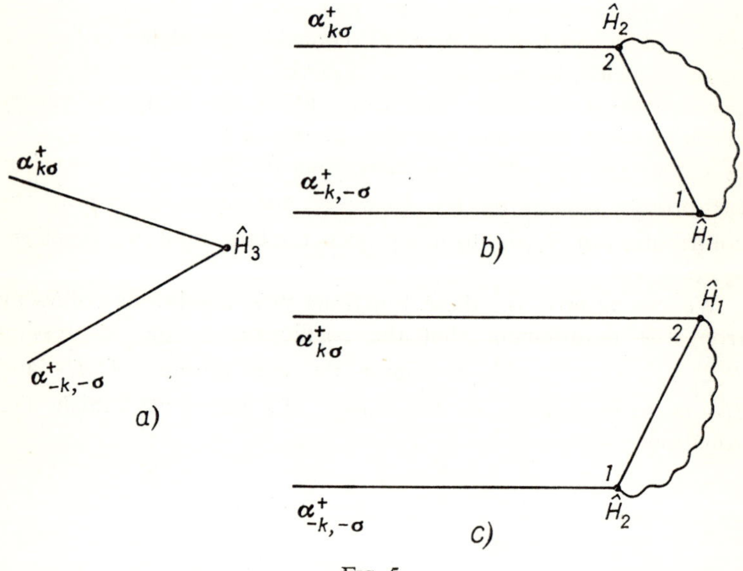

FIG. 5

From (I.39) and (I.9) we have

$$\Phi(p) = -u(p)v(p), \quad F(p) = v^2(p), \quad u^2(p)-v^2(p) = 1-2F(p).$$

We see that the eqn. (I.53) has the solution

$$u(k) \equiv \Theta_G(k) = \begin{cases} 1, \\ 0, \end{cases} \quad v(k) \equiv \Theta_F(k) = \begin{cases} 0, & E(k) > E_F, \\ 1, & E(k) < E_F, \end{cases} \tag{I.54}$$

which corresponds to the "trivial" solution (I.37b). This solution is consistent with the condition (I.39).

Let us now approximate (I.53) by substituting the trivial solution (I.54) in the expression for $\varepsilon_1(k)$. We then have

$$u_k^2 - v_k^2 = \begin{cases} 1, & E(k) > E_F \approx \lambda, \\ -1, & E(k) < E_F \approx \lambda. \end{cases} \tag{I.55}$$

Therefore, for both $k < k_F$ and $k > k_F$ the expression for $\varepsilon_1(k)$ is positive

$$\varepsilon_1(k) = \big(E(k) - \lambda\big)\,(u_k^2 - v_k^2) = |E(k) - \lambda| = |\varepsilon(k)|. \tag{I.56}$$

In the weak interaction limit we can put in (I.35) and (I.53)

$$E(p) - \lambda \approx \tilde{E}(p) - \lambda \approx \tilde{E}_1(p) - \lambda \approx \zeta(p).$$

The compensation equation (I.53) can be written in the form

$$2\zeta(p)\Phi(p) - \frac{(1-2F(p))}{V} \sum_{p'} g^2 \frac{\omega(p-p')}{\omega(p-p') + |E(p)-\lambda| + |E(p')-\lambda|}\Phi(p') = 0.$$

In order that the compensation equations corresponding to the Hamiltonians (I.31b) and (I.48) lead to the same results one must put

$$I(p_1, p_2; p_2', p_1') = -g^2 \frac{\omega(p_1'-p_1)\delta(p_1+p_2-p_1'-p_2')}{\omega(p_1'-p_1) + |E(p_1)-\lambda| + |E(p_1')-\lambda|} \tag{I.57}$$

for $p \approx p_F$ (Bogoliubov, Tolmachev and Shirkov, 1958). Hence, if $E - \lambda \ll \omega$, a good approximation of this expression is

$$I(p_1, p_2; p_4, p_3) = \begin{cases} -g^2 & \text{for} \quad p \approx p_i \approx p_F, \\ 0 & \text{for} \quad p \approx p_i \neq p_F, \end{cases} \tag{I.58}$$

Now, having I given by (I.58), we can try to solve (I.38). If $p_i \approx p_F$ then $\tilde{E}(p_i) \approx \tilde{E}(p_F)$. If we take into account that in the effective attraction mechanism the most important processes are one-phonon processes changing the energy of quasi-particles by $\pm\omega$ (ω is the mean energy of the phonon) then $\tilde{E}(p)$ will be contained in the spherical layer $\tilde{E}(p_F) - \omega \leqslant \tilde{E}(p) \leqslant \tilde{E}(p_F) + \omega$. We assume the formula $\lambda = \tilde{E}(p_F)$ which will be proved later.

Substituting (I.58) into (I.38) we see that $C(p)$ does not depend on p. Hence we look for solutions of the form

$$C(p) = C \quad \text{for} \quad p \quad \text{for which} \quad \lambda-\omega \leqslant E(p) \leqslant \lambda+\omega,$$
$$C(p) = 0 \quad \text{otherwise.}$$

Equation (I.38) now takes the form

$$1 = \frac{g^2}{2V}\sum_p \frac{1}{\sqrt{(\tilde{E}(p)-\lambda)^2+C^2}}. \tag{I.59}$$

We replace the summation by integration over $\boldsymbol{p}$

$$\frac{1}{V}\sum_p \to \frac{1}{(2\pi)^3}\int \mathrm{d}^3\boldsymbol{p}. \tag{I.60}$$

For a thin layer in the neighbourhood of the Fermi surface (if the integral depends only on $p = |\boldsymbol{p}|$) we have $\int \mathrm{d}^3p = 4\pi p_F^2 \int \mathrm{d}p$. Hence

$$\frac{g^2}{2}\frac{p_F^2}{(2\pi)^2}\int_{p_F-\Delta}^{p_F+\Delta} \frac{\mathrm{d}p}{\sqrt{(\tilde{E}(p)-\lambda)^2+C^2}} = 1, \tag{I.61}$$

$\Delta = \dfrac{\omega}{E'}$ (see below). For $p \approx p_F$ we have

$$\tilde{E}(p) \cong \lambda + \left(\frac{\mathrm{d}\tilde{E}}{\mathrm{d}p}\right)_{p=p_F}(p-p_F) = \lambda+\tilde{E}'(p-p_F). \tag{I.62}$$

We introduce in (I.61) a new integration parameter by means of the formula

$$\mathrm{d}p = \frac{1}{\tilde{E}'}\,\mathrm{d}\tilde{E}. \tag{I.63}$$

On the one hand, $\lambda-\omega \leqslant \tilde{E} \leqslant \lambda+\omega$, on the other hand, $p_F-\Delta \leqslant p \leqslant p_F+\Delta$. Taking into account (I.62) we have $\lambda-\tilde{E}'\Delta \leqslant \tilde{E} \leqslant \lambda+\tilde{E}'\Delta$ and we see that $\Delta = \dfrac{\omega}{\tilde{E}'}$. Finally, we can write (I.61) in the form

$$\frac{2g^2p_F^2}{(2\pi)^2\tilde{E}'}\int_0^{\frac{\omega}{c}} \frac{\mathrm{d}x}{\sqrt{x^2+1}} = 1. \tag{I.64}$$

From here

$$C = -\frac{2\omega e^{\frac{1}{\varrho}}}{1-e^{\frac{2}{\varrho}}} \cong 2\omega e^{-\frac{1}{\varrho}} \tag{I.65}$$

where

$$\varrho = g^2 \frac{\mathrm{d}n}{\mathrm{d}E}, \quad \frac{\mathrm{d}n}{\mathrm{d}E} = \frac{1}{2\pi_2}\left(\frac{p^2}{\frac{\mathrm{d}\tilde{E}}{\mathrm{d}p}}\right)_{p=p_F} = \frac{1}{V}\left\{\frac{V}{(2\pi)^3}\frac{4\pi p^2 \mathrm{d}p}{\mathrm{d}\tilde{E}}\right\}_{p=p_F}. \tag{I.66}$$

We assume that the density of states is constant in the small interval $(-\omega, \omega)$ around the Fermi surface. Moreover, we make an approximation by putting in the expression for $\tilde{E}$, resulting from (I.35), the "trivial" solution for $F(p)$ given by (I.37b).

Then $\mathrm{d}n/\mathrm{d}\tilde{E}$ is the density of single-particle states (in the energy interval $\tilde{E}$, $\tilde{E}+\mathrm{d}\tilde{E}$, per unit volume) in the normal state, for one spin orientation, evaluated at the Fermi surface. In the BCS notation, the energy interaction element is denoted by $\tilde{V}$ (we use a tilde in order to distinguish it from the volume V) and the density of states by $N(0)$. The relation between this notation and the one used here is the following:

$$\frac{g^2}{V} = \tilde{V}, \quad \frac{\mathrm{d}n}{\mathrm{d}\tilde{E}} = \frac{N(0)}{V} \quad \text{and} \quad g^2\frac{\mathrm{d}n}{\mathrm{d}\tilde{E}} = N(0)\tilde{V}.$$

The approximation in (I.65) is valid for small ϱ (weak interaction limit).

Thus the quasi-particle energy spectrum is given by formula (I.46), where C is the constant described by (I.65). We see that this spectrum cannot be found by means of a perturbation theory method because if we start with non-interacting particles ($g^2 = 0$) we have $\varrho = 0$ as the zero-order approximation. But the function $\exp(-1/\varrho)$ has an essential singularity at $\varrho = 0$ and cannot be expanded in a series in the neighbourhood of this point.

Taking

$$\hat{H} \cong \langle \hat{H} \rangle_0 + \sum_{p,\sigma} \Omega(p)\alpha^{+}_{p\sigma}\alpha_{p\sigma} \tag{I.67}$$

consider the difference between the vacuum state ($E_0 = \langle \hat{H} \rangle_0$, $|0\rangle = |C\rangle$) and the first excited one-particle state (E_1, $|1\rangle = \alpha^{+}_{p\sigma}|C\rangle$)

for which we have

$$\hat{H}|1\rangle = E_1|1\rangle = \left(E_0+\sqrt{(\tilde{E}(p)-\tilde{E}(p_F))^2+C^2}\right)|1\rangle. \quad \text{(I.68)}$$

Hence

$$E_1-E_0 = \sqrt{(\tilde{E}(p)-\tilde{E}(p_F))^2+C^2} = \Omega(p) > 0. \quad \text{(I.69)}$$

We see that $\Omega(p)$ is always positive and has its minimum, different from zero, at $p = p_F$

$$\Omega(p_F) = C. \quad \text{(I.70)}$$

So an energy gap C exists between the ground state and the first excited state of the one-particle spectrum. However, since quasi-particles with one-half spin must appear in pairs, the energy gap is in fact equal to $2C$ (see also Chapter III, § 2, formula (III.29)).

If a system is characterized by an energy spectrum with an energy gap, it cannot be excited by an amount of energy less than the energy gap. From this follows the great stability of the current-carrying state characterizing superconductivity.

In the current-carrying state the energy of the elementary excitation can be equal to zero. There exists, however, a critical drift velocity u_c such that for velocities smaller than u_c the energy gap between the ground state and the excited state exists.

The group velocity for the wave packet is given by

$$\boldsymbol{v}_g = \frac{\partial \Omega(p)}{\partial \boldsymbol{p}}.$$

We assume that the whole system moves with a drift velocity $\boldsymbol{u}$. Then from the point of view of the system at rest, the group velocity of the wave packet is

$$\boldsymbol{v}'_g = \boldsymbol{v}_g + \boldsymbol{u} = \frac{\partial \Omega'(p')}{\partial \boldsymbol{p}'} = \frac{\partial \Omega'(p)}{\partial \boldsymbol{p}}$$

where $\boldsymbol{p}' = \boldsymbol{p}+m\boldsymbol{u}$. After integration of this equation we have

$$\Omega'(p) = \Omega(p)+(\boldsymbol{u}\cdot\boldsymbol{p}).$$

The appearance of the term $(\boldsymbol{u}\cdot\boldsymbol{p})$ can decrease the energy of the quasi-particle by $-|\boldsymbol{u}|\,|\boldsymbol{p}|$. We define the critical drift velocity $\boldsymbol{u}_c$ as the velocity for which Ω' vanishes:

$$|\boldsymbol{u}_c| = \min_p \frac{\Omega(p)}{|\boldsymbol{p}|} = \frac{C}{|p_F|}.$$

For $|\boldsymbol{u}| < |\boldsymbol{u}_c|$ the energy of an elementary excitation is greater than zero and the creation of an elementary excitation is accompanied by an increase in the energy of the system. It is energetically unfavourable. We see that for the systems described by a spectrum with an energy gap, $u_c \neq 0$ is always true. Therefore, such a system can persist in a current-carrying state which is a superconducting state. For example, for $C = 10^{-3}$ eV, $E_F = 5$ eV, we have $u_c = 100$ m/sec. For the case of an ideal Fermi gas $\Omega(p) = p^2/2m - p_F^2/2m$, hence

$$\min_{(p)} \frac{\Omega(p)}{|\boldsymbol{p}|} = \frac{\Omega(p_F)}{|p_F|} = 0$$

and we could not have superconductivity.

Now we shall demonstrate that $\tilde{E}(p_F) = \lambda$. Let us take $\lambda \equiv \tilde{E}(p_0)$ where p_0 is arbitrary. The total particle number operator has the form

$$\hat{N} = \sum_f a_f^+ a_f. \tag{I.71}$$

For the ground state $|C\rangle$ we have

$$\langle \hat{N} \rangle_0 = \sum_f a_f^+ a_f = \sum_f F(f, f). \tag{I.72}$$

For the special case $f = (p, \sigma)$ and solution (I.37a) we obtain formula (I.16) in the form

$$\langle \hat{N} \rangle_0 = 2 \sum_p F(p) = 2 \sum_p \frac{1}{2} \left\{ 1 - \frac{\tilde{E}(p) - \lambda}{\sqrt{(\tilde{E}(p) - \lambda)^2 + C^2}} \right\} = N. \tag{I.73}$$

We see that $F(p) = \langle n_p \rangle_0$ gives, for the ground state, the mean values of the numbers of particles in the state $\boldsymbol{p}$. We note that $C = 0$

(i) for $\tilde{E}(p) < \lambda - \omega$, i.e. for $\tilde{E}(p) - \lambda < -\omega < 0$,

(ii) for $\tilde{E}(p) > \lambda + \omega$, i.e. $\tilde{E}(p) - \lambda > \omega > 0$.

In case (i) $\zeta(p) < 0$, in case (ii) $\zeta(p) > 0$. Hence, in case (i) $(p < p_0 - \Delta) F(p) = 1$; on the other hand, in case (ii) $(p > p_0 + \Delta)$ $F(p) = 0$. Therefore

$$N = 2 \sum_{\substack{p \\ \tilde{E}(p) < \lambda - \omega}} 1 + 2 \sum_{\substack{p \\ \lambda - \omega \leqslant \tilde{E}(p) \leqslant \lambda + \omega}} F(p) \tag{I.74}$$

$$= 2\sum_{\substack{p\\ \tilde{E}(p)<\lambda-\omega}} 1 + \sum_{\substack{p\\ \lambda-\omega\leqslant\tilde{E}(p)\leqslant\lambda+\omega}} 1 - \frac{Vp_0^2}{2\pi^2}\int_{-\omega}^{\omega}\frac{\tilde{E}\,\mathrm{d}\tilde{E}}{\sqrt{\tilde{E}^2+C^2}}.$$

The integral of the odd function in (I.74) vanishes. The second sum can (for a thin layer) be written in the form $2\sum_{\lambda-\omega\leqslant\tilde{E}(p)\leqslant\lambda} 1$.

Hence we see that

$$N = 2\sum_{\substack{p\\ \tilde{E}(p)\leqslant\tilde{E}(p_0)\equiv\lambda}} 1 \quad . \tag{I.75}$$

Thus, $p_0 = p_F$ and $\lambda = \tilde{E}(p_F)$. Now it is clear why we take the trivial solution in the form (I.37b) and not simply $F(p) = 1$ (which is also a solution of (I.33, 34)). $F(p)$ gives us the probability of occupation of a Bloch state $\boldsymbol{p}$ in the superconducting state. For the trivial solution (normal state) $\langle n_p\rangle_0$ has a jump at the Fermi surface, while for the solution (I.37a) (superconducting state) $\langle n_p\rangle_0$ has no jump and is smeared out in the neighbourhood of the Fermi surface.

Now let us calculate the difference between the mean energy for the non-trivial and the trivial solutions. Formula (I.11a) takes for (I.37a,b) the form

$$\begin{aligned}\langle\hat{H}\rangle_0 &= \sum_p [E(p)-\lambda+\zeta(p)]F(p)\\ &\quad+\frac{1}{V}\sum_{p,p'} I(\boldsymbol{p},-\boldsymbol{p};\boldsymbol{p},-\boldsymbol{p}')\Phi(p)\Phi(p') \qquad \text{(I.76)}\\ &\cong 2\sum_p \zeta(p)F(p) - \sum_p \frac{C^2}{\sqrt{\zeta^2+C^2}}.\end{aligned}$$

It is convenient to use the identity

$$F(p)-\Theta(p)_F = -\frac{\Theta_F}{2}\left(1+\frac{\zeta}{\Omega}\right)+\frac{1-\Theta_F}{2}\left(1-\frac{\zeta}{\Omega}\right). \tag{I.77}$$

With the help of this identity we obtain

$$\begin{aligned}\langle\hat{H}\rangle_0^{(s)}-\langle\hat{H}\rangle_0^{(n)} \equiv \Delta\bar{E} = \sum_p \Big\{(1-\Theta_F(p))\left[\zeta(p)-\frac{\zeta^2(p)}{\Omega(p)}-\frac{C^2}{2\Omega(p)}\right]\\ -\Theta_F(p)\left[\zeta(p)+\frac{\zeta^2(p)}{\Omega(p)}+\frac{C^2}{2\Omega(p)}\right]\Big\}. \qquad (1.78)\end{aligned}$$

Using (I.37a) we eliminate C^2 from (I.78), then we replace the sum by an integral and finally we change the integration variable p ($p \to \tilde{E} \to \zeta$) obtaining

$$\frac{\Delta \tilde{E}}{V} = -\frac{p_F^2}{2\pi^2}\frac{1}{\tilde{E}'}\int_0^{\omega}\left(1-\frac{\zeta}{\sqrt{\zeta^2+C^2}}\right)^2\sqrt{\zeta^2+C^2}\,\mathrm{d}\zeta. \qquad \text{(I.79)}$$

We introduce a new integration variable $\zeta/C = \sinh x$ and then take the upper limit of the integral as infinity ($\omega/C = \infty$). After some calculations we finally have

$$\frac{1}{V}\left(\langle \hat{H}\rangle_0^{(s)} - \langle \hat{H}\rangle_0^{(n)}\right) = -\frac{\mathrm{d}n}{\mathrm{d}E}\,2\omega^2 e^{-2/\varrho^2} < 0. \qquad \text{(I.80)}$$

Hence the mean energy of the ground state is less for the "superconducting" solution than for the "normal" one. For this reason, the state described by the superconducting solution is energetically more favourable than the normal one. However, it does not follow from (I.80) that the "superconducting" solution (I.37a) corresponds to a minimum of the variational problem. For (I.37a) to be the minimum, the second variation of W (formula (I.19)) must be positive.

Valatin (1958) found the condition for the existence of a stable solution for the generalized problem from the instability of the H–F solution. This condition was used (Bogoliubov, 1958c; Soloviev, 1958) to obtain the criterion for the existence of the superfluidity of nuclear matter.

§ 4 Stability of the solution describing the superconducting state

In order to verify whether the solutions (I.22) or (I.23) correspond to a minimum of (I.27), we must examine the sign of the second variation of $W[F, \Phi]$ given by Galasiewicz (1963a)

$$\begin{aligned}\delta^2 W = 2 &\sum_{f_1,f_2,f_1',f_2'} M(f_1',f_1;f_2,f_2')\delta F^*(f_1',f_1)\delta F(f_2,f_2')\\ +2 &\sum_{f_1,f_2,f_1',f_2'} N(f_1,f_2;f_1',f_2')\delta\Phi^*(f_1,f_2)\delta\Phi(f_1',f_2')\\ +2 &\sum_{f_1,f_2,f_1',f_2'} \mu^*(f_1',f_1)\delta_{f_2 f_2'}\delta F^*(f_1,f_2)\delta\Phi(f_1',f_2')\end{aligned} \qquad \text{(I.81)}$$

$$+2 \sum_{f_1, f_2, f_1', f_2'} \mu(f_1, f_1')\delta_{f_2, f_2'}\delta\Phi^*(f_1, f_2)\delta F(f_1', f_2')$$

where

$$\begin{aligned} M(f_1', f_1; f_2, f_2') &= M^*(f_2, f_2'; f_1', f_1) = \tfrac{1}{2}[U(f_1, f_2; f_2', f_1') \\ &\quad - U(f_1, f_2; f_1', f_2')] + \lambda(f_1, f_2')\delta_{f_2 f_1'}, \\ N(f_1, f_2; f_1', f_2') &= N^*(f_1', f_2'; f_1, f_2) \\ &= \tfrac{1}{2} U(f_1, f_2; f_2', f_1') + \lambda(f_2, f_2')\delta_{f_1 f_1'}. \end{aligned} \tag{I.82}$$

In order to ascertain whether the expression (I.81) is a positive quadratic form in δF, $\delta\Phi$, we must determine the sign of the eigenvalues E of the equations

$$\begin{aligned} &2\sum_{f_1, f_1'}[U(f_4, f_1; f_1', f_3) - U(f_4, f_1; f_3, f_1') + \lambda(f_4, f_1')\delta_{f_3 f_1} \\ &\quad + \lambda(f_1, f_3)\delta_{f_1' f_4}]\delta F(f_1, f_1') + 2\sum_{f_1, f_2}\mu^*(f_1, f_3)\delta_{f_4 f_2}\delta\Phi(f_1, f_2) \\ &\quad + 2\sum_{f_1, f_2}\mu(f_1, f_3)\delta_{f_2 f_3}\delta\Phi^*(f_1, f_2) = E\delta F(f_3, f_4), \end{aligned} \tag{I.83a}$$

$$\begin{aligned} &2\sum_{f_1, f_2}[U(f_3, f_4; f_1, f_2) + \lambda(f_4, f_2)\delta_{f_3 f_1} - \lambda(f_3, f_2)\delta_{f_4 f_1}]\delta\Phi(f_1, f_2) \\ &\quad + \sum_{f_1, f_2}[\mu(f_3, f_1)\delta_{f_4 f_2} - \mu(f_4, f_1)\delta_{f_3 f_2}]\delta F(f_1, f_2) \\ &\quad = E\delta\Phi(f_3, f_4). \end{aligned} \tag{I.83b}$$

If E is positive, (I.81) is a positive quadratic form and the mean value $\langle \hat{H} \rangle_0$ has a minimum. The general eqns. (I.83a,b) can be used in the examination of the "trivial" solution (I.37b) which is a solution of the H–F variational principle (see eqns. (I.27–29) and of the "superconducting" solution (I.37a) which is a solution of the generalized variational principle (see eqns. (I.17), (I.18a,b), (I.23).

Let us examine what restrictions are imposed on $\delta F(f_1, f_2)$ by the subsidiary conditions (I.29). After variation (I.29) yields

$$[\Theta_F(p_1) + \Theta_F(p_2) - 1]\delta F(f_1, f_2) = 0. \tag{I.84}$$

We see that $\delta F(f_1, f_2) \neq 0$ either if $p_1 < p_F < p_2$ or if $p_2 < p_F < p_1$.

Now we want to obtain the Lagrange multipliers $\lambda(f, f') =$

$\lambda(p)\,\delta(f-f')$ for the generalized and H–F variational principles. From (I.21a) we have

$$\zeta(p)-[1-2F(p)]\,\lambda(p) = 0. \tag{I.85}$$

On the other hand, from (I.33) and (I.37a) we obtain

$$1-2F(p) = \frac{\zeta(p)}{\Omega(p)}. \tag{I.86}$$

Thus, for the generalized solution (energy gap $C \neq 0$), λ_C has the form (see Valatin (1961))

$$\lambda_c(p) = \sqrt{\zeta^2(p)+C^2} = \Omega(p). \tag{I.87}$$

For the H–F solution ($C = 0$), one has

$$\lambda_0(p) = \begin{cases} +\zeta_0(p) > 0 & \text{for} \quad p > p, \\ -\zeta_0(p) > 0 & \text{for} \quad p \leqslant p. \end{cases} \tag{I.88}$$

From (I.88) we obtain

$$\lambda_0(p_1)+\lambda_0(p_2) = |\zeta(p_1)|+|\zeta(p_2)| > 0. \tag{I.89}$$

In order to stress the property of U that

$$U(f_1,f_2;f_3,f_4) = -U(f_1,f_2;f_4,f_3) = -U(f_2,f_1;f_4,f_3),$$

we write U in the form

$$U(f_1,f_2;f_3,f_4) = \frac{1}{2V}\,\delta(p_1+p_2-p_3-p_4)$$
$$\times [I(p_1,p_2;p_3,p_4)\delta_{\sigma_1\sigma_4}\delta_{\sigma_2\sigma_3}-I(p_1,p_2;p_4,p_3)\delta_{\sigma_1\sigma_3}\delta_{\sigma_2\sigma_4}]$$

and then use (I.58). In this case, for the $H-F$ solution eqns. (I.83) have the form ($g^2/V = \tilde{g}^2$, $E \to 2E$)

$$-\tilde{g}^2\sum_{p_1,p_2} \delta F(p_1,-\sigma|p_2,-\sigma)+[\lambda_0(p_3)+\lambda_0(p_4)]\,\delta F(p_3,\sigma|p_4,\sigma)$$
$$= E_F^0\,\delta F(p_3,\sigma|p_4,\sigma). \tag{I.90}$$

On the other hand, for the generalized solution eqns. (I.83) give

$$-\tilde{g}^2\sum_{p_1,p_2} \delta F(p_1,-\sigma|p_2,-\sigma)+[\lambda_c(p_3)+\lambda_c(p_4)]\,\delta F(p_3,\sigma|p_4,\sigma)$$
$$= E_F^c\,\delta F(p_3,\sigma|p_4,\sigma), \tag{I.91a}$$

$$\tilde{g}^2\sum_{p_1,p_2} \delta F(p_1,\sigma|p_2,-\sigma)+[\lambda_c(p_3)+\lambda_c(p_4)]\,\delta F(p_3,\sigma|p_4,\sigma)$$
$$= E_F^c(+,-)\,\delta F(p_3,\sigma|p_4,-\sigma), \tag{I.91b}$$

$$\tilde{g}^2 \sum_{p_1, p_2} \delta\Phi(p_1, \sigma|p_2, -\sigma) + [\lambda_c(p_3)+\lambda_c(p_4)]\,\delta\Phi(p_3, \sigma|p_4, -\sigma)$$

$$= E_\Phi^C\,\delta\Phi(p_3, \sigma|p_4, -\sigma), \quad \text{(I.91c)}$$

$$[\lambda_c(p_3)+\lambda_c(p_4)]\,\delta\Phi(p_3, \sigma|p_4, \sigma) = E_\Phi^C(\sigma, \sigma)\,\delta\Phi(p_3, \sigma|p_4, \sigma). \quad \text{(I.91d)}$$

In (I.90) and (I.91a–d) we can omit the spin dependence, since there is no preferred direction. In (I.91c) it is best to retain the spin indices since the functions Φ are antisymmetric in $f = (\boldsymbol{p}, \sigma)$.

For the more detailed analysis of (I.91) it will be necessary to use the "gap equation" (I.59) which follows from the equation of compensation of dangerous diagrams for pairs of quasi-particles with opposite momenta. Therefore we must now confine ourselves to the case $\boldsymbol{p}_4 = -\boldsymbol{p}_3$ and the eqn. (I.91a) has the form

$$-\tilde{g}^2 \sum_{p_1} \delta F(p_1, -\sigma|-p_1, -\sigma)$$

$$= [E_F^C - 2\lambda_C(p)]\,\delta F(p, \sigma|-p, \sigma). \quad \text{(I.92)}$$

Denoting in (I.92) $\sum_p \delta F = D$ we have

$$\delta F(p, \sigma|-p, \sigma) = \frac{\tilde{g}^2 D}{E_F^C - 2\lambda_C(p)}, \quad \text{(I.93)}$$

and finally

$$\sum_p \frac{1}{E_F^C - 2\lambda_C(p)} \equiv \tilde{\Phi}^C(E_F^C) = -\frac{1}{\tilde{g}^2}. \quad \text{(I.94)}$$

We call now the solution of this eigenvalue equation for the limit $C = 0$ the H–F solution. Note that according to the condition (I.84) the real H–F solution has to be found for $|\boldsymbol{p}_3| \neq |\boldsymbol{p}_4|$, i.e. from eqn. (I.90).

In order to have the discrete set of one-particle energies $|\zeta(p)| = \lambda_0(p)$ we consider a large but finite box. The eigenvalue problem is now formally similar to the problem of one pair considered in Cooper's paper (Cooper, 1956; see also Schrieffer, 1964). The formula (I.94) gives the following shape of the curve $\tilde{\Phi}^{(0)}(E)$: for $E \to -\infty$ the function $\tilde{\Phi}^{(0)}$ tends to -0; for $E \to 2\lambda_0^{(n)}(p)$, if $E < 2\lambda_0^{(n)}(p)$, $\tilde{\Phi}^{(0)}$ tends to $-\infty$ and if $E > 2\lambda_0^{(n)}(p)$, $\tilde{\Phi}^{(0)}$ tends to $+\infty$. The lowest value of $2\lambda_0^{(n)}$ is $2\lambda_0(p_F) = 0$. Thus, for $E \to \pm 0$ we have $\tilde{\Phi}^{(0)} \to \pm\infty$, respectively. The plot of the function $\tilde{\Phi}^{(0)}$ is given in Fig. 6.

The eigenvalues E_i (roots of (I.94)) are given by the intersections of $\tilde{\Phi}$ with a straight line parallel to the E-axis which lies below the axis. We see that one of the roots is negative. This gives, in

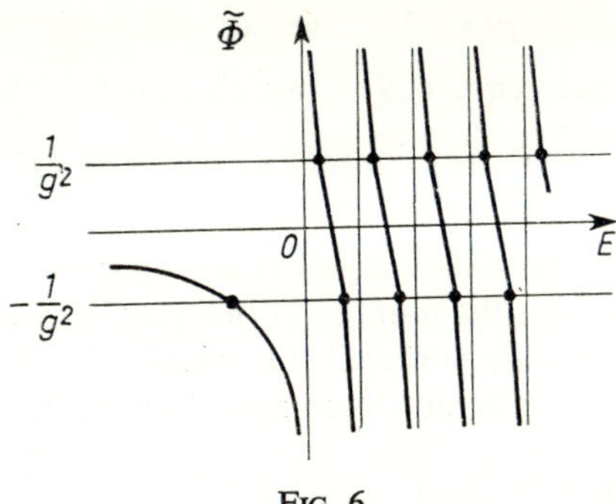

FIG. 6

the Cooper theory, a bound state, that is a Cooper pair. In our case (Galasiewicz, 1969a) this root gives the instability of the H–F solution.

For the superconducting solution $C \neq 0$ and due to the existence of the energy gap, the lowest value of $2\lambda_c^{(n)}$ is now $2\lambda_c(p_F) = 2C$. Therefore the function $\tilde{\Phi}^{(c)}$ reaches the values $\pm\infty$ first in the vicinity of $E = 2C > 0$. We see that the plot of $\tilde{\Phi}^{(c)} = \tilde{\Phi}^{(c)}(E)$ is shifted by $2C$ to the right with respect to that of Fig. 6. In this case it seems possible to have intersections of $\tilde{\Phi}^{(c)}$ with the straight horizontal line $-1/\tilde{g}^2$ for non-negative values of E. Consider the eqn. (I.59) which can be written in the form

$$\sum_p \frac{1}{2\lambda_c(p)} = \frac{1}{\tilde{g}^2}. \tag{I.95}$$

We see that this equation is identical to (I.94) taken for $E_F^c = 0$. So according to the "gap equation" we have the first intersection of $\tilde{\Phi}^{(c)}$ with the line $-1/\tilde{g}^2$ on the axis $E = 0$. The existence of the energy gap insures the stability of the solution describing the superconducting state.

Note that in the case under consideration the eqns. (I.91a,b) have no negative roots because the straight line lies above the E-axis.

In order to be able to carry through the calculations, we consider the special case, formally similar to that of Mottelson (1959), in which we can take

$$\begin{aligned} \zeta_0(p) \approx 0, \quad &\text{hence} \quad \lambda_0 = 0, \\ \zeta_c(p) \approx 0, \quad &\text{hence} \quad \lambda_c = C. \end{aligned} \tag{I.96}$$

First we investigate eqn. (I.91a) which now has the form

$$-\tilde{g}^2 \sum_{p_1, p_2} \delta F(p_1, p_2) + 2C\, \delta F(p_3, p_4) = E_F^c \delta F(p_3, p_4). \tag{I.97}$$

We may find the eigenvalues E_F^c by equating to zero a determinant of order n^2, where n is the number of one-particle states in a narrow layer near the Fermi surface, for which the matrix elements $I(p_1, p_2; p_3, p_4)$ are different from zero (see (I.58))

$$\begin{vmatrix} 1-\tilde{E}, & 1, & 1, & \ldots, & 1 \\ 1, & 1-\tilde{E}, & 1, & \ldots, & 1 \\ 1, & 1, & 1-\tilde{E}, & \ldots, & 1 \\ \ldots & \ldots & \ldots & \ldots & \ldots \\ 1, & 1, & 1, & \ldots, & 1-\tilde{E} \end{vmatrix} \equiv \tilde{E}^{n^2-1} \begin{vmatrix} W_{n^2-2} & \begin{matrix} 0, & 0 \\ \vdots & \vdots \\ 0, & 0 \\ 1, & 0 \end{matrix} \\ \begin{matrix} 0, \ldots, 0, 1, -1, \\ 0, \ldots, 0, 0, -1, \end{matrix} & \begin{matrix} 1 \\ 1-\tilde{E} \end{matrix} \end{vmatrix} = 0. \tag{I.98}$$

Here W_s is a determinant of order s

$$W_s = \begin{vmatrix} -2, & 1, & 0, & 0, \ldots \\ 1, & -2, & 1, & 0, \ldots \\ 0, & 1, & -2, & 1, \ldots \\ 0, & 0, & 1, & -2, \ldots \\ \ldots & \ldots & \ldots & \ldots \ldots \end{vmatrix} = (-1)^s(s+1), \tag{I.99}$$

and

$$\tilde{E} = -\frac{E_F^c - 2C}{\tilde{g}^2}. \tag{I.100}$$

Equation (I.98) has the (n^2-1)-fold root

$$E_{F,\, n^2-1}^c = 2C \tag{I.101}$$

and a single root given by

$$1-\tilde{E} = -\frac{W_{n^2-2}}{W_{n^2-2} + W_{n^2-3}} = 1-n^2. \tag{I.102}$$

From this equation we obtain

$$E_{F,1}^c = -n^2\tilde{g}^2 + 2C. \tag{I.103}$$

Thus (I.90) has the eigenvalues

$$E^0_{F,n^2-1} = 0, \quad E^0_{F,1} = -n^2\tilde{g}^2. \tag{I.104}$$

The H–F solution is unstable, i.e. $E^0_{F,1} < 0$, if the interaction is attractive, $I = -\tilde{g}^2 < 0$.

Equations (I.91a–d) have the eigenvalues

$$E^c_{F,n^2-1} = 2C > 0, \qquad E^c_{F,1} = -n^2\tilde{g}^2+2C, \tag{I.105a}$$

$$E^c_{F,n^2-1}(+,-) = 2C > 0, \quad E^c_{F,1}(+,-) = n^2\tilde{g}^2+2C > 0, \tag{I.105b}$$

$$E^c_{\Phi,n^2-1} = 2C > 0, \qquad E^c_{\Phi,1} = n^2\tilde{g}^2+2C > 0, \tag{I.105c}$$

$$E^c_{\Phi,n^2} = 2C > 0. \tag{I.105d}$$

While $E^0_{F,1}$ is negative, $E^c_{F,1}$ (given by (I.105a)) can be positive for suitable values of $\tilde{g}^2$, since $2C > 0$.

CHAPTER II

Thermodynamics

§ 1 The mean value of the Hamiltonian

Now let us consider a system at temperatures $\Theta \neq 0$. For this purpose we will use the thermodynamic Wick theorem and we now define, instead of (I.7), the new contractions

$$\overline{\hat{A}\hat{B}} = \langle \hat{A}\hat{B} \rangle = \frac{\mathrm{Tr}\{\hat{A}\hat{B}D\}}{\mathrm{Tr}\{D\}} \tag{II.1}$$

where D is the unnormalized density matrix (see Matsubara, 1955; Thouless, 1957; Bloch and de Dominicis, 1958; Gaudin, 1960).

We obtain

$$\overline{a^+_{k_1 s_1} a_{k_2 s_2}} = \langle a^+_{k_1 s_1} a_{k_2 s_2} \rangle = a^+_{k_1 s_1} a_{k_2 s_2} - N(a^+_{k_1 s_1} a_{k_2 s_2}). \tag{II.2}$$

Here, N denotes the normal product in α, α^+ operators, after performing the Bogoliubov transformation. From (II.2) we see that

$$\langle N(a^+_{k_1 s_1} a_{k_2 s_2}) \rangle = 0.$$

The functions F and Φ now have the form

$$\overline{a^+_{f_1} a_{f_2}} = F(f_1, f_2) = \frac{\mathrm{Tr}\{a^+_{f_1} a_{f_2} D\}}{\mathrm{Tr}\{D\}} = \sum_\nu \{v^*_{f_1 \nu} v_{f_2 \nu}(1-\bar{n}_\nu) + u^*_{f_1 \nu} u_{f_2 \nu} \bar{n}_\nu\}, \tag{II.3}$$

$$\overline{a_{f_1} a_{f_2}} = \Phi(f_1, f_2) = \frac{\mathrm{Tr}\{a_{f_1} a_{f_2} D\}}{\mathrm{Tr}\{D\}} = \sum_\nu \{u_{f_1 \nu} v_{f_2 \nu}(1-\bar{n}_\nu) + v_{f_1 \nu} u_{f_2 \nu} \bar{n}_\nu\}, \quad \bar{n}_\nu = \langle \alpha^+_\nu \alpha_\nu \rangle.$$

For the case of the special Bogoliubov transformation ($f = \boldsymbol{k}, s$)

we obtain

$$\overbrace{a^{+}_{k_1 s_1} a_{k_2 s_2}} = F(f_1, f_2) = \delta(f_1 - f_2) F(k_1, s_1)$$
$$= \delta(\boldsymbol{k}_1 - \boldsymbol{k}_2)\,\delta(s_1 - s_2)\,[u(k_1)^2 \bar{n}_{k_1 s_1} + v(k_1)^2 (1 - \bar{n}_{-k_1, -s_1})],$$

$$\overbrace{a_{k_1 s_1} a_{k_2 s_2}} = \Phi(f_1, f_2) = \delta(f_1 + f_2)\Phi(k_1, s_1) \tag{II.4}$$
$$= \delta(\boldsymbol{k}_1 + \boldsymbol{k}_2)\,\delta(s_1 + s_2) u(k_1) v(k_1, s_1)\,[\bar{n}_{k_1 s_1} + \bar{n}_{k_2 s_2} - 1],$$
$$v(k, +) = -v(k, -) = v(k),$$
$$\Phi(k) = u(k) v(k)\,(\bar{n}_{ks} + \bar{n}_{-k, -s} - 1).$$

The normal products have the form

$$N(a^{+}_{k_1 s_1} a_{k_2 s_2}) = u(k_1) u(k_2)\,\alpha^{+}_{k_1 s_1} \alpha_{k_2 s_2}$$
$$- v(k_1, s_1) v(k_2, s_2)\,\alpha^{+}_{-k_2, -s_2} \alpha_{-k_1, -s_1}$$
$$+ u(k_1) v(k_2, s_2)\,\alpha^{+}_{k_1 s_1} \alpha^{+}_{-k_2, -s_2}$$
$$+ v(k_1, s_1) u(k_2)\,\alpha_{-k_1, -s_1} \alpha_{k_2 s_2}, \tag{II.5}$$
$$N(a_{k_1 s_1} a_{k_2 s_2}) = v(k_1, s_1) u(k_2)\,\alpha^{+}_{-k_1, -s_1} \alpha_{k_2 s_2}$$
$$- u(k_1) v(k_2, s_2)\,\alpha^{+}_{-k_2, s_2} \alpha_{k_1 s_1}$$
$$+ u(k_1) u(k_2)\,\alpha_{k_1 s_1} \alpha_{k_2 s_2}$$
$$+ v(k_1, s_1) v(k_2, s_2)\,\alpha^{+}_{-k_1, -s_1} \alpha^{+}_{-k_2, -s_2}$$

Now we write the Hamiltonian (I.31), with I defined by (I.58), in terms of the operators α, α^{+} and instead of contractions ⌐ (I.9), we write in (I.7a,b) the suitable contractions ⌐. After using (II.5) we can write the Hamiltonian of our system in the form

$$\hat{H} = \langle \hat{H} \rangle + \sum_{k,s} \Omega(k, s)\,\alpha^{+}_{ks} \alpha_{ks} + \sum_{k,s} R^{*}_{-k, -s;\, k, s}\,\alpha_{-k, -s} \alpha_{k, s}$$
$$+ \sum_{k,s} R_{-k, -s;\, k, s}\,\alpha^{+}_{k, s} \alpha^{+}_{-k, s} + \sum_{(k,s)} N(a^{+} a^{+} a a) \tag{II.6}$$

where now

$$\langle \hat{H} \rangle = \frac{1}{2} \sum_{k,\sigma} [\varepsilon_k - \lambda + \zeta_k(\sigma)]\, F(k, \sigma) - \frac{g^2}{V} \sum_{k_1, k_2} \Phi(k_1) \Phi(k_2), \tag{II.7}$$

$$\Omega(k, s) = \zeta_k(s)\,[u^2(k) - v^2(k)] - u(k) v(k, s) \frac{g^2}{V} \sum_{p} \Phi(p, s), \tag{II.8}$$

$$R^{+}_{-k,s;k,s} = \zeta_k(s)u(k)v(k,s) + [u^2(k) - v^2(k)]\frac{g^2}{2V}\sum_p \Phi(p,s), \tag{II.9}$$

$$\zeta_k(\sigma) = \tilde{\varepsilon}_k(\sigma) - \lambda \approx \varepsilon_k - \lambda, \tag{II.10}$$

$$\tilde{\varepsilon}_k(\sigma) = \varepsilon_k - \frac{g^2}{V}\sum_p F(p, -\sigma) \approx \varepsilon_k.$$

In the absence of an external field we have $\bar{n}_{k+} = \bar{n}_{k-} = \bar{n}_k$. Hence $\zeta_k(+) = \zeta_k(-) = \zeta_k$.

In the expression for $\langle\hat{H}\rangle$, given by (II.7), there appears a sum of all possible products of $\overline{a^+a}$, $\overline{aa}$, $\overline{a^+a^+}$. According to the Wick–Bloch–de Dominicis thermodynamic theorem (see, for example, Landau, 1956), $\langle\hat{H}\rangle$ is a mean value of the operator $\hat{H}$. The definition of the mean value is given in (II.1). The mean energy of the system for $\Theta \neq 0$ is equal to $\bar{E} = \langle\hat{H}\rangle + \lambda\langle\hat{N}\rangle$ since we take the original Hamiltonian in the form (I.1). The mean energy is identified with the internal energy of the system. It depends on the temperature since the $\bar{n}_k$ are functions of temperature.

It can be shown in the general case that $\langle\hat{H}\rangle$ is given by the formula

$$\langle\hat{H}\rangle = \tfrac{1}{2}\sum_{f,f'}[T(f,f') + \zeta(f,f')] + \tfrac{1}{2}\sum_{(f)} U(f_1,f_2;f_2',f_1') \times \Phi^*(f_1,f_2)\Phi(f_1',f_2') \tag{II.11}$$

where F and Φ are now given by (II.3) and the other quantities by (I.1) and (I.11d).

Let us calculate the derivative

$$\frac{\delta\langle\hat{H}\rangle}{\delta\bar{n}_\nu} = \frac{1}{2}\sum_{f,f'}\left\{\frac{\delta\zeta(f,f')}{\delta\bar{n}_\nu}F(f,f') + [T(f,f') + \zeta(f,f')]\frac{\delta F(f,f')}{\delta\bar{n}_\nu}\right\} + \frac{1}{2}\sum_{(f)} U(f_1,f_2;f_2',f_1') \times\left[\frac{\delta\Phi^*(f_1,f_2)}{\delta\bar{n}_\nu}\Phi(f_1',f_2') + \Phi^*(f_1,f_2)\frac{\delta\Phi(f_1',f_2')}{\delta\bar{n}_\nu}\right]. \tag{II.12}$$

We find $\frac{\delta F}{\delta \bar{n}_\nu}$ and $\frac{\delta \Phi}{\delta \bar{n}_\nu}$ from (II.3) and finally obtain

$$\left(\frac{\delta \langle \hat{H} \rangle}{\delta \bar{n}_\nu}\right)_{\bar{n}_\nu = 0} = \Omega(\nu, \nu) = \Omega(\nu) \tag{II.13}$$

where Ω is given by (I.11b). Formula (II.13) is identical to that for the energy of elementary excitations given by Landau (1956) and with the formula for the energies of elementary excitations $\Omega(\nu) = \langle C|\alpha_\nu \hat{H} \alpha_\nu^+|C\rangle$, considered by Bogoliubov, Tolmachev and Shirkov (1958). Formula (II.13) gives the proper interpretation of the energy $\Omega(p)$ (formula (I.46)) which is the energy of the excitation of the system if one additional electron is added to the system (see, for example, Schrieffer, 1964). Until now the coefficients $\{u, v\}$ have satisfied only one condition

$$u^2(k) + v^2(k) = 1. \tag{II.14}$$

Now we impose an additional condition: the diagonalization of the Hamiltonian (II.6) (by neglecting the last term), which corresponds to the principle of the compensation of dangerous diagrams for $\Theta = 0$ (see Khalatnikov and Abrikosov, 1958)

$$R^*_{-k, -s; k, s} = 0. \tag{II.15}$$

We write

$$C = -\frac{g^2}{2V} \sum_k \Phi(k). \tag{II.16}$$

C is now a function of temperature $C = C(\Theta)$. Taking $\bar{n}_{k+} = \bar{n}_{k-}$ we have a system of equations

$$2\zeta_k u(k) v(k) - C[u^2(k) - v^2(k)] = 0, \tag{II.17}$$

$$u^2(k) + v^2(k) = 1.$$

Hence

$$v^2(k) = \frac{1}{2}\left\{1 - \frac{\zeta_k}{\sqrt{\zeta_k^2 + C^2}}\right\}, \qquad u^2(k) = \frac{1}{2}\left\{1 + \frac{\zeta_k}{\sqrt{\zeta_k^2 + C^2}}\right\},$$

$$u(k) v(k) = \frac{1}{2} \frac{C}{\sqrt{\zeta_k^2 + C^2}}, \quad u^2(k) - v^2(k) = \frac{\zeta_k}{\sqrt{\zeta_k^2 + C^2}}. \tag{II.18}$$

From eqns. (II.9, 4, 16, 18) we have

$$\Omega(k, s) = \sqrt{\zeta^2 + C^2} = \Omega(k). \tag{II.19}$$

In § 3 we will prove that

$$\bar{n}_k = \frac{1}{e^{\frac{\Omega(k)}{\Theta}}+1}. \tag{II.20}$$

Taking into account (II.4, 18) we have

$$\langle\hat{H}\rangle = \sum_k \left[\zeta_k + \left(\frac{C^2}{2\Omega(k)} - \Omega(k)\right)(1-2\bar{n}_k)\right]. \tag{II.21}$$

In order to know the temperature dependence of $\langle\hat{H}\rangle$ we must have $C = C(\Theta)$.

§ 2 Temperature dependence of the energy gap

Equations (II.16, 4, 18) give us the equation for the energy gap when $\Theta \neq 0$:

$$C(\Theta) = \frac{g^2}{2V}\sum_{\substack{k \\ k\sim k_F}} \frac{1}{2}\frac{C(\Theta)}{\sqrt{\zeta_k^2+C^2(\Theta)}}(1-2\bar{n}_k). \tag{II.22}$$

From (II.20) it follows that

$$1-2\bar{n}_k = \tanh\frac{\Omega(k)}{2\Theta}. \tag{II.23}$$

We assume, for a narrow interval of summation over $\boldsymbol{k}$, when $k \sim k_F$, that C does not depend on k. Hence we have

$$\frac{g^2}{2V}\sum_{\substack{k \\ k\sim k_F}} \frac{\tanh\dfrac{\sqrt{\zeta_k^2+C^2}}{2\Theta}}{\sqrt{\zeta_k^2+C(\Theta)^2}} = 1. \tag{II.24}$$

Changing, similarly as in formula (I.59), the summation to an integration we have

$$1 = \varrho\int_0^{\frac{\omega}{C(\Theta)}} \frac{\mathrm{d}x}{\sqrt{x^2+1}} - 2\varrho\int_0^{\omega} \frac{\mathrm{d}\zeta}{\Omega(e^{\frac{\Omega}{\Theta}}+1)}, \tag{II.25}$$

$$\varrho = g^2\frac{1}{2\pi^2}\left(\frac{k^2}{\dfrac{\mathrm{d}\tilde{\varepsilon}}{\mathrm{d}k}}\right)_{k=k_F}.$$

The first term in (II.25) diverges as $\omega \to \infty$. But for $\Theta = 0$ we have

$$1 = \varrho \int_0^{\frac{\omega}{C(0)}} \frac{\mathrm{d}x}{\sqrt{x^2+1}}. \tag{II.26}$$

We substitute the right-hand side of this formula into the left-hand side of (II.25) and use the formula

$$\int_0^{\frac{\omega}{C}} \frac{\mathrm{d}x}{\sqrt{x^2+1}} \approx \ln \frac{2\omega}{C}. \tag{II.27}$$

We have then

$$\ln \frac{C(0)}{C(\Theta)} = 2 \int_0^{\infty} \frac{1}{\Omega(e^{\frac{\Omega}{\Theta}}+1)} \mathrm{d}\zeta. \tag{II.28}$$

This formula does not contain divergent terms when $\omega \to \infty$. The integral on the right-hand side is convergent and therefore we can write ∞ for the upper limit. Now, as was done by Khalatnikov and Abrikosov (1958), we express the right-hand side of (II.28) by means of Bessel functions. We change from ζ to the new variables Ω and then to φ:

$$\mathrm{d}\zeta = \frac{\Omega \, \mathrm{d}\Omega}{\sqrt{\Omega^2 - C^2}}, \quad \Omega = C \cosh \varphi, \quad \mathrm{d}\Omega = C \sinh \varphi \, \mathrm{d}\varphi. \tag{II.29}$$

Instead of (II.28), we now have

$$\ln \frac{C(0)}{C(\Theta)} = 2 \int_0^{\infty} \frac{\mathrm{d}\varphi}{e^{\beta \cosh \varphi}+1}, \quad \beta = \frac{C}{\Theta}. \tag{II.30}$$

The function under the integration sign can be written in the form

$$\frac{1}{1+e^{\beta \cosh \varphi}} = e^{-\beta \cosh \varphi} \frac{1}{1-(-e^{-\beta \cosh \varphi})} = \sum_{m/1}^{\infty} (-1)^{m+1} e^{-m\beta \cosh \varphi}. \tag{II.31}$$

However, we know that

$$\int_0^{\infty} e^{-\alpha\varphi - m\beta \cosh \varphi} \mathrm{d}\varphi = K_\alpha(m\beta), \quad K_\alpha(m\beta) = K_{-\alpha}(m\beta) \tag{II.32}$$

(see, for example, Gradstein and Ryzhik, 1963), where K_α is a Bessel function. Finally, we can write (II.28) in the form

$$\ln \frac{C(0)}{C(\Theta)} = 2\sum_{m/1}^{\infty} (-1)^{m+1} K_0 \left(\frac{(mC(\Theta)}{\Theta}\right). \tag{II.33}$$

Formula (II.33) allows us to find the temperature dependence of the energy gap in the two extreme cases: (i) for $\Theta \sim 0°$, (ii) for $\Theta \sim \Theta_c$ where Θ_c is a transition temperature. For (i) $\Theta \to 0$, the argument of the Bessel function goes to infinity. In this case we can use the asymptotic formula

$$\lim_{z\to\infty} K_0(z) = \sqrt{\frac{\pi}{2z}}\, e^{-z}. \tag{II.34}$$

As the exponential term is small, the first term for $m = 1$ is dominant. For $\Theta \sim 0$ we have $C(\Theta) \sim C(0)$ and we can expand the left-hand side of (II.33) in a series about $C(0)$, and on the right-hand side use (II.34). We obtain

$$\ln\left(1+\frac{C(0)-C(\Theta)}{C(\Theta)}\right) = \sqrt{\frac{2\pi\Theta}{C(\Theta)}}\, e^{-\frac{C(\Theta)}{\Theta}}. \tag{II.35}$$

Hence

$$C(\Theta) = C(0) - \sqrt{2\pi\Theta C(0)}\, e^{-\frac{C(0)}{\Theta}}. \tag{II.36}$$

On the right-hand side we put $C(0)$ instead of $C(\Theta)$. For (ii), $\Theta \to \Theta_c$, we have $C(\Theta) \to 0$, and $C(\Theta_c) = 0$. Since the argument of K_0 has a factor m, which can take values from 1 to ∞, we cannot use an expression for small C. As was done by Khalatnikov and Abrikosov (1958), we wish to change the right-hand side of (II.33) using the appropriate formulae given by Gradstein and Ryzhik (1963):

$$\begin{gathered} 2\sum_{m/1}^{\infty} (-1)^{m+1} K_0 \left(\frac{mC(\Theta)}{\Theta}\right) = -\ln \gamma \frac{C(\Theta)}{4\pi\Theta} \\ -2\pi \sum_{l/1}^{\infty} \left\{ \frac{1}{(2l-1)^2\pi^2 + \left(\frac{C}{\Theta}\right)^2} - \frac{1}{2\pi l} \right\}, \quad \ln\gamma = 0.577, \\ \sum_{l/1}^{\infty} (-1)^{l+1} \frac{1}{l} = \sum_{l/1}^{\infty} \frac{1}{2l-1} - \frac{1}{2l} = \ln 2; \end{gathered} \tag{II.37}$$

moreover,

$$2\sum_{l/1}^{\infty}\frac{1}{(2l-1)^p}=\sum_{l/1}^{\infty}\frac{1}{l^p}(1+(-1)^{l+1})=(2-2^{1-p})\zeta(p)$$

(II.38)

where $\ln\gamma$ is Euler's constant, $\zeta(p)$ is a Riemann zeta-function. Using (II.37) we have

$$\ln\frac{C(0)}{C(\Theta)}=\ln\frac{\pi\Theta}{\gamma C(\Theta)}+2\pi\sum_{/1}^{\infty}\left\{\frac{1}{(2l-1)\pi}-\frac{1}{\sqrt{\left(\frac{C}{\Theta}\right)^2+(2l-1)^2\pi^2}}\right\}. \quad \text{(II.39)}$$

Formula (II.39) allows us to find the critical temperature Θ_c. For Θ_c we have $C(\Theta_c)=0$. In this case,

$$\ln C(0)=\ln\frac{\pi\Theta_c}{\gamma},$$
$$\Theta_c=\frac{\gamma C(0)}{\pi}=\frac{2\omega e^{-2\varrho}}{1.76}. \quad \text{(II.40)}$$

Formula (II.33) is rewritten in the form (II.39). Now the sum on the right-hand side can be expanded in a power series in the small parameter C/Θ. Finally, we have

$$\left[(2l-1)^2\pi^2+\left(\frac{C}{\Theta}\right)^2\right]^{-1/2}=\frac{1}{(2l-1)\pi}\left[1-\frac{1}{2}\frac{1}{\pi^2(2l-1)^2}\left(\frac{C}{\Theta}\right)^2+\frac{3}{8}\frac{1}{\pi^4(2l-1)^4}\left(\frac{C}{\Theta}\right)^4++\right]. \quad \text{(II.41)}$$

Using (II.38, 40) we obtain

$$\ln\frac{\Theta}{\Theta_c}=\frac{7\zeta(3)}{8\pi^2}\left(\frac{C(\Theta)}{\Theta}\right)^2+\frac{93}{128\pi^4}\zeta(5)\left(\frac{C(\Theta)}{\Theta}\right)^4. \quad \text{(II.42)}$$

In first order in $\Theta_c-\Theta$ equation (II.42) gives

$$\ln\left(1+\frac{\Theta-\Theta_c}{\Theta_c}\right)\approx\frac{\Theta-\Theta_c}{\Theta_c}=-\frac{1}{(3.06)^2}\left(\frac{C}{\Theta}\right)^2\approx\frac{1}{(3.06)^2}\left(\frac{C}{\Theta_c}\right)^2.$$

(II.43)

Hence

$$\frac{C(\Theta)}{\Theta_c} = \sqrt{\frac{8\pi}{7\zeta(3)}}\sqrt{\frac{\Theta_c-\Theta}{\Theta_c}} = 3.06\sqrt{\frac{\Theta_c-\Theta}{\Theta_c}}. \tag{II.44}$$

The dependence $C(\Theta)/\Theta_c$ on Θ/Θ_c has the form shown in Fig. 7.

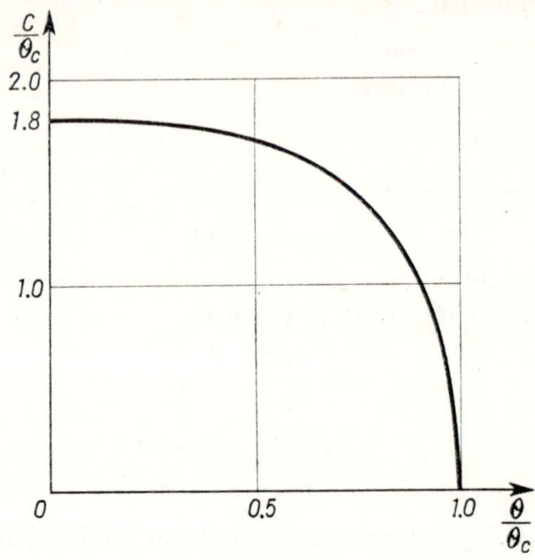

FIG. 7

In formula (II.40) we have a factor ω, which is of the order of the maximum Debye frequency. This frequency is proportional to $1/\sqrt{M}$ where M is the mass of the ion in the crystal lattice. We also have $\omega \sim 1/\sqrt{M}$ and hence

$$\Theta_c \sim \frac{1}{\sqrt{M}}. \tag{II.45}$$

(II.45) gives the isotope effect, i.e. the dependence of the transition temperature on the ionic mass.

§ 3 Thermodynamic potential

We will now demonstrate, using the method given by Bogoliubov, Zubarev and Tserkovnikov (1957), that for the Hamiltonian

(I.31) with the interaction given by (I.58) (in the above-mentioned reference the interaction has a more general form) one can obtain for $V \to \infty$ an asymptotically exact expression for the thermodynamic potential

$$\psi = F - \lambda\langle\hat{N}\rangle = -\Theta \ln \mathrm{Tr}\, e^{-\frac{\hat{H}}{\Theta}} \tag{II.46}$$

where $F = E - \Theta S$ is the free energy and S the entropy.

We define an operator

$$\hat{B}_k = N(a_{k+}a_{-k-}) = a_{k+}a_{k-} - \underbrace{a_{k+}a_{-k,-}}$$

$$= a_{k+}a_{-k,-} + u(k)v(k) \tag{II.47}$$

(see (I.8)) where $\{u, v\}$ are the coefficients of the transformation

$$a_{k\sigma} = u(k)\beta_{k\sigma} + v(k, \sigma)\beta^{+}_{-k,-\sigma} \tag{II.48}$$

connected by means of (I.39).

The Hamiltonian (I.31) now has the form

$$\hat{H} = \hat{H}^{(0)} + \hat{H}',$$

$$\hat{H}^{(0)} = \sum_k V_k + \sum_k \hat{H}_k, \qquad \hat{H}' = -\frac{g^2}{2V}\sum_{k,k} \hat{B}^{+}_{k}\hat{B}_k, \tag{II.49}$$

where

$$\tilde{V}_k = 2\zeta_k v(k)^2 - u(k)v(k)\frac{g^2}{2V}\sum_{k'} u(k')v(k'),$$

$$\hat{H}_k = \tilde{\Omega}(k)(\beta^{+}_{k+}\beta_{k+} + \beta^{+}_{-k,-}\beta_{-k,-}) + \tilde{\Lambda}_k(\beta^{+}_{k+}\beta^{+}_{-k,-} + \beta_{-k,-}\beta_{k,+}),$$

$$\tilde{\Omega}(k) = \zeta_k[u(k)^2 - v(k)^2] + u(k)v(k)\frac{g^2}{V}\sum_{k'} u(k')v(k'), \tag{II.50}$$

$$\tilde{\Lambda}_k = 2\zeta_k u(k)v(k) - [u(k)^2 - v(k)^2]\frac{g^2}{2V}\sum_{k'} u(k')v(k'),$$

$$\varepsilon(k) - \lambda = \zeta_k.$$

Now we wish to prove that as $V \to \infty$, we have

$$\ln \mathrm{Tr}\exp(-\hat{H}/\Theta) = \ln \mathrm{Tr}\exp(-\hat{H}^{(0)}/\Theta).$$

Using thermodynamic perturbation theory we have

$$\frac{\operatorname{Tr} e^{-\frac{\hat{H}}{\Theta}}}{\operatorname{Tr} e^{-\frac{\hat{H}^{(0)}}{\Theta}}} = 1 + \sum_{n>1} (-1)^n \int_0^{\frac{1}{\Theta}} \mathrm{d}t_1 \int_0^{t_1} \mathrm{d}t_2 \ldots \int_0^{t_{n-1}} \mathrm{d}t_n$$

$$\times \frac{\operatorname{Tr}\{e^{-\frac{\hat{H}^{(0)}}{\Theta}} H'(t_1) \ldots H'(t_n)\}}{\operatorname{Tr}\{e^{-\frac{\hat{H}^{(0)}}{\Theta}}\}}, \qquad \text{(II.51)}$$

$$H'(t) = e^{\hat{H}^{(0)}t} H' e^{-\hat{H}^{(0)}t}.$$

This formula can be written in the form

$$\ln \operatorname{Tr} e^{-\frac{\hat{H}}{\Theta}} - \ln \operatorname{Tr} e^{-\frac{\hat{H}^{(0)}}{\Theta}}$$

$$= \ln\Big\{1 + \sum_{n>1} (-1)^n \int_0^{\frac{1}{\Theta}} \mathrm{d}t_1 \int_0^{t_1} \mathrm{d}t_2 \ldots \int_0^{t_{n-1}} \mathrm{d}t_n \mathfrak{A}_n\Big\} \qquad \text{(II.52)}$$

where

$$\mathfrak{A}_n = \left(\frac{g^2}{2V}\right)^n \sum_{\substack{k_1, \ldots, k_n \\ k'_1, \ldots, k'_n}} \frac{\operatorname{Tr}\{e^{-\frac{\hat{H}^{(0)}}{\Theta}} \tilde{\hat{B}}_{k_1}(t_1) \hat{B}_{k'_1}(t_1) \ldots \tilde{\hat{B}}_{k_n}(t_n) \hat{B}_{k'_n}(t_n)\}}{\operatorname{Tr}\{e^{-\frac{\hat{H}^{(0)}}{\Theta}}\}}$$

$$\hat{B}_k(t) = e^{\hat{H}^{(0)}t} \hat{B}_k e^{-\hat{H}^{(0)}t} = e^{\hat{H}_k t} \hat{B}_k e^{-\hat{H}_k t}, \quad \tilde{\hat{B}}_k(t) = e^{\hat{H}_k t} \hat{B}_k^+ e^{\hat{H}_k t}. \qquad \text{(II.53)}$$

Taking the logarithm on both sides of (II.52) we obtain, on the left-hand side, logarithms of exponential functions. The exponents of these functions are proportional to the energy. Energy is an extensive quantity proportional to the volume, hence the two terms on the left-hand side of (II.52) go to infinity as $V \to \infty$. After dividing by V, the left-hand side of (II.52) is finite. We wish to demonstrate that the right-hand side of (II.52) is a finite quantity as $V \to \infty$, and therefore vanishes if divided by V. For this purpose we must prove that $\mathfrak{A}_n$ is independent of V. This will be achieved by assuming that for all k

$$\operatorname{Tr}\{e^{-\frac{\hat{H}_k}{\Theta}} \hat{B}_k\} = 0. \qquad \text{(II.54)}$$

We will demonstrate later that this formula can in fact be satisfied after diagonalizing $\hat{H}_k$ (II.50).

We remark that the operators $\hat{H}_k$, $\hat{B}_k$, $\hat{B}_k^+$ commute with each other for different k. If in some term of the sum $\mathfrak{A}_n$, among the products of the operators $\hat{B}$, $\hat{\tilde{B}}$, only one operator $\hat{B}_q$ (or $\hat{\tilde{B}}_q$) appears, such that q is different from other indices, we can rearrange the operators in order to exclude the factor of the type (II.54), i.e.

$$\mathrm{Tr}\{e^{-\frac{\hat{H}_q}{\Theta}}\hat{B}_q(t)\} = \mathrm{Tr}\{e^{-\frac{\hat{H}_q}{\Theta}}\hat{B}_q\} = 0$$

or

$$\mathrm{Tr}\{e^{-\frac{\hat{H}_q}{\Theta}}\hat{\tilde{B}}_q(t)\} = \mathrm{Tr}\{e^{-\frac{\hat{H}_q}{\Theta}}\hat{B}_q^+\} = 0.$$

Hence, in order to have a non-zero term in $\mathfrak{A}_n$, at least n of the $2n$ indices $k_1, \ldots k_n, \ldots k'_1, \ldots k'_n$ must be different.

If we have exactly n different indices, then each must appear twice. Hence in $\mathfrak{A}_n$ we have a sum over n indices $\sum\limits_{p'_1 \cdots p_n}$ which is of the order V^n; $(1/V)\sum\limits_p$ is a finite quantity, i.e. $\sum\limits_p \sim V$. The factor $1/V^n$ in $\mathfrak{A}_n$ is compensated, and all expressions are independent of V. In a second extreme case all indices $k_1, \ldots k'_n$ are equal to each other. In this case, in $\mathfrak{A}_n$ there is a single sum which gives a term proportional to V/V^n vanishing in the limit as $V \to \infty$.

Therefore, we have proved that with the condition (II.54), $\mathfrak{A}_n$ is independent of volume and hence that the right-hand side of (II.52) does not depend on V and can be neglected in the limit as $V \to \infty$, in comparison with the left-hand side which goes to infinity. We also obtain the asymptotic formula as $V \to \infty$

$$\ln \mathrm{Tr}\, e^{-\frac{\hat{H}}{\Theta}} = \ln e^{-\frac{\hat{H}^{(0)}}{\Theta}}. \tag{II.55}$$

The proof of (II.55) given here is incomplete since we have not ascertained whether the series on the right-hand side of (II.52) is convergent. A rigorous proof of (II.55) is given by Bogoliubov, Zubarev and Tserkovnikov (1960).

Formula (II.54) is an equation additional to (I.39) determining $\{u, v\}$. After diagonalizing $\hat{H}_k$, in (II.54) we have only the diagonal part of $\hat{B}_k$ expressed in terms of $\{u, v\}$. In order to use (II.54) we introduce the new Fermi operators

$$\beta_{k\sigma} = \lambda(k)\alpha_{k\sigma} - \mu(k, \sigma)\alpha^{+}_{-k, -\sigma} \tag{II.56}$$

where

$$\lambda^2(k) + \mu^2(k) = 1, \quad \mu(k) = \mu(k, +) = -\mu(k, -).$$

The expression for (a, a^+) in terms of (α, α^+) is a result of two transformations (II.48, 56) and can be written

$$\begin{aligned} a_{k\sigma} &= \tilde{u}(k)\,\alpha_{k\sigma} + \tilde{v}(k,\sigma)\,\alpha^+_{-k,-\sigma}, \\ \tilde{u}(k) &= u(k)\,\lambda(k) + v(k)\,\mu(k), \\ \tilde{v}(k,\sigma) &= v(k,\sigma)\;\lambda(k) - u(k)\,\mu(k,\sigma), \\ \tilde{u}^2(k) + \tilde{v}^2(k) &= 1. \end{aligned} \tag{II.57}$$

The operator $\tilde{B}_k$ defined by (II.47) can be expressed in the form

$$\begin{aligned} \hat{B}_k = u(k)v(k) - \tilde{u}(k)\tilde{v}(k)\,[1 + \alpha^+_{k+}\,\alpha_{k+} + \alpha^+_{-k,-}\,\alpha_{-k,-}] \\ - \tilde{u}^2(k)\,\alpha_{-k,-}\,\alpha_{k+} + \tilde{v}^2(k)\alpha^+_{k+}\,\alpha^+_{-k,-}. \end{aligned} \tag{II.58}$$

We put

$$C = \frac{g^2}{2V}\sum_k u(k)v(k) \tag{II.59}$$

and write

$$\begin{aligned} u(k) &= \cos\varphi_k, \quad & v(k) &= \sin\varphi_k, \\ \lambda(k) &= \cos\chi_k, \quad & \mu(k) &= \sin\chi_k. \end{aligned} \tag{II.60}$$

Then

$$\begin{aligned} u^2(k) - v^2(k) &= \cos 2\varphi_k, \quad & 2u(k)v(k) &= \sin 2\varphi_k, \\ \lambda^2(k) - \mu^2(k) &= \cos 2\chi_k, \quad & 2\lambda(k)\mu(k) &= \sin 2\chi_k. \end{aligned} \tag{II.61}$$

From (II.57) it follows that

$$\tilde{u}(k) = \cos(\varphi_k - \chi_k), \qquad \tilde{v}(k) = \sin(\varphi_k - \chi_k). \tag{II.62}$$

With the help of (II.59) and (II.61) the quantities $\tilde{\Omega}(k)$, $\tilde{\Lambda}(k)$ are expressed by the equation

$$\begin{pmatrix} \tilde{\Omega}(k) \\ \tilde{\Lambda}(k) \end{pmatrix} = \begin{pmatrix} \cos 2\varphi_k, & \sin 2\varphi_k \\ \sin 2\varphi_k, & -\cos 2\varphi_k \end{pmatrix} \begin{pmatrix} \zeta_k \\ C \end{pmatrix}. \tag{II.63}$$

After the transformation (II.56) we can write

$$\begin{aligned} \hat{H}_k = V_k + \Omega(k)\,(\alpha^+_{k+}\,\alpha_{k+} + \alpha^+_{-k,-}\,\alpha_{-k,-}) \\ + \Lambda(k)(\alpha^+_{k+}\,\alpha^+_{-k,-} + \alpha_{-k,-}\,\alpha_{k+}) \end{aligned} \tag{II.64}$$

where

$$\begin{aligned} V_k &= \zeta_k[u(k)^2 - v(k)^2] + 2u(k)v(k)C - \Omega(k), \\ \begin{pmatrix} \Omega(k) \\ -\Lambda(k) \end{pmatrix} &= \begin{pmatrix} \cos 2\chi_k, & \sin 2\chi_k \\ \sin 2\chi_k, & -\cos 2\chi_k \end{pmatrix} \begin{pmatrix} \Omega(k) \\ \tilde{\Lambda}(k) \end{pmatrix}. \end{aligned} \tag{II.65}$$

From (II.63,65) we have

$$\begin{pmatrix} \Omega(k) \\ -\Lambda(k) \end{pmatrix} = \begin{pmatrix} \cos 2(\varphi_k-\chi_k), & \sin 2(\varphi_k-\chi_k) \\ -\sin 2(\varphi_k-\chi_k), & \cos 2(\varphi_k-\chi_k) \end{pmatrix} \begin{pmatrix} \zeta_k \\ C \end{pmatrix}, \qquad \text{(II.66)}$$

and

$$\Omega(k)^2+\Lambda(k)^2 = \zeta_k^2+C^2. \qquad \text{(II.67)}$$

The condition for the diagonalization of $\hat{H}_k$ has the form (we use (II.66,62))

$$\Lambda(k) = 0 = \zeta_k 2\tilde{u}(k)\tilde{v}(k)-C[\tilde{u}(k)^2-\tilde{v}(k)^2]. \qquad \text{(II.68)}$$

Hence

$$\Omega(k) = \sqrt{\zeta_k^2+C^2}. \qquad \text{(II.69)}$$

Equations (II.57,58), similarly to (II.17), give the relations

$$\tilde{u}(k)^2 = \frac{1}{2}\left\{1-\frac{\zeta_k}{\Omega(k)}\right\}, \quad \tilde{v}(k)^2 = \left\{1+\frac{\zeta_k}{\Omega(k)}\right\},$$
$$\tilde{u}(k)\tilde{v}(k) = \frac{1}{2}\frac{C}{\Omega(k)}. \qquad \text{(II.70)}$$

Our aim is to calculate $u(k)$, $v(k)$ by means of (II.54). As $\hat{H}_k$ is diagonal, we are interested only in the diagonal part of $\hat{B}_k$. We use (II.70) and let $\alpha_{k\sigma}^{+}\alpha_{k\sigma} = n_{k\sigma}$. From (II.58) we have

$$(\hat{B}_k)_{\text{diag}} = u(k)v(k)-\frac{1}{2}\frac{C}{\Omega(k)}[1-(n_{k+}+n_{k-})]. \qquad \text{(II.71)}$$

The condition (II.54) can be rewritten

$$\operatorname{Tr}\{e^{-\frac{\hat{H}_k}{\Theta}}\hat{B}_k\} = e^{-\frac{V_k}{\Theta}}\Bigg\{u(k)v(k)-\frac{C}{2\Omega(k)}\sum_{n_{k\sigma}=0,1} e^{-\frac{\Omega(k)}{\Theta}(n_{k+}+n_{k-})}$$
$$+\frac{C}{2\Omega(k)}\sum_{n_{k\sigma}=0,1}[n_{k+}+n_{k-}]e^{-\frac{\Omega(k)}{\Theta}(n_{k+}+n_{k-})}\Bigg\} = 0. \qquad \text{(II.72)}$$

In order to have a common factor $\sum_{n_{k\sigma}} e^{-\frac{\Omega(k)}{\Theta}(n_{k+}+n_{k-})}$, we replace $n_{k\sigma}$ in the square bracket by $\bar{n}_{k\sigma}$. We have

$$u(k)v(k) = \frac{C}{2\Omega(k)}(1-\bar{n}_{k+}-\bar{n}_{k-}). \qquad \text{(II.73)}$$

For $\bar{n}_{k+} = \bar{n}_{k-}$ we obtain from (II.59) an equation for $C(\Theta)$ identical to (II.22).

For the thermodynamic potential we have the expression

$$\psi = -\Theta \ln \operatorname{Tr} e^{-\frac{\hat{H}^{(0)}}{\Theta}} \tag{II.74}$$

where

$$\hat{H}^{(0)} = \sum_k [\zeta_k + Cu(k)v(k) - \Omega(k)] + \sum_{k,\sigma} \Omega(k) n_{k\sigma}. \tag{II.75}$$

The first term in (II.75) can be expressed in terms of $\tilde{V}_k$, V_k given by (II.50,65). We see that for the calculation of ψ it is sufficient to have u, v given by (II.73).

From (II.75) and (II.73) we finally obtain

$$\psi = \sum_k [\zeta_k + u(k)v(k)C - \Omega(k)] - \Theta \sum_k \ln \sum_{n_{k\sigma}=0,1} e^{\frac{\Omega(k)}{\Theta}(n_{k+}+n_{k-})}$$
$$= \sum_k \left[\zeta_k + \frac{C^2}{2\Omega(k)}(1-\bar{n}_{k+}-\bar{n}_{k-}) - \Omega(k) - 2\Theta \ln\left(1+e^{-\frac{\Omega(k)}{\Theta}}\right)\right]. \tag{II.76}$$

Now we wish to calculate $\bar{n}_{k\sigma} = \langle n_{k\sigma}\rangle$ and to demonstrate that $\bar{n}_{k\sigma} = n_k$. From the definition of the mean value

$$\bar{n}_{k+} = \frac{\operatorname{Tr}\{n_{k+} e^{-\frac{\hat{H}^{(0)}}{\Theta}}\}}{\operatorname{Tr}\{e^{-\frac{\hat{H}^{(0)}}{\Theta}}\}} = \frac{\sum_{n_{k+}=0,1} \{n_{k+} e^{-\frac{\Omega(k)n_{k+}}{\Theta}}\}}{\sum_{n_{k+}=0,1} e^{-\frac{\Omega(k)n_{k+}}{\Theta}}}$$
$$= \frac{1}{e^{\frac{\Omega(k)}{\Theta}}+1} = \bar{n}_k. \tag{II.77}$$

Using (II.77) (see also (II.20)) we have

$$\ln\left(1+e^{-\frac{\Omega(k)}{\Theta}}\right) = -\frac{\Omega(k)}{\Theta} - \ln \bar{n}_k.$$

For ψ we now write the expression

$$\psi = \sum_k \left\{\zeta_k - \left[\Omega(k) - \frac{C^2}{2\Omega(k)}\right](1-2\bar{n}_k)\right\}$$
$$+\Theta 2 \sum_k \left[\ln \bar{n}_k + (1-\bar{n}_k)\frac{\Omega(k)}{\Theta}\right]. \tag{II.78}$$

By means of (II.77) it is easy to prove that

$$\ln \bar{n}_k - (\bar{n}_k - 1)\frac{\Omega(k)}{\Theta} = \bar{n}_k \ln \bar{n}_k + (1-\bar{n}_k)\ln(1-\bar{n}_k). \qquad \text{(II.79)}$$

The first term in (II.78) is equal to $\langle \hat{H} \rangle$ (see (II.21). Hence, from (II.21) and (II.79) we have

$$\psi = \langle \hat{H} \rangle + \Theta 2 \sum_k [\bar{n}_k \ln \bar{n}_k + (1-\bar{n})\ln(1-\bar{n}_k)]. \qquad \text{(II.80)}$$

From (II.78,77,69) we see that ψ is a function of C^2. For $C \neq 0$ we have a superconducting state and write $\psi(C^2) = \psi_s$. For $C = 0$ we have a normal state and write $\psi(0) = \psi_n$. In order to examine the function $\psi = \psi(C^2)$, we consider the derivative $\psi' = \frac{\partial \psi}{\partial C^2}$. We have

$$\psi' = -\frac{C^2}{4\Theta} \sum_k f\left(\frac{\Omega(k)}{\Theta}\right) \leqslant 0 \qquad \text{(II.81)}$$

where

$$f(x) = \frac{\sinh x - x}{2x^3 \cosh^2 \frac{x}{2}} = -\frac{1}{x}\frac{\mathrm{d}}{\mathrm{d}x}\left(\frac{1}{x}\tanh\frac{x}{2}\right) > 0. \qquad \text{(II.82)}$$

We see that ψ is a decreasing function of C^2 and has its maximum value for $C = 0$, i.e. for the normal state. Therefore we have

$$\psi_s < \psi_n. \qquad \text{(II.83)}$$

Thus, the transition to the superconducting state takes place at the temperature $\Theta = \Theta_c$ given by the formula (II.40). Below this temperature, $C \neq 0$.

Consider ψ_s near $C = 0$, i.e. for $\Theta \approx \Theta_c$, by expanding it in powers of C^2. From (II.81) we see that $(\psi')_0 = 0$. If we use the formula

$$\tanh\frac{x}{2} = 4x \sum_{l/1}^{\infty} \frac{1}{(2l-1)^2\pi^2 + x^2} \qquad \text{(II.84)}$$

we obtain

$$(\psi'')_0 = -\frac{2}{\Theta^3} \sum_k \sum_l \frac{1}{\left[(2l-1)^2\pi^2 + \left(\frac{\zeta_k}{\Theta}\right)^2\right]^2}. \qquad \text{(II.85)}$$

Integrating instead of summing we have

$$(\psi'')_0 = -\frac{1}{\Theta^2}\frac{dn}{dE}\frac{1}{\pi^2}\sum_{l/1}^{\infty}\frac{1}{(2l-1)^3} = -\frac{1}{(\pi\Theta)^2}\frac{dn}{dE}\frac{7}{8}\zeta(3). \tag{II.86}$$

The next differentiation gives

$$(\psi''')_0 = \frac{1}{(\pi\Theta)^4}\frac{dn}{dE}\frac{93}{32}\zeta(5), \quad \zeta(5) = 1.04 \approx 1.$$

Finally, we write the expansion in the form

$$\psi_s - \psi_n = -\frac{\Theta^2}{\pi^2}\frac{dn}{dE}\frac{7}{16}\zeta(3)\left(\frac{C}{\Theta}\right)^2 + \frac{\Theta^2}{\pi^4}\frac{dn}{dE}\frac{31}{64}\left(\frac{C}{\Theta}\right)^6. \tag{II.87}$$

Now we will examine what kind of transition takes place at $\Theta = \Theta_c$. For this purpose we will consider the entropy and the temperature dependence of the specific heat.

§ 4 Entropy and specific heat

We have the following expression for the thermodynamic potential (II.46):

$$\psi = \bar{E} - \lambda\langle\hat{N}\rangle - \Theta S = \langle\hat{H}\rangle - \Theta S. \tag{II.88}$$

From (II.80) we see that the entropy of our system is given by the following expression:

$$S = -2\sum_k [\bar{n}_k \ln \bar{n}_k + (1-\bar{n}_k)\ln(1-\bar{n}_k)] \tag{II.89}$$

where the $\bar{n}_k$ are given by (II.77) and are connected with quasi-particles with energy spectrum $\Omega(p)$.

Now we wish, as was done by Khalatnikov and Abrikosov (1958), to find the dependence $S = S(\Theta)$ for $\Theta \approx 0$. We must express (II.89) in terms of Bessel functions. In (II.89) we integrate rather than sum. As in (II.29), we introduce a new variable Ω and have

$$d\zeta = \left[\frac{d}{d\Omega}\sqrt{\Omega^2 - C^2}\right] d\Omega.$$

For S we obtain

$$S = -4\frac{dn}{dE}\left[\ln\bar{n} - \frac{\bar{n}-1}{\Theta}\Omega\right]_0^{\omega} - 4\frac{dn}{dE}\int_C^{\omega}\Omega\sqrt{\Omega^2 - C^2}\frac{d\bar{n}}{d\Omega}\, d\Omega. \tag{II.90}$$

We have used eqns. (II.79) and the identities

$$\frac{d}{d\Omega}\ln\bar{n} = \frac{1}{\bar{n}}\frac{d\bar{n}}{d\Omega} = \frac{1}{\Theta}(\bar{n}-1). \tag{II.91}$$

In (II.90) we have excluded the terms that are divergent for $\Omega = \omega \to \infty$. In the limit we have

$$\bar{n} \approx e^{-\frac{\Omega}{\Theta}}, \quad \ln\bar{n} \approx -\frac{\Omega}{\Theta}, \quad 1-\bar{n} \approx 1. \tag{II.92}$$

We see that the upper limit in (II.90) can be replaced by ∞ since the terms in the bracket vanish and the integral, due to the factor $\frac{d\bar{n}}{d\Omega}$, is convergent. So we have

$$S = -4\frac{dn}{dE}\int_{C}^{\infty}\Omega\sqrt{\Omega^2-C^2}\frac{d\bar{n}}{d\Omega}\,d\Omega. \tag{II.93}$$

As in (II.29), we introduce a new variable and after integration by parts we have

$$S = 2\frac{dn}{dE}\frac{C^2}{\Theta}\int_{c}^{\infty}\frac{\cosh 2\varphi}{\exp\frac{C\cosh\varphi}{\Theta}+1}\,d\varphi. \tag{II.94}$$

The function under the integral sign can be written in the form

$$\frac{1}{2}\sum_{n/1}^{\infty}(-1)^{n+1}(e^{2\varphi}+e^{-2\varphi})e^{-\frac{mC}{\Theta}\cosh\varphi}. \tag{II.95}$$

From (II.32)

$$S = \frac{2C(\Theta)^2}{\Theta}\frac{dn}{dE}\sum_{m/1}^{\infty}(-1)^{m+1}K_2\left(\frac{mC(\Theta)}{\Theta}\right). \tag{II.96}$$

Formula (II.96) allows us to find the temperature dependence of S for $\Theta \sim 0$. We use the asymptotic formula for Bessel functions (II.34) and obtain

$$S = \frac{dn}{dE}\sqrt{\frac{2\pi C(0)^3}{\Theta}}\,e^{-\frac{C(0)}{\Theta}}. \tag{II.97}$$

On the right-hand side we write $C(\Theta) = C(0)$.

From (II.97) and the formula $c = \Theta\left(\frac{\partial S}{\partial\Theta}\right)_V$ we obtain the temperature dependence of the specific heat, in the superconducting state, for $\Theta \sim 0$

$$c_s = \sqrt{2\pi}\,\frac{dn}{dE}\,C(0)\left(\frac{C(0)}{\Theta}\right)^{3/2} e^{-\frac{C(0)}{\Theta}}. \tag{II.98}$$

We see that we have obtained the exponential temperature dependence of the specific heat which is observed in experiment.

Putting $C = 0$ in (II.93) and integrating by parts we obtain an expression for the entropy for the normal state (at low temperatures)

$$S_n = 8\,\frac{dn}{dE}\,\frac{1}{\Theta}\int_0^\infty \frac{\Omega}{e^{\frac{\Omega}{\Theta}}+1}\,d\Omega = \frac{2}{3}\,\pi^2\frac{dn}{dE}\,\Theta. \tag{II.99}$$

Hence, for the specific heat we have

$$c_n = \frac{2}{3}\,\pi^2\frac{dn}{dE}\,\Theta. \tag{II.100}$$

Using (II.100) we can rewrite (II.98) in the form

$$\frac{c_s(\Theta)}{c_n(\Theta_c)} = \frac{3}{\gamma}\sqrt{\frac{2}{\pi}}\left(\frac{C(0)}{\Theta}\right)^{3/2} e^{-\frac{C(0)}{\Theta}} \tag{II.101}$$

where Θ_c is given by (II.40).

Now we are interested in the temperature dependence of the entropy and the specific heat, in the superconducting state, for $\Theta \sim \Theta_c$. In this region of temperatures $C \sim 0$ and we can use formula (II.87) which gives $\psi_s - \psi_n$ as a function of C^2. The temperature dependence of C, for $\Theta \sim \Theta_c$, is given by (II.44).

We note that $\psi_s = F_s - \lambda\langle\hat{N}\rangle$, $\psi_n = F_n - \lambda\langle\hat{N}\rangle$. Hence

$$\psi_s - \psi_n = F_s - F_n. \tag{II.102}$$

Therefore, from (II.87) and the formula $S = -\left(\frac{\partial F}{\partial \Theta}\right)_V$ we obtain the temperature dependence of $S_s - S_n$ for $\Theta \sim \Theta_c$. From (II.88) and (II.44) we have

$$\psi_s - \psi_n = -\frac{dn}{dE}\,\frac{4\pi^2}{7\zeta(3)}\,\Theta_c^2\left(\frac{\Theta_c - \Theta}{\Theta}\right)^2, \quad \zeta(3) = 1.2,$$

$$\frac{\Theta_c - \Theta}{\Theta} = \frac{\Theta_c - \Theta}{\Theta_c}\left[1 + \frac{\Theta_c - \Theta}{\Theta_c} + +\right]. \tag{II.103}$$

The entropy is given by the formula

$$S_s - S_n = -\frac{\mathrm{d}n}{\mathrm{d}E}\frac{8\pi^2}{7\zeta(3)}\Theta_c\left(\frac{\Theta_c - \Theta}{\Theta_c}\right), \tag{II.104}$$

and the specific heat by

$$c_s - c_n = \frac{\mathrm{d}n}{\mathrm{d}E}\frac{8\pi^2}{7\zeta(3)}\Theta. \tag{II.105}$$

We see that $c_s(\Theta_c) \neq c_n(\Theta_c)$, i.e. the specific heat has a jump at $\Theta = \Theta_c$. This is characteristic of second-order phase transitions. The functions ψ and S do not have a jump at $\Theta = \Theta_c$.

Using (II.100) we have

$$\frac{c_s(\Theta_c)}{c_n(\Theta_c)} = 1 + \frac{16}{21}\zeta(3) = 1.63.$$

If in (II.103) we also consider higher powers of $\Theta_c - \Theta$ (see (II.87)), we must express C^2 by means of the formula

$$C^2 = \frac{8\pi^2}{7\zeta(3)}\Theta^2\frac{\Theta_c - \Theta}{\Theta_c}.$$

In this case, we obtain for the entropy

$$S_s - S_n = \frac{2}{3}\pi^2\frac{\mathrm{d}n}{\mathrm{d}E}\Theta\left[-\frac{2}{3\pi^2}\left(\frac{C(\Theta)}{\Theta}\right)^2 + \frac{143}{32}\frac{\zeta(3)}{\pi^4}\left(\frac{C(\Theta)}{\Theta}\right)^4\right].$$

These more exact calculations give a jump in the specific heat by a factor 2.43 and the temperature dependence of the specific heat in the form

$$\frac{c_s(\Theta)}{c_n(\Theta_c)} = 2.43 - 10.6\left(\frac{\Theta_c - \Theta}{\Theta_c}\right). \tag{II.106}$$

§ 5 Critical magnetic field

We find the critical magnetic field from the relation

$$\frac{H_c^2}{8\pi} = F_n - F_s = \psi_n - \psi_s \tag{II.107}$$

(see (II.102)). For temperatures $\Theta \sim \Theta_c$ we find the temperature dependence of H_c from (II.103). Now we are interested in an expression for $\psi_s = \langle \hat{H} \rangle - \Theta S_s$ from which we can obtain the temperature dependence of H_c for $\Theta \sim 0$. For this reason we wish to transform the expression for the entropy given by

(II.89), by considering (II.79) (see Khalatnikov and Abrikosov, 1958). After integration by parts we have

$$\sum_k \ln \bar{n}_k = a + \frac{1}{\Theta} \sum_k \left(\frac{C^2}{2\Omega(k)} - \Omega(k) \right) (\bar{n}_k - 1) \quad \text{(II.108)}$$

where

$$a = -2 \frac{dn}{dE} \omega \ln(1 + e^{\frac{1}{\Theta}\sqrt{\omega^2+C^2}}) \approx -2 \frac{dn}{dE} \frac{\omega}{\Theta} \sqrt{\omega^2 + C^2},$$

$$\sum_k \left(\frac{C^2}{2\sqrt{\zeta_k^2 + C^2}} - \sqrt{\zeta_k^2 + C^2} \right)$$

$$= 2 \frac{dn}{dE} \int_0^{\omega} \left(\frac{C^2}{2\sqrt{y^2+C^2}} - \sqrt{y^2+C^2} \right) dy$$

$$= -\frac{dn}{dE} \omega \sqrt{\omega^2 + C^2} = \frac{2}{\Theta} a. \quad \text{(II.109)}$$

Hence the entropy in the superconducting state is given by

$$S_s = -\frac{4}{\Theta} \sum_k \left(\frac{C^2}{2\Omega(k)} - \Omega(k) \right) \bar{n}_k. \quad \text{(II.110)}$$

By means of (II.110) we can rewrite formula (II.21) in the form

$$\langle \hat{H} \rangle = \sum_k \left[\zeta_k + \left(\frac{C^2}{2\sqrt{\zeta_k^2 + C^2}} - \sqrt{\zeta_k^2 + C^2} \right) \right] + \frac{1}{2} S_s \Theta. \quad \text{(II.111)}$$

Using (II.109) we obtain

$$\psi_s = \sum_k \zeta_k - \frac{dn}{dE} \omega \sqrt{\omega^2 + C^2} - \frac{1}{2} S_s \Theta. \quad \text{(II.112)}$$

The first sum can be written in the form

$$\sum_k \zeta_k = 2 \sum_{\zeta_k < 0} \zeta_k + \sum_k |\zeta_k| = b + \frac{dn}{dE} \omega^2. \quad \text{(II.113)}$$

The constant b does not depend on the gap and is common to the normal and superconducting state, so that it can be omitted. Finally, for ψ_s we have

$$\psi_s = \omega^2 \frac{\mathrm{d}n}{\mathrm{d}E}\left(1+\sqrt{1+\left(\frac{C}{\omega}\right)^2}\right)-\frac{1}{2}S_s\Theta$$

$$= -\frac{1}{2}\left(\frac{\mathrm{d}n}{\mathrm{d}E}C^2+S\Theta\right). \qquad \text{(II.114)}$$

For the normal state ($C = 0$)

$$\psi_n = -\tfrac{1}{2}S_n\Theta. \qquad \text{(II.115)}$$

Hence

$$\frac{H_c^2}{8\pi} = \frac{1}{2}\frac{\mathrm{d}n}{\mathrm{d}E}C^2+\frac{1}{2}(S_s-S_n)\Theta. \qquad \text{(II.116)}$$

For $\Theta = 0$ we have

$$\frac{H_{c0}^2}{8\pi} = \frac{1}{2}\frac{\mathrm{d}n}{\mathrm{d}E}C^2(0). \qquad \text{(II.117)}$$

From (II.40) it follows that

$$H_{c0} = 2\Theta_c\frac{\pi}{\gamma}\sqrt{\pi\frac{\mathrm{d}n}{\mathrm{d}E}}, \qquad \frac{\mathrm{d}n}{\mathrm{d}E} = \left(\frac{H_{c0}}{\Theta_c}\right)^2\frac{\gamma^2}{4\pi^2}. \qquad \text{(II.118)}$$

For $\Theta \sim 0$ the entropy S_s decreases exponentially and S_n is proportional to Θ (II.99), therefore $S_n \gg S_s$. For this reason, as we see from (II.36), we can in (II.116) take $C(\Theta) = C(0) = \frac{\pi}{\gamma}\Theta_c$ and obtain

$$\frac{H_c^2}{8\pi} = \left(\frac{H_{c0}}{\Theta_c}\right)^2\frac{\gamma^2}{4\pi^3}\left[\frac{1}{2}\left(\frac{\pi}{\gamma}\Theta_c\right)^2-\frac{\pi^2}{3}\Theta^2\right]. \qquad \text{(II.119)}$$

Finally we can write

$$H_c = H_{c0}\left[1-\frac{\gamma^2}{3}\left(\frac{\Theta}{\Theta_c}\right)^2\right], \qquad \gamma^2 = (1.78)^2 = 3.06. \qquad \text{(II.120)}$$

For temperatures $\Theta \sim \Theta_c$ we use (II.103). After eliminating $\mathrm{d}n/\mathrm{d}E$ we have

$$H_c(\Theta) = H_{c0}\gamma\sqrt{\frac{8}{7\zeta(3)}}\left(\frac{\Theta_c-\Theta}{\Theta_c}\right) \approx 1.73H_{c0}\left(\frac{\Theta_c-\Theta}{\Theta_c}\right). \qquad \text{(II.121)}$$

The experimental formula for the total temperature region is

$$H_c = H_{c0}\left(1-\left(\frac{\Theta}{\Theta_c}\right)^2\right) = H_{c0}\left(\frac{\Theta_c+\Theta}{\Theta_c}\right)\left(\frac{\Theta_c-\Theta}{\Theta_c}\right). \qquad \text{(II.122)}$$

For $\Theta \sim \Theta_c$ we can write this formula

$$H_c = 2H_{c0}\left(\frac{\Theta_c - \Theta}{\Theta_c}\right). \tag{II.123}$$

We see that for $\Theta \sim 0$ (II.120) gives a result very much in agreement with experimental formula and that for $\Theta \sim \Theta_c$, likewise, (II.121) is in agreement with the formula (II.123).

CHAPTER III

Collective oscillations

§ 1 The approximate second quantization Hamiltonian

We now wish to consider the collective oscillations of the system, first in the case of uncharged particles, and then in the case of charged particles.

For the problem of one-particle excitations, with energy $\Omega(p)$, the last term in the Hamiltonian (I.10) is not important and one can omit it. However, for a description of collective oscillations this term plays a rather important role. This term is given by the following expression (Galasiewicz, 1960a):

$$\sum_{(f)} U(f_1, f_2; f_2', f_1') N(a_{f_1}^+ a_{f_2}^+ a_{f_2'} a_{f_1'}) = \hat{H}_2 + \hat{H}_4 + \hat{H}_3 \quad \text{(III.1)}$$

where

$$\hat{H}_2 = \sum_{(f)} U(f_1, f_2; f_2', f_1') \sum_{\substack{\mu,\nu \\ \varrho,\delta}} \{u_{f_1\mu}^* u_{f_2\nu}^* u_{f_2'\varrho} u_{f_1'\delta} \alpha_\mu^+ \alpha_\nu^+ \alpha_\varrho \alpha_\delta$$
$$+ v_{f_1\mu}^* v_{f_2\nu}^* v_{f_2'\varrho} v_{f_1'\delta} \alpha_\varrho^+ \alpha_\delta^+ \alpha_\mu \alpha_\nu - u_{f_1\mu}^* v_{f_2\nu}^* v_{f_2'\varrho} u_{f_1'\delta} \alpha_\mu^+ \alpha_\varrho^+ \alpha_\nu \alpha_\delta$$
$$- v_{f_1\mu}^* u_{f_2\nu}^* u_{f_2'\varrho} v_{f_1'\delta} \alpha_\nu^+ \alpha_\delta^+ \alpha_\mu \alpha_\varrho + u_{f_1\mu}^* v_{f_2\nu}^* u_{f_2'\varrho} v_{f_1'\delta} \alpha_\mu^+ \alpha_\delta^+ \alpha_\nu \alpha_\varrho$$
$$+ v_{f_1\mu}^* u_{f_2\nu}^* v_{f_2'\varrho} u_{f_1'\delta} \alpha_\nu^+ \alpha_\varrho^+ \alpha_\mu \alpha_\delta\}, \quad \text{(III.2)}$$

$$\hat{H}_4 = \frac{1}{2} \sum_{(f)} \sum_{\substack{\mu,\nu, \\ \varrho,\delta}} \{u_{f_1\mu}^* u_{f_2\nu}^* v_{f_2'\varrho} v_{f_1'\delta} \alpha_\mu^+ \alpha_\nu^+ \alpha_\varrho^+ \alpha_\delta^+$$
$$+ v_{f_1\mu}^* v_{f_2\nu}^* u_{f_2'\varrho} u_{f_1'\delta} \alpha_\mu \alpha_\nu \alpha_\varrho \alpha_\delta\},$$

$$\hat{H}_3 = \frac{1}{2} \sum_{(f)} U(f_1, f_2; f_2', f_1') \sum_{\substack{\mu,\nu, \\ \varrho,\delta}} N\{u_{f_1\mu}^* v_{f_2\nu}^* v_{f_2'\varrho} v_{f_1'\delta} \alpha_\mu^+ \alpha_\nu^+ \alpha_\varrho \alpha_\delta^+$$
$$+ u_{f_1\mu}^* v_{f_2\nu}^* v_{f_2'\varrho} v_{f_1'\delta} \alpha_\mu^+ \alpha_\nu \alpha_\varrho^+ \alpha_\delta^+ + v_{f_1\mu}^* v_{f_2\nu}^* v_{f_2'\varrho} v_{f_1'\delta} \alpha_\mu \alpha_\nu^+ \alpha_\varrho^+ \alpha_\delta^+$$
$$+ u_{f_1\mu}^* u_{f_2\nu}^* v_{f_2'\varrho} v_{f_1'\delta} \alpha_\mu^+ \alpha_\nu^+ \alpha_\varrho^+ \alpha_\delta + v_{f_1\mu}^* v_{f_2\nu}^* v_{f_2'\varrho} u_{f_1'\delta} \alpha_\mu \alpha_\nu \alpha_\varrho^+ \alpha_\delta$$

$$+u^*_{f_1\mu}v^*_{f_2\nu}u_{f'_2\varrho}u_{f'_1\delta}\alpha^+_\mu\alpha_\nu\alpha_\varrho\alpha_\delta+v^*_{f_1\mu}u^*_{f_2\nu}u_{f'_2\varrho}u_{f'_1\delta}\alpha_\mu\alpha^+_\nu\alpha_\varrho\alpha_\delta$$

$$+v^*_{f_1\mu}v^*_{f_2\nu}v_{f'_2\varrho}v_{f'_1\delta}\alpha_\mu\alpha_\nu\alpha_\varrho\alpha^+_\delta\}.$$

Now we would like to obtain, from the Hamiltonian (I.10), the approximate second quantization Hamiltonian (asq Hamiltonian). As in the particular case (Bogoliubov, Tolmachev and Shirkov, 1958), we make use of the results of Gell-Mann and other authors (Gell-Mann and Brueckner, 1957; Sawada, 1957; Sawada *et al.*, 1957; Brout, 1957). This means that in order to obtain the fundamental approximation, we may restrict ourselves to a summation of diagrams of the form of the complex given in Fig. 8.

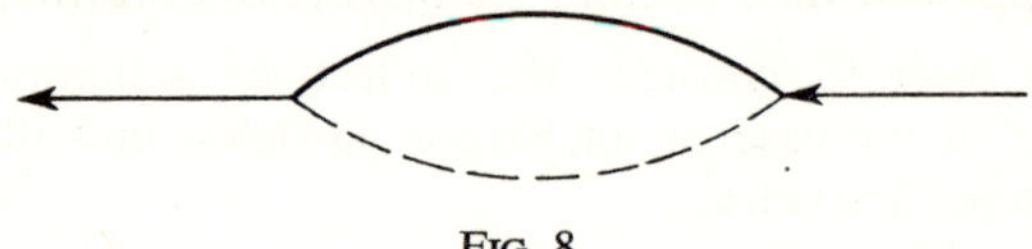

FIG. 8

All more complicated diagrams, built from elementary complexes, are of the form represented in Fig. 9 (see also Bogoliubov, Tolmachev and Shirkov (1958)).

These diagrams contain the vertex parts of Fig. 10. In Figs. 8–10 solid lines denote particles (energy greater than λ) and dashed lines, holes (energy smaller than λ).

In the exact Hamiltonian the terms grouped in $\hat{H}_2$ and $\hat{H}_4$ correspond to these vertex parts. Therefore, in the approximate Hamiltonian we omit the contribution from $\hat{H}_3$.

The second term of $\hat{H}_4$ corresponds to the vertex parts *a*, the first term of $\hat{H}_4$ to the vertex parts *b* and the terms of $\hat{H}_2$ correspond to the vertex parts *c*. In order to obtain the asq Hamiltonian we consider only $\hat{H}_2$ and $\hat{H}_4$. Now, instead of the products $\alpha_{\nu_1}\alpha_{\nu_2}$ of Fermi amplitudes, describing the "particle–hole" complexes, we introduce the Bose amplitudes $\beta_{\nu_1\nu_2}$ ($\beta_{\nu_1\nu_2}=-\beta_{\nu_2\nu_1}$). Hence from the operators $\hat{H}_2$, $\hat{H}_4$ we must extract the pairs of operators $\alpha_{\nu_1}\alpha_{\nu_2}$, $(\alpha_{\nu_3}\alpha_{\nu_4})^+$ and substitute for them the Bose amplitudes $\beta_{\nu_1\nu_2}$, $\beta^+_{\nu_3\nu_4}$. So we must find in $\hat{H}_2$, $\hat{H}_4$ the coefficients of the products $(\alpha_{\nu_1}\alpha_{\nu_2})(\alpha_{\nu_3}\alpha_{\nu_4})$ and $(\alpha_{\nu_3}\alpha_{\nu_4})^+(\alpha_{\nu_1}\alpha_{\nu_2})$. These coefficients are given by the formulae

$$\tfrac{1}{4}\langle\alpha^+_{\nu_4}\alpha^+_{\nu_3}\alpha^+_{\nu_2}\alpha^+_{\nu_1}(\hat{H}_2+\hat{H}_4)\rangle_0, \quad \tfrac{1}{4}\langle\alpha_{\nu_3}\alpha_{\nu_4}(\hat{H}_2+\hat{H}_4)\alpha^+_{\nu_2}\alpha^+_{\nu_1}\rangle_0.$$

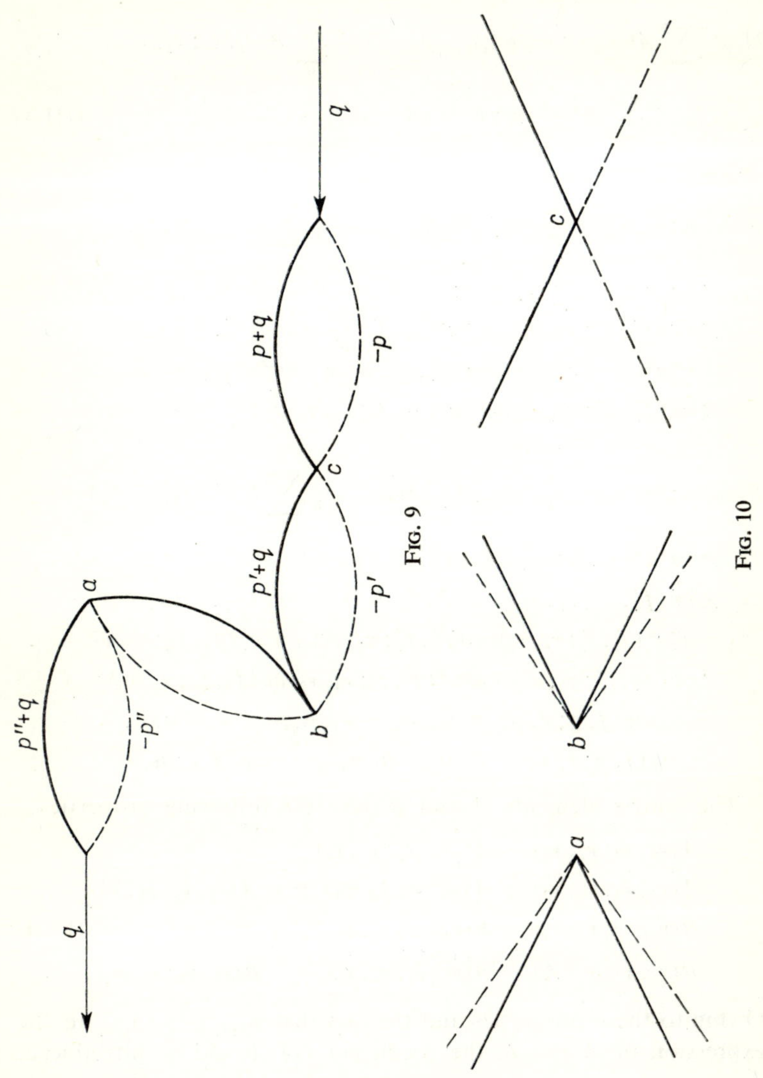

FIG. 9

FIG. 10

The expectation value corresponds to the vacuum state defined by (I.6). From $\hat{H}_2+\hat{H}_4$ we finally obtain

$$\hat{\tilde{H}}_2=\sum_{(\nu)}A(\nu_3,\nu_4;\nu_1,\nu_2)\beta^+_{\nu_3\nu_4}\beta_{\nu_1\nu_2}+\tfrac{1}{2}\sum_{(\nu)}B(\nu_1,\nu_2;\nu_3,\nu_4)\beta^+_{\nu_3\nu_4}\beta^+_{\nu_1\nu_2}$$
$$+\tfrac{1}{2}\sum_{(\nu)}B^*(\nu_1,\nu_2;\nu_3,\nu_4)\beta_{\nu_1\nu_2}\beta_{\nu_3\nu_4} \quad \text{(III.3)}$$

where

$$A(\nu_3,\nu_4;\nu_1,\nu_2)=\tfrac{1}{4}\langle\beta_{\nu_3\nu_4}\hat{H}_2\beta^+_{\nu_1\nu_2}\rangle_0=\tfrac{1}{8}\sum_{(f)}U(f_1,f_2;f_2',f_1')$$
$$\times\{u^*u^*(f_1,f_2;\nu_4,\nu_3)uu(f_1',f_2';\nu_2,\nu_1)+v^*v^*(f_1,f_2;\nu_1,\nu_2)$$
$$\times vv(f_1',f_2';\nu_3,\nu_4)+v^*u^*(f_1,f_2;\nu_1,\nu_3)vu(f_1',f_2';\nu_4,\nu_2)$$
$$+v^*u^*(f_1,f_2;\nu_2,\nu_4)vu(f_1',f_2';\nu_3,\nu_1)+v^*u^*(f_1,f_2;\nu_2,\nu_3)$$
$$\times uv(f_1',f_2';\nu_1,\nu_4)+v^*u^*(f_1,f_2;\nu_1,\nu_4)uv(f_1',f_2';\nu_2,\nu_3)\}, \quad \text{(III.5)}$$

$$B(\nu_1,\nu_2;\nu_3,\nu_4)=\tfrac{1}{4}\langle\beta_{\nu_3\nu_4}\beta_{\nu_1\nu_2}\hat{H}_4\rangle_0=\tfrac{1}{8}\sum_{(f)}U(f_1,f_2;f_2',f_1')$$
$$\times\{u^*u^*(f_1,f_2;\nu_2,\nu_1)vv(f_1',f_2';\nu_3,\nu_4)+u^*u^*(f_1,f_2;\nu_4,\nu_3)$$
$$\times vv(f_1',f_2';\nu_1,\nu_2)+u^*u^*(f_1,f_2;\nu_3,\nu_2)vv(f_1',f_2';\nu_1,\nu_4)$$
$$+u^*u^*(f_1,f_2;\nu_4,\nu_1)vv(f_1',f_2';\nu_2,\nu_3)+u^*u^*(f_1,f_2;\nu_1,\nu_3)$$
$$\times vv(f_1',f_2';\nu_2,\nu_4)+u^*u^*(f_1,f_2;\nu_4,\nu_2)vv(f_1',f_2';\nu_1,\nu_3)\},$$
$$ww(f,g;\nu,\mu)=w_{f\nu}w_{g\mu}-w_{f\mu}w_{g\nu}, \qquad w=u,v,$$
$$vu(f,g;\nu,\mu)=v_{f\nu}u_{g\mu}-u_{f\mu}v_{g\nu}=-uv(f,g;\mu,\nu).$$

The matrix elements A and B have the following properties:

$$A(\nu_3,\nu_4;\nu_1,\nu_2)=A^*(\nu_1,\nu_2;\nu_3,\nu_4),$$
$$A(\nu_3,\nu_4;\nu_1,\nu_2)=A(\nu_4,\nu_3;\nu_2,\nu_1)=-A(\nu_4,\nu_3;\nu_1,\nu_2),$$
$$B(\nu_1,\nu_2;\nu_3,\nu_4)=B(\nu_3,\nu_4;\nu_1,\nu_2), \quad \text{(III.4)}$$
$$B(\nu_1,\nu_2;\nu_3,\nu_4)=B(\nu_2,\nu_1;\nu_4,\nu_3)=-B(\nu_2,\nu_1;\nu_3,\nu_4).$$

Owing to these properties and the fact that $\beta_{\nu_1\nu_2}=-\beta_{\nu_2\nu_1}$ in the expression of A and B, the coefficient 1/4 should be introduced.

Now we must add to $\hat{H}_2$ the term $\hat{H}_1$, the self-energy of the "particle–hole" complex. In order to obtain the correct energy denominators, i.e. the same as in the correct Hamiltonian, it is necessary to choose

$$\hat{\tilde{H}}_1 = \frac{1}{2} \sum_{\nu_1, \nu_2} [\Omega(\nu_1) + \Omega(\nu_2)] \beta^+_{\nu_1\nu_2} \beta_{\nu_1\nu_2} \tag{III.6}$$

where $\Omega(\nu) = \Omega(\nu, \nu)$ and is given by (I.11b).

Thus the complete Hamiltonian in the approximate second quantization scheme has the form

$$\hat{\tilde{H}} = \frac{1}{2} \sum_{\nu_1, \nu_2} [\Omega(\nu_1) + \Omega(\nu_2)] \beta^+_{\nu_1\nu_2} \beta_{\nu_1\nu_2} + \sum_{(\nu)} A(\nu_3, \nu_4; \nu_1, \nu_2) \beta^+_{\nu_3\nu_4} \beta_{\nu_1\nu_2}$$
$$+ \frac{1}{2} \sum_{(\nu)} B(\nu_3, \nu_4; \nu_1, \nu_2) \beta^+_{\nu_3\nu_4} \beta^+_{\nu_1\nu_2} + \frac{1}{2} \sum_{(\nu)} B^*(\nu_3, \nu_4; \nu_1, \nu_2) \beta_{\nu_3\nu_4} \beta_{\nu_1\nu_2}. \tag{III.7}$$

This Hamiltonian is a quadratic form in the Bose operators β.

§ 2 Diagonalization of the asq Hamiltonian and collective oscillations

Bogoliubov (1947a) and Tyablikov (1965) proved that the problem of diagonalizing the quadratic form described by (III.7) may be reduced to solving a system of linear, homogeneous equations in the c-number functions $\xi(\nu_1, \nu_2)$, $\eta(\nu_1, \nu_2)$ (see (III. 10)):

$$E\xi(\nu_1, \nu_2) = [\Omega(\nu_1) + \Omega(\nu_2)] \xi(\nu_1, \nu_2) + 2 \sum_{\nu_3, \nu_4} A(\nu_1, \nu_2; \nu_3, \nu_4)$$
$$\times \xi(\nu_3, \nu_4) + 2 \sum_{\nu_3, \nu_4} B(\nu_1, \nu_2; \nu_3, \nu_4) \eta(\nu_3, \nu_4), \tag{III.8}$$
$$-E\eta(\nu_1, \nu_2) = [\Omega(\nu_1) + \Omega(\nu_2)] \eta(\nu_1, \nu_2) + 2 \sum_{\nu_3, \nu_4} A^*(\nu_1, \nu_2; \nu_3, \nu_4)$$
$$\times \eta(\nu_3, \nu_4) + 2 \sum_{\nu_3, \nu_4} B^*(\nu_1, \nu_2; \nu_3, \nu_4) \xi(\nu_3, \nu_4),$$

with the normalization condition

$$\sum_{\nu_1, \nu_2} \{|\xi(\nu_1, \nu_2)| - |\eta(\nu_1, \nu_2)|^2\} = 1. \tag{III.9}$$

Equations (III.8) lead to a secular equation for determining E. When the E_n are the roots of this equation and ξ_n, η_n are the corresponding functions, we can diagonalize the Hamiltonian (III.7) through the canonical transformation

$$\beta_{\nu_1\nu_2} = \sum_n \{\xi_n(\nu_1, \nu_2) b_n + \eta_n(\nu_1, \nu_2) b^+_n\} \tag{III.10}$$

where b_n, b_n^+ are new Bose amplitudes. We obtain (III.7) in the form

$$\hat{\tilde{H}} = -\sum_n E_n \sum_{\nu_1,\nu_2} \eta^*(\nu_1,\nu_2)\eta(\nu_1,\nu_2) + \sum_n E_n b_n^+ b_n. \qquad \text{(III.11)}$$

The functions ξ_n, η_n must obey the following normalization condition

$$\sum_n \{|\xi_n|^2 - |\eta_n|^2\} = 1. \qquad \text{(III.12)}$$

If in (III.8,4) we write $2A = X$, $2B = -Y$, these formulae will be identical to the formulae obtained by Bogoliubov (1959) by means of the self-consistent field method. The normalization condition for ξ, η gives the positive sign of E. This is the advantage of the asq method over the self-consistent field method.

We assume that the matrices A and B are real. For the types of interaction investigated, this assumption leads to the requirement, that the functions $\{u, v\}$ be real. After these assumptions we define new functions

$$\vartheta(\nu_1,\nu_2) = \xi(\nu_1,\nu_2) + \eta(\nu_1,\nu_2), \qquad \Theta(\nu_1,\nu_2) = \xi(\nu_1,\nu_2) - \eta(\nu_1,\nu_2). \qquad \text{(III.13)}$$

For Θ, ϑ, we have the equations

$$\begin{aligned} E\Theta(\nu_1,\nu_2) &= [\Omega(\nu_1)+\Omega(\nu_2)]\vartheta(\nu_1,\nu_2) + 2\sum_{\nu_3,\nu_4}[A(\nu_1,\nu_2;\nu_3,\nu_4) \\ &\quad + B(\nu_1,\nu_2;\nu_3,\nu_4)]\vartheta(\nu_3,\nu_4), \\ E\vartheta(\nu_1,\nu_2) &= [\Omega(\nu_1)+\Omega(\nu_2)]\Theta(\nu_1,\nu_2) + 2\sum_{\nu_3,\nu_4}[A(\nu_1,\nu_2;\nu_3,\nu_4) \\ &\quad - B(\nu_1,\nu_2;\nu_3,\nu_4)]\Theta(\nu_3,\nu_4). \end{aligned} \qquad \text{(III.14)}$$

Let us consider eqns. (III.14) in the case of a simple form of the transformation (I.3) given by (I.39,40). For $T(f,f')$, $U(f_1,f_2;f_2',f_1')$ we take the expressions (I.30).

The non-vanishing matrix elements of the Hamiltonian (III.7) are of the form

$$\begin{aligned} A_{-+,-+}(p_1,p_2;p_1',p_2') &= A_{-+,+-}(\ldots) = A_{-+}(\ldots), \\ B_{-+,-+}(p_1,p_2;p_1',p_2') &= B_{-+,+-}(\ldots) = B_{-+}(\ldots), \\ A_{++.++}(p_1,p_2;p_1',p_2') &= A_{--,--}(\ldots) = A^{\pm}(\ldots), \\ B_{++,--}(p_1,p_2;p_1',p_2') &= B_{--,++}(\ldots) = B^{-+}(\ldots). \end{aligned} \qquad \text{(III.15)}$$

The spectrum of collective oscillations is also divided into two branches. For the first, the oscillations take place for pairs of particles having opposite spins (elementary excitations with spin 0), for the second, for pairs of particles having parallel spins (elementary excitations with spin ± 1). Let us note that the compensation equations (I.41) are, according to (I.39,40), equations of compensation for pairs of particles having opposite spins. Compensation equations for pairs of particles with parallel spins have been considered by various authors (Galasiewicz, 1960b; Fisher, 1960; Anderson and Morel, 1961; Balian and Werthamer, 1963; Czerwonko, 1966a) in order to obtain, in the superconducting state, a magneticsusceptibility different from zero.

Collective oscillations and other problems connected with the first branch were investigated by Bogoliubov (1959) by means of the self-consistent field method; here we use the asq method.

The Hamiltonian (III.7) is, by virtue of (III.15), a sum of two independent Hamiltonians

$$\hat{\hat{H}} = \hat{\hat{H}}(-+, -+) + \hat{\hat{H}}(--, ++). \tag{III.16}$$

Consider now the first Hamiltonian which has the form

$$\begin{aligned}\hat{\hat{H}}(-+, -+) = &\sum_{p_1, p_2} [\Omega(p_1) + \Omega(p_2)] \beta^{+}_{-+}(p_1, p_2) \beta_{-+}(p_1, p_2)\\ &+ \sum_{(p)} A_{-+}(p_1, p_2; p_1', p_2') \beta^{+}_{-+}(p_1, p_2) \beta_{-+}(p_1', p_2')\\ &+ \sum_{(p)} B_{-+}(p_1, p_2; p_1', p_2') [\beta^{+}_{-+}(p_1, p_2) \beta^{+}_{-+}(-p_2', -p_1')\\ &+ \beta_{-+}(p_1, p_2) \beta_{-+}(-p_2', -p_1')],\end{aligned} \tag{III.17}$$

where

$$\begin{aligned}A_{-+}(p_1, p_2; p_1', p_2') = &\frac{1}{V} \delta(p_1 + p_2 - p_1' - p_2') \{ I(p_1, p_2; p_1', p_2')\\ &\times [u(p_1) v(p_2) u(p_1') v(p_2') + v(p_1) v(p_2) v(p_1') v(p_2')]\\ &+ I(p_1, -p_1'; p_2', -p_2) [u(p_1) v(p_2) v(p_1') u(p_2')\\ &+ v(p_1) u(p_2) u(p_1') v(p_2')] + [I(-p_1, p_2'; -p_1', p_2)\\ &- I(p_2', -p_1; -p_1', p_2)] [v(p_1) u(p_2) v(p_1') u(p_2')\\ &+ u(p_1) v(p_2) u(p_1') v(p_2')]\},\end{aligned} \tag{III.18}$$

$$B_{-+}(p_1, p_2; p_1', p_2') = \frac{1}{V} \delta(p_1 + p_2 - p_1' - p_2') \{ -I(p_1, p_2; p_2', p_1')$$

$$\times [u(p_1)u(p_2)v(p_1')v(p_2')+v(p_1)v(p_2)u(p_1')u(p_2')]$$
$$+I(p_1, -p_1'; p_2', -p_2)\,[v(p_1)u(p_2)v(p_1')u(p_2')$$
$$+u(p_1)v(p_2)u(p_1')v(p_2')]+[I(-p_1, p_2'; -p_1', p_2)$$
$$-I(p_2', -p_1; -p_1', p_2)]\,[u(p_1)v(p_2)v(p_1')u(p_2')$$
$$+v(p_1)u(p_2)u(p_1')v(p_2')]\}.$$

The Hamiltonian (III.17) describes the collective oscillations of pairs of particles having opposite spins. In order to diagonalize this Hamiltonian we must solve the following equations:

$$E\xi_{-+}(p_1, p_2) = [\Omega(p_1)+\Omega(p_2)]\,\xi_{-+}(p_1, p_2)$$
$$+2\sum_{p_1', p_2'} A_{-+}(p_1, p_2; p_1', p_2')\,\xi_{-+}(p_1', p_2')$$
$$-2\sum_{p_1', p_2'} B_{-+}(p_1, p_2; p_1', p_2')\,\eta_{-+}(-p_2', -p_1'), \tag{III.19}$$
$$-E\eta_{-+}(-p_2, -p_1) = [\Omega(p_1)+\Omega(p_2)]\,\eta_{-+}(-p_2, -p_1)$$
$$+2\sum_{p_1', p_2'} A_{-+}(p_1, p_2; p_1', p_2')\,\eta_{-+}(-p_2', -p_1')$$
$$-2\sum_{p_1', p_2'} B_{-+}(p_1, p_2; p_1', p_2')\,\xi_{-+}(p_1', p_2').$$

From (III.18) we see that eqns. (III.19) connect only functions with fixed $\boldsymbol{p}_1+\boldsymbol{p}_2$. Therefore we may put

$$\boldsymbol{p}_1 = \boldsymbol{p}, \quad \boldsymbol{p}_2 = -\boldsymbol{p}+\boldsymbol{q} \tag{III.20}$$

and define new functions

$$\begin{aligned}\Theta_q^{(-+)}(p) &= \xi_{-+}(p_1, p_2)-\eta(-p_2, -p_1) = \Theta_q(p),\\ \vartheta_q^{(-+)}(p) &= \xi_{-+}(p_1, p_2)+\eta(-p_2, -p_1) = \vartheta_q(p).\end{aligned} \tag{III.21}$$

For these functions we have the equations

$$L_q(\Theta) = E\vartheta, \quad M_q(\vartheta) = E\Theta \tag{III.22}$$

where

$$\begin{aligned}L_q(\Theta) &= [\Omega(p)+\Omega(p-q)]\,\Theta(p)+\frac{1}{V}\sum_{p'} Q_q(p, p')\,\Theta(p'),\\ M_q(\vartheta) &= [\Omega(p)+\Omega(p-q)]\,\vartheta(p)+\frac{1}{V}\sum_{p'} R_q(p, p')\,\vartheta(p').\end{aligned} \tag{III.23}$$

For the kernels Q_q, R_q we have the expressions

$$\begin{aligned} Q_q(p,p') &= \mathscr{I}_q(p,p')U_q(p)U_q(p')+I_q(p,p')W_q(p)W_q(p'), \\ R_q(p,p') &= \mathscr{I}_q(p,p')V_q(p)V_q(p')+G_q(p,p')\chi_q(p)\chi_q(p'), \end{aligned} \tag{III.24}$$

$$\begin{aligned} \mathscr{I}_q(p,p') &= I(p,-p+q;-p'+q,p'), \\ I_q(p,p') &= I(p,p'-q;p',p-q)-I(p,p'-q;p-q,p') \\ &\quad -I(p,-p';-p'+q,p-q), \\ G_q(p,p') &= I(p,p'-q;p',p-q)-I(p,p'-q;p-q,p') \\ &\quad +I(p,-p';-p'+q;p-q), \end{aligned} \tag{III.25}$$

$$\begin{aligned} U_q(p) &= u(p)u(p-q)+v(p)v(p-q), \qquad U_0(p) = 1, \\ W_q(p) &= v(p)u(p-q)-u(p)v(p-q), \qquad W_0(p) = 0, \\ V_q(p) &= u(p)u(p-q)-v(p)v(p-q), \qquad V_0(p) = \frac{\zeta_p}{\Omega(p)}, \\ \chi_q(p) &= v(p)u(p-q)+u(p)v(p-q), \qquad \chi_0(p) = \frac{C(p)}{\Omega(p)}. \end{aligned} \tag{III.26}$$

In equations (III.23) we now put the solution in the form

$$\Theta(p) = S_1\,\delta(p-p_0), \qquad \vartheta(p) = S_2\,\delta(p-p_0) \tag{III.27}$$

(plane waves in the r-representation), where S_1, S_2 are constants, and p_0 is an arbitrary, but fixed, momentum. Then we let $V \to \infty$ and neglect in (III.23) terms of the order $1/V$. The solution (III.27) would be a good one if S_1, S_2 were connected by the equations

$$\begin{aligned} S_1[\Omega(p_0)+\Omega(p_0-q)] &= ES_2, \\ S_2[\Omega(p_0)+\Omega(p_0-q)] &= ES_1. \end{aligned} \tag{III.28}$$

After eliminating S_1, S_2 we have

$$E = \Omega(p_0)+\Omega(p_0-q). \tag{III.29}$$

As we remarked above, the positive sign of E is a consequence of the normalization condition (III.12).

Thus, examining the collective oscillations we are convinced of the existence of a continuous spectrum separated from the ground state by a gap $2C$ (see Chapter I, § 3, formulae (I.67–70)). For a given q the energy is a continuous function of p_0.

Let us now consider collective oscillations for a system of uncharged particles. When the Coulomb interaction is absent, we may assume that $\mathscr{I}_q$, I_q and G_q take finite values for all q.

For $q = 0$ (see III.26) we have

$$Q_0(p, p') = \mathscr{I}_0(p, p'),$$
$$R_0(p, p') = \frac{\mathscr{I}_0(p, p')}{\Omega(p)\Omega(p')}[\zeta(p)\zeta(p')+C(p)C(p')]. \tag{III.30}$$

It is easy to prove that

$$\Theta = u(p)v(p) = \Phi(p) \tag{III.31}$$

is the solution of

$$L_0(\Theta) = 0 \tag{III.32}$$

which is identical with the "gap" equation (I.38).

If we have the inhomogeneous equation

$$L_0(\Theta) = f(p), \tag{III.33}$$

a solution exists only if

$$\sum_p f(p)u(p)v(p) = 0. \tag{III.34}$$

We see that for $q = 0$, eqns. (III.22) have the solution

$$\Theta = u(p)v(p), \quad \vartheta = 0, \quad E = 0. \tag{III.35}$$

We will take this as the zero-order solution and look for the solution for small $|q|$ in the form of a power series

$$\Theta = u(p)v(p)+|q|\Theta_1(p, e)+|q|^2\Theta_2(p, e)++$$
$$\vartheta = |q|\vartheta_1(p, e), \tag{III.36}$$
$$E = |q|E_1, \quad e = \frac{\boldsymbol{q}}{|\boldsymbol{q}|}.$$

After substituting (III.36) into (III.22) we have for the first and second order in $|q|$:

$$L_0(\Theta_1) = -e[\nabla_q L_q(uv)]_{q=0}, \tag{III.37}$$
$$M_0(\vartheta_1) = E_1 u(p)v(p), \tag{III.38}$$
$$L_0(\Theta_2) = E_1\vartheta_1 - e[\nabla_q L_q(\Theta_1)]_{q=0} - \tfrac{1}{2}\{e\nabla_q[e\nabla_q L_q(uv)]\}_{q=0}. \tag{III.39}$$

The function $f(p)$ on the right-hand side of (III.37, 33) is a sum of terms of the form

$$\left[\frac{\partial}{\partial q}g(p-q)\right]_0 = -\frac{\partial}{\partial p}g(p) = \tilde{f}(p) = -\tilde{f}(-p).$$

Therefore, in this case, $f(p) = -f(-p)$, and (III.34) is fulfilled automatically.

In order that the solution of (III.39) exist, the condition (III.34) must be satisfied. This gives

$$E_1 \sum_p \vartheta_1(p, e) u(p) v(p) = \sum_p u(p) v(p) \{ [e\nabla_q L_q(\Theta)]_{q=0} + \tfrac{1}{2} [e\nabla_q(e\nabla_q L_q(uv))] \}_{q=0}. \tag{III.40}$$

From (III.38) we see that $\vartheta_1 \sim E_1$. After calculations (for which we take $E(p) = p^2/2m$) we find $E_1^2 = \frac{1}{3} v_F^2$. The factor $v_F = p_F/m$ is the velocity of a particle near the Fermi surface. Thus, we have

$$E = \frac{v_F}{\sqrt{3}} |q| \tag{III.41}$$

describing the oscillations of a quasi-acoustical character. The momenta $|q|$ are restricted to those values for which the levels E lie below the continuous spectrum (in the energy gap).

Consider now a system of electrons. In this case, the kernel G_q has a singularity due to the Coulomb interaction and is given by $G_q = \dfrac{8\pi e^2}{|q|^2} + G_q'$.

We split off from $M_q(\vartheta)$ the part which is singular for $q = 0$

$$M_q(\vartheta) = M_q'(\vartheta) + \chi_q(p) \frac{\psi}{2} \tag{III.42}$$

where

$$\frac{|q|^2}{16\pi e^2} \psi = \frac{1}{V} \sum_{p'} \vartheta(p') \chi_q(p').$$

We now have the system of equations

$$\begin{aligned} L_q(\Theta) &= E\vartheta, \\ M_q'(\vartheta) + \chi_q(p) \frac{\psi}{2} &= E\Theta, \\ \frac{1}{V} \sum_{p'} \vartheta(p') \chi_q(p') &= \frac{|q|^2}{16\pi e^2} \psi. \end{aligned} \tag{III.43}$$

For $q = 0$ this system has a solution (for arbitrary E)

$$\Theta = u(p) v(p), \quad \vartheta = 0, \quad E = \psi.$$

Starting from this solution we put, for $q \neq 0$,

$$\begin{aligned} \Theta &= u(p)\,v(p)+|q|\,\Theta_1(p, e)+|q|^2\Theta_2(p, e), \\ \vartheta &= |q|\vartheta_1(p, e), \\ E &= E_0+|q|E_1, \qquad \psi = E_0+|q|\psi_1. \end{aligned} \tag{III.44}$$

From (III.43) we find

$$L_0(\Theta_1) = E_0\vartheta_1-[e\nabla_q L_q(uv)]_{q=0}, \tag{III.45}$$

$$M_0'(\vartheta_1) = u(p)\,v(p)[E_1-\psi_1]+E_0\Theta_1+\tfrac{1}{2}\,E_0 e\nabla_p(u(p)\,v(p)), \tag{III.46}$$

$$\frac{1}{V}\sum_{p'}\vartheta_1(p')u(p')v(p') = 0, \tag{III.47}$$

$$L_0(\Theta_2) = E_0\vartheta_2+E_1\vartheta_1-[e\nabla_q L_q(\Theta_1)]_{q=0}-\tfrac{1}{2}\,[e\nabla_q(e\nabla_q L_q(uv))]_{q=0}, \tag{III.48}$$

$$\frac{2}{V}\sum_{p}\vartheta_2(p)u(p)v(p) = \frac{E_0}{16\pi e^2}+\frac{1}{V}\sum_{p}\vartheta_1(p)e\nabla_p(u(p)v(p)). \tag{III.49}$$

If

$$E_1-\psi_1 = 0,$$

we see from (III.44,46) that $\Theta_1(p) = -\Theta_1(-p)$, $\vartheta_1(p) = -\vartheta_1(-p)$ and hence condition (III.47) is fulfilled automatically.

To solve (III.48) we use formulae (III.34) and (III.49) and find

$$\begin{aligned} &\sum_{p} u(p)v(p)\{[e\nabla_q L_q(\Theta_1)]_{q=0}+\tfrac{1}{2}[e\nabla_q(e\nabla_q L_q(uv))]_{q=0}\} \\ &\qquad = E_0\sum_{p}\vartheta_2(p)u(p)v(p) \\ &\qquad = VE_0\left\{\frac{E_0}{32\pi e^2}+\frac{1}{2V}\sum_{p}\vartheta_1(p)e\nabla_p(u(p)v(p))\right\}. \end{aligned} \tag{III.50}$$

The term standing on the left-hand side of this equation is identical to the term in (III.40) and for this reason is different from zero; therefore E_0 must be different from zero. We wish to calculate E_0 for the spherically symmetric case. We take $E(p) = p^2/2m$ and assume that

$$I_{\text{ph}}(p_1, p_2; p_2', p_1') = -g^2(p_1-p_1'), \qquad p_1+p_2 = p_1'+p_2'. \tag{III.51}$$

From these assumptions we have the identity

$$L_q(\chi_q) = \left\{ \frac{(p-q)^2}{2m} - \frac{p^2}{2m} \right\} W_q(p). \tag{III.52}$$

As the kernel $Q_q(p, p')$ (III.24) is a real and symmetric function of p, p', the operator L_q defined by (III.23) is a hermitian operator and therefore we have the relation

$$\sum_p L_q(\Theta)\chi_q = \sum_p L_q(\chi_q)\Theta. \tag{III.53}$$

From this relation it follows that

$$E\frac{1}{V}\sum_p \vartheta(p)\chi_q(p) = \frac{1}{V}\sum_p \Theta(p)\left\{\frac{(p-q)^2}{2m} - \frac{p^2}{2m}\right\}W_q(p). \tag{III.54}$$

Introducing (III.36) and the last equation of (III.43) into (III.54) we have

$$E_0^2 = 16\pi e^2 \frac{1}{V}\sum_p u(p)v(p)\frac{pe}{m}\{v(p)e\nabla_p u(p) - u(p)e\nabla_p v(p)\}. \tag{III.55}$$

After introducing $u(p)$, $v(p)$ from (II.18), for $\Theta = 0$, we obtain for E_0 (Bogoliubov, 1959; see also Anderson, 1958):

$$E_0 = \sqrt{\frac{4e^2}{3\pi}\frac{p_F^2}{m}}. \tag{III.56}$$

Formula (III.56) gives the energy of plasma oscillations. These oscillations also take place in the normal state. We see that quasi-acoustic oscillations do not appear in the case of charged particles.

Now let us consider the collective oscillations of the second branch. These are described by the second Hamiltonian in (III.17):

$$\begin{aligned}\hat{\tilde{H}}(--, ++) &= \tfrac{1}{2}\sum_{p_1, p_2}[\Omega(p_1)+\Omega(p_2)]\sum_\sigma \beta^+_{\sigma\sigma}(p_1, p_2)\beta_{\sigma\sigma}(p_1, p_2)\\ &+ \sum_{(p)}\sum_\sigma A^\sigma(p_1, p_2; p_1', p_2')\beta^+_{\sigma\sigma}(p_1, p_2)\beta_{\sigma\sigma}(p_1', p_2')\\ &+ \sum_{(p)} B^{-+}(p_1, p_2; p_1', p_2')\,[\beta^+_{++}(p_1, p_2)\beta^+_{--}(-p_1', -p_2')\\ &+ \beta_{++}(p_1, p_2)\beta_{--}(-p_1, -p_2')]\end{aligned} \tag{III.57}$$

where

$$A^{\pm}(p_1, p_2; p_3, p_4) = \frac{1}{4V}\delta(p_1+p_2-p_1'-p_2')\{[I(p_2, p_1; p_1', p_2')$$
$$-I(p_1, p_2; p_1', p_2')]\times[u(p_1)u(p_2)u(p_1')u(p_2')+v(p_1)v(p_2)v(p_1')v(p_2')]$$
$$+I(p_2, -p_2'; -p_1, p_1')[v(p_1)u(p_2)u(p_1')v(p_2')+u(p_1)v(p_2)v(p_1')u(p_2')]$$
$$-I(p_2, -p_1'; -p_1, p_2')[v(p_1)u(p_2)v(p_1')u(p_2')+u(p_1)v(p_2)u(p_1')v(p_2')]\}, \tag{III.58}$$

$$B^{-+}(p_1, p_2; p_3, p_4) = \frac{1}{4V}\delta(p_1+p_2-p_1'-p_2')\{[(Ip_1, p_2; p_1', p_2')$$
$$-I(p_2, p_1; p_1', p_2')]\times[u(p_1)u(p_2)v(p_1')v(p_2')+v(p_1)v(p_2)u(p_1')u(p_2')]$$
$$+I(p_2-p_2'; -p_1, p_1')[v(p_1)u(p_2)v(p_1')u(p_2')+u(p_1)v(p_2)u(p_1')v(p_2')]$$
$$-I(p_2, -p_1', -p_1, p_2')[v(p_1)u(p_2)u(p_1')v(p_2')$$
$$+u(p_1)v(p_2)v(p_1')u(p_2')]\}.$$

The Hamiltonian (III.57) describes the elementary excitations with spin ± 1, which interact with each other. Therefore the equations for the functions $\xi_{\sigma\sigma}$, $\eta_{\sigma\sigma}$ do not split into the equations for two independent branches. We define new functions

$$\xi_{--}(p_1, p_2)-\eta_{++}(-p_2, -p_1) = \tilde{\Theta}_q^{(-+)}(p) = \tilde{\Theta}_q(p),$$
$$\xi_{--}(p_1, p_2)+\eta_{++}(-p_2, -p_1) = \tilde{\vartheta}_q^{(-+)}(p) = \tilde{\vartheta}_q(p),$$
$$p_1+p_2 = p_1'+p_2' = q. \tag{III.59}$$

For these functions we have

$$E\tilde{\Theta}_q(p) = [\Omega(p)+\Omega(p-q)]\tilde{\vartheta}_q(p)$$
$$+2\sum_{p'}[A(p, -q+p; p', -p'+q)$$
$$+B(p, -p+q; p', -p'+q)]\tilde{\vartheta}_q(p'), \tag{III.60}$$

$$E\tilde{\vartheta}_q(p) = [\Omega(p)+\Omega(p-q)]\tilde{\Theta}_q(p)$$
$$+2\sum_{p'}[A(p, -p+q; p', -p'+q)$$
$$+B(p, -p+q; p', -p'+q)]\tilde{\Theta}_q(p')$$

where

$$A+B = \frac{1}{4V}\{I_q^{(1)}(V_q(p)V_q(p')+(I-I_q^{(2)})\chi_q(p)\chi_q(p')\},$$

$$A-B = \frac{1}{4V}\{I_q^{(1)}U_q(p)U_q(p')+(I-I_q^{(2)})W_q(p)W_q(p')\},$$

$$\text{(III.61)}$$

$$I_q^{(1)}(p,p') = I(-p+q,p;p',-p'+q)-I(p,-p+q;p',-p'+q),$$

$$I_q^{(2)}(p,p') = I(-p+q,p'-q;-p,p')-I(-p+q,-p';-p,-p'+q)$$

Note that the functions ξ, η, and consequently the functions Θ, ϑ, are odd functions of p_1, p_2. Thus, for $q = 0$ we must have

$$\Theta_0(p) = \Theta(\boldsymbol{p}) = -\Theta(-\boldsymbol{p}), \quad \vartheta_0(p) = \vartheta(\boldsymbol{p}) = -\vartheta(-\boldsymbol{p}). \quad \text{(III.62)}$$

If we restrict ourselves to the case where the interaction may be replaced by a constant inside a thin layer near the Fermi surface ($E_F \pm \omega$) and by zero outside it, we see that

$$A = B = 0.$$

The Hamiltonian (III.57) was also obtained by Yosida (1959). In that paper it was stated that in the case of an interaction which is constant in a thin layer near the Fermi surface, we do not obtain collective oscillations of particles with parallel spins. In this case the temperature dependence of the paramagnetic susceptibility obtained by Yosida (1958) is not altered and remains inconsistent with Reif's experiment (Reif, 1957; see also Yosida, 1958). Various attempts have been made to explain the results of Reif's experiment, based on the pairing with parallel spins (Galasiewicz, 1960b; Fisher, 1960; Anderson and Morel, 1961; Balian and Werthamer, 1963; Czerwonko, 1966a).

We now consider the case when the interaction I is effective only near the Fermi surface and has the form (III.51).

For $q = 0$ we obtain from (III.60), (III.61), (III.26)

$$E\vartheta(\boldsymbol{p}) = 2\Omega(p)\Theta(\boldsymbol{p}) - \frac{1}{V}\sum_{p'} g^2(|\boldsymbol{p}'-\boldsymbol{p}|)\Theta(\boldsymbol{p}),$$

$$\text{(III.63)}$$

$$E\Theta(\boldsymbol{p}) = 2\Omega(p)\vartheta(\boldsymbol{p}) + \frac{1}{V}\sum_{p'} g^2(|\boldsymbol{p}'-\boldsymbol{p}|)\frac{C(p)C(p')-\zeta_p\zeta_{p'}}{\Omega(p)\Omega(p')}.$$

These equations are identical with the equations obtained by Bogoliubov, Tolmachev and Shirkov (1958; see also Bogoliubov

(1959)) by considering collective oscillations for the first branch. However, in this case the solution must be an odd function of $\boldsymbol{p}$, while in the former case it had to be an even function of $\boldsymbol{p}$.

We change to spherical polars taking for the polar axis the direction of the vector $\boldsymbol{p}$. In this case

$$g^2(|\boldsymbol{p}'-\boldsymbol{p}|) \cong g^2(p, p', p_F\sqrt{2(1-\cos\alpha)}), \qquad \begin{aligned} \vartheta(\boldsymbol{p}) &= \vartheta(p, \cos\alpha), \\ \Theta(\boldsymbol{p}) &= \Theta(p, \cos\alpha), \end{aligned} \tag{III.64}$$

where α is the angle between the vector $\boldsymbol{p}$ and $\boldsymbol{p}'$. If the angle α takes on the value π, the functions Θ, ϑ change sign. In the expressions for Θ, ϑ we have only terms containing odd spherical harmonics

$$\Theta = \sum_{n/0}^{\infty} a_{2n+1}(p)P_{2n+1}(\cos\alpha), \qquad \vartheta = \sum_{n/0}^{\infty} b_{2n+1}(p)P_{2n+1}(\cos\alpha), \tag{III.65}$$

$$g^2 = \sum_{n/0}^{\infty} g_n(p, p')P_n(\cos\alpha).$$

Putting (III.65) into (III.63) we obtain the equations for the separate terms of the expansions. We consider the equations for the first term

$$Eb_1(\zeta) = 2\Omega(\zeta)a_1(\zeta) - \frac{1}{3}\frac{\mathrm{d}n}{\mathrm{d}E}\int_{-\infty}^{+\infty} g_1(\zeta, \zeta')a_1(\zeta)\mathrm{d}\zeta',$$

$$Ea_1(\zeta) = 2\Omega(\zeta)b_1(\zeta) + \frac{1}{3}\frac{\mathrm{d}n}{\mathrm{d}E}\int_{-\infty}^{+\infty} \times \frac{C(\zeta)C(\zeta')-\zeta\zeta'}{\Omega(\zeta)\Omega(\zeta')}\, b_1(\zeta)g_1(\zeta, \zeta')\mathrm{d}\zeta'. \tag{III.66}$$

Consider the case when C is a constant which is different from zero only near the Fermi surface. We define

$$C_1(\zeta) = \frac{1}{3}\frac{\mathrm{d}n}{\mathrm{d}E}\int_{-\infty}^{+\infty} g_1(\zeta, \zeta')a_1(\zeta')\mathrm{d}\zeta',$$

$$C_2(\zeta) = \frac{1}{3}\frac{\mathrm{d}n}{\mathrm{d}E}\int_{-\infty}^{+\infty} \frac{\zeta' g_1(\zeta, \zeta')}{\Omega(\zeta')}\, b_1(\zeta')\mathrm{d}\zeta', \tag{III.67}$$

$$C_3(\zeta) = \frac{1}{3}C\frac{\mathrm{d}n}{\mathrm{d}E}\int_{-\infty}^{+\infty} \frac{g_1(\zeta, \zeta')}{\Omega(\zeta')}\, b_1(\zeta')\mathrm{d}\zeta'.$$

From (III.66,67) we have

$$a_1(\zeta) = \frac{1}{4\Omega(\zeta)^2 - E^2}\left[2\Omega(\zeta)C_1(\zeta) - \frac{CE}{\Omega(\zeta)}C_3(\zeta) + \frac{\zeta E}{\Omega(\zeta)}C_2(\zeta)\right], \tag{III.68}$$

$$b_1(\zeta) = \frac{1}{4\Omega(\zeta)^2 - E^2}\left[EC_1(\zeta) - 2CC_3(\zeta) + 2\zeta C_2(\zeta)\right].$$

The first and second terms on the right-hand side of (III.68) are even functions of ζ, the third one is an odd function. We have the following relation

$$EC_3(\zeta) = 2CC_1(\zeta) - C\frac{1}{3}\frac{dn}{dE}\int_{-\infty}^{+\infty}\frac{g_1(\zeta, \zeta')}{\Omega(\zeta')}C_1(\zeta')d\zeta'$$

and the following equations from which to determine E:

$$C_1(\zeta) = \frac{1}{3}\frac{dn}{dE}\int_{-\infty}^{+\infty}\frac{2g_1(\zeta, \zeta')\zeta'^2}{\Omega(\zeta')[4\Omega(\zeta')^2 - E^2]}C_1(\zeta')d\zeta'$$
$$+C^2\left(\frac{1}{3}\frac{dn}{dE}\right)^2\int_{-\infty}^{+\infty}\frac{g_1(\zeta, \zeta')}{\Omega(\zeta')[4\Omega^2(\zeta') - E^2]}$$
$$\times\left[\int_{-\infty}^{+\infty}\frac{g_1(\zeta', \zeta'')}{\Omega(\zeta'')}C_1(\zeta'')d\zeta''\right]d\zeta', \tag{III.69}$$
$$C_2(\zeta) = \frac{1}{3}\frac{dn}{dE}\int_{-\infty}^{+\infty}\frac{2g_1\zeta, \zeta')\zeta'^2}{\Omega(\zeta')[4\Omega(\zeta)^2 - E^2]}C_2(\zeta')d\zeta'.$$

For $g_1(\zeta, \zeta') = g_1(0, 0) = g_1(p_F)$ we can take C_1 and C_2 as constant, and eqns. (III.69) lead to expressions

$$\left\{\tilde{\varrho}\ln\frac{2\omega}{C} - \tilde{\varrho}\frac{\sqrt{1-\varepsilon^2}}{\varepsilon}\arctan\frac{\varepsilon}{\sqrt{1-\varepsilon^2}} - 1\right\} = 0,$$

$$\left\{\left(\tilde{\varrho}\ln\frac{2\omega}{C} - 1 + \varepsilon^2\right)\left(\frac{\tilde{\varrho}}{\varepsilon\sqrt{1-\varepsilon}}\arctan\frac{\varepsilon}{\sqrt{1-\varepsilon^2}} + 1\right) - \varepsilon^2\right\} = 0,$$

$$\tilde{\varrho} = \frac{1}{3}g_1(p_F)\frac{dn}{dE}, \quad \varepsilon = \frac{E}{2C}. \tag{III.70}$$

From (III.70) we see that the stability conditions for the collective excitations of the second branch, in the superconducting state, are the same as the stability conditions for the first branch for transverse waves obtained by Bogoliubov, Tolmachev and Shirkov (1958). For $\tilde{\varrho}$ in the interval

$$-1 < \tilde{\varrho} < \frac{1}{\ln \frac{2\omega}{C}}, \tag{III.71}$$

equation (III.70) has a single root.

Thus, the interaction (III.51), leads to spin-wave-like elementary excitations. These excitations are stable if the interaction is not too strong.

Let us now consider the collective excitations in the normal state. Substituting $C = 0$ in (III.66), we finally obtain

$$C_1(\zeta) = \frac{1}{3}\frac{\mathrm{d}n}{\mathrm{d}E}\int_0^{\infty} g_1(\zeta, \zeta')\frac{4\zeta'}{4\zeta' - E^2}C_1(\zeta')\mathrm{d}\zeta' \tag{III.72}$$

and hence the asymptotic formula

$$-E^2 \sim e^{-\frac{2}{g_1}}$$

which means that the normal state is unstable.

CHAPTER IV

Electrodynamics

§ 1 The asq Hamiltonian in the case of weak external fields

Let us consider a dynamical system of Fermi particles in a weak external field. We assume that this field gives rise only to a variation of $E(f,f')$ in the one-particle Hamiltonian. Thus, rather than (I.1), we have

$$\hat{H}+\delta\hat{H} = \hat{H}+\sum_{f,f'} \delta E(f,f') a_f^+ a_{f'}. \qquad \text{(IV.1)}$$

Applying the transformation (I.3) to the additional term in (IV.1) and—as above—changing to the Bose amplitudes $\beta_{\mu\nu}$ we obtain in the approximate Hamiltonian (III.7) the additional term $\delta\hat{\hat{H}}$

$$\hat{\hat{H}}' = \hat{\hat{H}}+\delta\hat{\hat{H}} = \hat{\hat{H}}+\tfrac{1}{2}\sum_{f,f'} \delta E(f,f') \sum_{\nu_1,\nu_2} (u^*_{f\nu_2} v_{f'\nu_1} - u^*_{f\nu_1} v_{f'\nu_2}) \beta^+_{\nu_1\nu_2}$$

$$+\tfrac{1}{2}\sum_{f,f'} \delta E^*(f,f') \sum_{\nu_1,\nu_2} (u_{f\nu_2} v^*_{f'\nu_1} - u_{f\nu_1} v^*_{f'\nu_2}) \beta_{\nu_1\nu_2}. \qquad \text{(IV.2)}$$

We see that the asq Hamiltonian now contains terms, linear in the operators $\beta_{\mu\nu}$, which lead to forced collective oscillations. (In the Hamiltonian (III.7) we do not have terms linear in the operators $\beta_{\nu\mu}$ due to the compensation relations (I.12).)

To remove the terms linear in $\beta_{\mu\nu}$ from the Hamiltonian (IV.2) we can use the method of "translation" of the Bose amplitudes

$$\beta_{\nu\mu} = \beta'_{\nu\mu}+C(\nu, \mu), \qquad \beta^+_{\nu\mu} = \beta^{+\prime}_{\nu\mu}+C^*(\nu, \mu) \qquad \text{(IV.3)}$$

where C and C^* are c-numbers. They are determined from the condition that the linear forms

$$\frac{\partial \hat{\tilde{H}}'}{\partial \beta'^{+}_{\nu\mu}} = 0, \qquad \frac{\partial \hat{\tilde{H}}'}{\partial \beta'_{\nu\mu}} = 0 \tag{IV.4}$$

vanish.

We now write the Hamiltonian (IV.2) for the case of the special transformation (I.39). In the approximation used, we have the sum of two terms, according to the spin of quasi-particles

$$\hat{\tilde{H}}' = \hat{\tilde{H}}'(-+, -+) + \hat{\tilde{H}}'(--, ++). \tag{IV.5}$$

We are interested in the perturbation terms, caused by weak external fields, of the form

$$-\frac{e}{2m}\{(\boldsymbol{p}\boldsymbol{A}(r)) + (\boldsymbol{A}(r)\boldsymbol{p})\} - \frac{e}{2m}\boldsymbol{\sigma}\boldsymbol{H}(r), \qquad \hbar = 1, c = 1 \tag{IV.6}$$

where $+e$, m are the electronic charge and mass, respectively, $\boldsymbol{H}$ is the magnetic field vector, $\boldsymbol{A}$ the vector potential of this field, $\boldsymbol{p}$ the momentum of a particle, and $\boldsymbol{\sigma}$ the spin vector (with Pauli matrices as components).

By means of the functions

$$\psi(r, s_z) = \frac{1}{\sqrt{V}} \sum_{p,\sigma} a_{p\sigma} e^{i(p\cdot r)} S_\sigma(s_z)$$

and using the formulae

$$\sigma_x S_\alpha = S_{-\alpha}, \qquad \sigma_y S_\alpha = iS_{-\alpha}\operatorname{sgn}\alpha, \qquad \sigma_z S_\alpha = S_\alpha \operatorname{sgn}\alpha$$

we write (IV.6) in the second quantization formalism

$$\delta\hat{H}_A = -\frac{e}{2m}\sum_{p_1+p_2=q} (p_2 - p_1)A(-p_1-p_2)\,[a^{+}_{-p_1+}a_{p_2+} + a^{+}_{-p_1-}a_{p_2-}], \tag{IV.7}$$

$$\begin{aligned}\delta\hat{H}_{\mathscr{H}} = -\frac{e}{2m}\sum_{p_1+p_2=q} & \mathscr{H}_{-1}(-p_1-p_2)a^{+}_{-p_1+}a_{p_2-} \\ & + \mathscr{H}_{1}(-p_1-p_2)a^{+}_{-p_1-}a_{p_2+} \\ & + \mathscr{H}_{0}(-p_1-p_2)\,[a^{+}_{-p_1+}a_{p_2+} - a^{+}_{-p_1-}a_{p_2-}],\end{aligned} \tag{IV.8}$$

$$\mathscr{H}_{-1} = \mathscr{H}_x - i\mathscr{H}_y, \qquad \mathscr{H}_1 = \mathscr{H}_x + i\mathscr{H}_y, \qquad \mathscr{H}_0 = \mathscr{H}_z.$$

(The Hamiltonian (IV.8) for $\delta\mathscr{H} = \delta\mathscr{H}_x$ has been used by Yosida (1958) to obtain the temperature dependence of the paramagnetic susceptibility in the BCS theory (Reif, 1957).)

Finally we obtain, in the approximate Hamiltonian (IV.5), the following terms linear in the β operators

$$\delta\hat{\tilde{H}}_A(-+) = -\frac{e}{2m}\sum_{p_1+p_2=q} vu(p_1, p_2)(p_2-p_1)A(-p_1-p_2) \times[\beta_{-+}(p_1, p_2)-\beta^+_{-+}(-p_2, -p_1)],$$

$$\delta\hat{\tilde{H}}_{\mathscr{H}_0}(-+) = -\frac{e}{2m}\sum_{p_1+p_2=q} vu(p_1, p_2)\mathscr{H}_0(-p_1-p_2) \times[\beta_{-+}(p_1, p_2)-\beta^+_{-+}(-p_2, -p_1)],$$

$$\delta\hat{\tilde{H}}_{\mathscr{H}_{\pm1}}(++) = \frac{e}{4m}\sum_{p_1+p_2=q} vu(p_1, p_2)\{\mathscr{H}_1\beta_{++}(p_1, p_2) -\mathscr{H}_{-1}\beta^+_{++}(-p_2, -p_1)\},$$

$$\delta\hat{\tilde{H}}_{\mathscr{H}_{\pm1}}(--) = -\frac{e}{4m}\sum_{p_1+p_2=q} vu(p_1, p_2)\{\mathscr{H}_{-1}\beta_{--}(p_1, p_2) -\mathscr{H}_1\beta^+_{++}(-p_2, -p_1)\},$$

$$vu(p_1, p_2) = v(p_1)u(p_2)-v(p_2)u(p_1). \tag{IV.9}$$

Thus the Hamiltonian (IV.5) is the sum of the Hamiltonians

$$\hat{\tilde{H}}'(-+, -+) = \hat{\tilde{H}}(-+, -+)+\delta\hat{\tilde{H}}_A(-+)+\delta\hat{\tilde{H}}_{\mathscr{H}_0}(-+) \tag{IV.10}$$

and

$$\hat{\tilde{H}}'(--, ++) = \hat{\tilde{H}}(--, ++)+\delta\hat{\tilde{H}}_{\mathscr{H}_{\pm1}}(++)+\delta\hat{\tilde{H}}_{\mathscr{H}_{\pm1}}(--). \tag{IV.11}$$

The Hamiltonian (IV.11) describes forced collective oscillations of pairs of particles with parallel spins. In the perturbation terms only transverse components of the magnetic field along the z-axis appear.

We obtain a very simple formula for (IV.11) if the interaction is replaced by a constant for $p_i \sim p_F$ and by zero elsewhere. In this case,

$$\hat{\tilde{H}}'(--, ++) = \tfrac{1}{2}\sum_{p_1, p_2, \sigma}\{[\Omega(p_1)+\Omega(p_2)]\beta^+_{\sigma\sigma}(p_1, p_2) \times\beta_{\sigma\sigma}(p_1, p_2)+\delta\hat{\tilde{H}}(\sigma, \sigma)\}. \tag{IV.12}$$

The perturbation terms in (IV.10,11), linear in β, are given by

(IV.9). In order to remove these terms in the quadratic forms (IV.10,11), we perform translations of the amplitudes:

$$\begin{aligned} \beta_{-+}(p_1, p_2) &= \beta'_{-+}(p_1, p_2) + C_{-+}(p_1, p_2), \\ \beta_{\sigma\sigma}(p_1, p_2) &= \beta'_{\sigma\sigma}(p_1, p_2) + C_{\sigma\sigma}(p_1, p_2), \\ C_{\sigma\sigma}(p_1, p_2) &= -C_{\sigma\sigma}(p_2, p_1). \end{aligned} \tag{IV.13}$$

For $C(p_1, p_2)$, according to (IV.4), we have the following equations:

$$\begin{aligned} &[\Omega(p_1)+\Omega(p_2)]\,[C_{-+}(p_1, p_2)-C^*_{-+}(-p_2, -p_1)] \\ &\quad + \sum_{p'_1, p'_2} [A_{-+}(p_1, p_2; p'_1, p'_2)-B_{-+}(p_1, p_2; p'_1, p'_2)] \\ &\quad \times [C_{-+}(p'_1, p'_2)-C^*_{-+}(-p'_2, -p'_1)] = \frac{e}{m} vu(p_1, p_2) \\ &\quad \times [(\mathbf{p}_2-\mathbf{p}_1)\mathbf{A}^*(\mathbf{p}_1+\mathbf{p}_2)+\tfrac{1}{2}\mathscr{H}^*_0(\mathbf{p}_1+\mathbf{p}_2)], \end{aligned} \tag{IV.14}$$

$$\begin{aligned} &[\Omega(p_1)+\Omega(p_2)]\,[C_{-+}(p_1, p_2)+C^*_{-+}(-p_2, -p_1)] \\ &\quad + \sum_{p'_1, p'_2} [A_{-+}(p_1, p_2; p'_1, p'_2)+B_{-+}(p_1, p_2; p'_1, p'_2)] \\ &\quad \times [C_{-+}(p'_1, p'_2)+C^*_{-+}(-p'_2, -p'_1)] = 0, \\ &[\Omega(p_1)+\Omega(p_2)]\,[C_{--}(p_1, p_2)+C^*_{++}(-p_2, -p_1)] \\ &\quad + 2\sum_{p'_1, p'_2} [A(p_1, p_2; p'_1, p'_2)-B(p_1, p_2; p'_1, p'_2)] \\ &\quad \times [C_{--}(p'_1, p'_2)+C^*_{++}(-p'_2, -p'_1)] \\ &\quad - vu(p_1, p_2)\frac{e}{2m}\mathscr{H}_{-1}(\mathbf{p}_1+\mathbf{p}_2) = 0, \\ &[\Omega(p_1)+\Omega(p_2)]\,[C_{--}(p_1, p_2)-C^*_{++}(-p_2, -p_1)] \\ &\quad + 2\sum_{p'_1, p'_2} [A(p_1, p;_2 p'_1, p'_2)+B(p_1, p_2; p'_1, p'_2)] \\ &\quad \times [C_{--}(p'_1, p'_2)-C^*_{++}(-p'_2, -p'_1)] = 0. \end{aligned} \tag{IV.15}$$

Equations (IV.14,15) connect functions $C(p_1, p_2)$, $C(p_1, p_2)$ corresponding to given $\mathbf{p}_1+\mathbf{p}_2 = \mathbf{q}$ only. If in (IV.14) we let

$$\begin{aligned} [C_{-+}(p_1, p_2)-C^*_{-+}(-p_2, -p_1)] &\to \Theta_q(p), \\ [C_{-+}(p_1, p_2)+C^*_{-+}(-p_2, -p_1)] &\to \vartheta_q(p) \end{aligned} \tag{IV.16}$$

and take $\mathscr{H}_0 = 0$ (or take $\mathbf{q}$ parallel to the z-axis), these equations will be identical with the eqns. (81) and (117) solved by Bogoliubov (1959). We shall apply the results of this section and of Bogoliubov's paper to the investigation of the electrodynamics of the superconducting state.

§ 2 Sum rule

Buckingham (1957) considered a system of electrons at absolute zero in the presence of a weak magnetic field. The magnetic field is described by the vector potential $A(r)$ (A does not depend on time). The system of electrons is in a state described by $|0\rangle$ with energy E_0. Here $|k\rangle$ represents a complete set of states with energy eigenvalues E_k.

Up to first order in A the perturbation expansion has Fourier components of the form

$$j_\mu(q) = -\frac{e^2\varrho_0}{m}\sum_{\nu/1}^{3}\left(\delta_{\nu\mu}-S_{\mu\nu}(\boldsymbol{q})\right)A_\nu(q) \tag{IV.17}$$

where ϱ_0 is the number of electrons per unit volume and

$$S_{\mu\nu}(\boldsymbol{q}) = -\frac{1}{2}\sum_{k\neq 0}\sum_{\nu/1}^{3}\left\{\frac{\langle 0|e^{i(q\cdot r)}\nabla_\mu+\nabla_\mu\, e^{i(q\cdot r)}|k\rangle}{(2m)(E_0-E_k)}\right.$$

$$\left.\times\frac{\langle k|e^{-i(q\cdot r)}\nabla_\nu+\nabla_\nu\, e^{-i(q\cdot r)}|0\rangle}{(2m)(E_0-E_k)}+\text{compl. conj.}\right\}. \tag{IV.18}$$

If $\hat{H}$ is the Hamiltonian of the system without perturbation, i.e. $\hat{H}|k\rangle = E_k|k\rangle$, the commutator of $\hat{H}$ and $\exp[i(\boldsymbol{q}\cdot\boldsymbol{r})]$ gives the following identity:

$$\sum_{\mu/1}^{3}\frac{q_\mu\langle k|e^{i(q\cdot r)}\nabla_\mu+\nabla_\mu e^{i(q\cdot r)}|k'\rangle}{2m(E_{k'}-E)} = \langle k|e^{i(q\cdot r)}|k'\rangle. \tag{IV.19}$$

Using this identity and the completeness of the set $|k\rangle$ we obtain the so-called Buckingham sum rule (Buckingham, 1957)

$$\sum_{\mu,\nu/1}^{3}\frac{q_\mu\, q_\nu}{q^2}\, S_{\mu\nu}(\boldsymbol{q}) = 1. \tag{IV.20}$$

As $S_{\mu\nu}(\boldsymbol{q})$ is a function of $\boldsymbol{q}$ only, it must be of the form

$$S_{\mu\nu}(\boldsymbol{q}) = a(q)\frac{q_\mu q_\nu}{q^2} + b(q)\,\delta_{\mu\nu} \tag{IV.21}$$

which gives the relation

$$\operatorname{Tr} S_{\mu\nu} = a + 3b. \tag{IV.22}$$

From (IV.20) we have

$$a + b = 1. \tag{IV.23}$$

These two equations give

$$\alpha = \tfrac{1}{2}(3 - \operatorname{Tr} S_{\mu\nu}), \qquad b = \tfrac{1}{2}(\operatorname{Tr} S_{\mu\nu} - 1). \tag{IV.24}$$

Hence,

$$\delta_{\mu\nu} - S_{\mu\nu} = \tfrac{1}{2}(\operatorname{Tr} S_{\mu\nu} - 3)\left(\frac{q_\mu q_\nu}{q^2} - \delta_{\mu\nu}\right). \tag{IV.25}$$

We see that the sum rule leads to the existence of the projection operator $\left(\frac{q_\mu q_\nu}{q} - \delta_{\mu\nu}\right)$ in the expression for the current density. This operator acting on a vector gives the projection of this vector on a direction perpendicular to $\boldsymbol{q}$. Since gauge transformations add a longitudinal part to $\boldsymbol{A}(\boldsymbol{q})$, parallel to $\boldsymbol{q}$, we see that (IV.20) gives the gauge invariance of the current.

§ 3 Gauge invariance of the current and the Meissner effect

Let us consider the response of the system to a weak external electromagnetic field (linear response of the system). We wish to obtain the current as a function of a vector potential $\boldsymbol{A}$ and the magnetic field $\mathscr{H} = \operatorname{curl} \boldsymbol{A}$ (Galasiewicz, 1960a) (both functions are independent of time).

Considering the part of the Hamiltonian given by ($c = 1$)

$$\hat{H} = \sum_{s_z} \int \psi^+(r, s_z)\left[\frac{1}{2m}(\boldsymbol{p} - e\boldsymbol{A}(r))^2 - \frac{e}{2m}(\sigma\cdot\mathscr{H}(r))\right]\psi(r, s_z)\,\mathrm{d}V, \tag{IV.26}$$

we obtain the current as the coefficient of the variation of $\boldsymbol{A}$ in the formula

$$\delta\hat{H} = -\sum_{s_z}\int(\boldsymbol{J}\cdot\delta\boldsymbol{A})\,\mathrm{d}V. \tag{IV.27}$$

Thus for the current we have

$$J = J_1(r)+J_2(r), \tag{IV.28}$$

$$J_1(r) = \sum_{s_z}\left\{\frac{ie}{m}\left[(\nabla\psi^+(r, s_z))\psi(r, s_z)-\psi^+(r, s_z)\nabla\psi(r, s_z)\right]\right.$$

$$\left.-\frac{e^2}{m}A(r)\psi^+(r, s_z)\psi(r, s_z)\right\}, \tag{IV.29}$$

$$J_2(r) = -\frac{e}{2m}\sum_{s_z}\operatorname{curl}\psi^+(r, s_z)\boldsymbol{\sigma}\psi(r, s_z). \tag{IV.30}$$

Consider now the Fourier representation of the currents J_1 and J_2:

$$J_1(r, -+) = \frac{e}{2m}\frac{1}{V}\sum_{p_1p_2}(p_2-p_1)e^{i((p_1+p_2)\cdot r)}$$

$$\times(a^+_{-p_1+}a_{p_2+}+a^+_{-p_1-}a_{p_2-})-\frac{e^2}{m}A(r)\Bigg[\varrho_0+\sum_{p_1+p_2=q\neq 0}e^{i((p_1+p_2)\cdot r)}$$

$$\times(a^+_{-p_1+}a_{p_2+}+a^+_{-p_1-}a_{p_2-})\Bigg], \tag{IV.31}$$

$$\varrho_0 = \frac{2}{V}\sum_p v^2(p),$$

$$J_{2\gamma}(r, -+) = \gamma\frac{e}{2m}\frac{1}{V}\sum_{p_1+p_2=q}e^{i((p_1+p_2)\cdot r)}$$

$$\times(p_1+p_2)_\gamma(a^+_{-p_1+}a_{p_2+}-a^+_{-p_1-}a_{-p_2-}), \tag{IV.32}$$

the index $\gamma = -1, 0, 1$,

$$J_{2\gamma}(r, ++--) = \gamma\frac{e}{m}\frac{1}{V}\sum_{p_1+p_2=q}e^{i((p_1+p_2)\cdot r)}$$

$$\times(p_1+p_2)_0 a^+_{-p_1\gamma}a_{p_2,-\gamma}, \tag{IV.33}$$

$$\gamma = -1, 1,$$

$$J_{20}(r, ++--) = -\frac{e}{2m}\frac{1}{V}\sum_{p_1+p_2=q}\sum_\gamma e^{i((p_1+p_2)\cdot r)}$$

$$\times\gamma(p_1+p_2)_\gamma a^+_{-p_1,-\gamma}a_{p_2\gamma}. \tag{IV.34}$$

In the same approximation in which we have obtained the

Hamiltonians (IV.10,11), we express the current $\boldsymbol{j}$ as a linear function of β, β^+. Then we switch to the translated Bose amplitudes $\beta+C$, β^++C^*. The functions C, C^* are given by eqns. (IV.14,15).

Let us consider the current expectation value

$$\boldsymbol{j}(+) = \langle \boldsymbol{J}(r)\rangle^0 \tag{IV.35}$$

where the symbol $\langle\ \rangle^0$ denotes a mean value in the vacuum state of the quasi-particles described by the operators β', β'^+. Taking into account that

$$\begin{aligned}\langle \beta_{\sigma_1\sigma_2}+C_{\sigma_1\sigma_2}\rangle^0 &= C_{\sigma_1\sigma_2}(p_1, p_2),\\ \langle \beta'_{\sigma_1\sigma_2}+C^*_{\sigma_1\sigma_2}\rangle^0 &= C^*_{\sigma_1\sigma_2}(p_1, p_2),\end{aligned} \tag{IV.36}$$

we obtain $\boldsymbol{j}$ as a function of C, C^*. From (IV.14,15) we see that the functions C, C^* are linear functions of A_α and $\mathscr{H}_\alpha$. Therefore, after solving these equations we can obtain the current as a linear function of A_α and $\mathscr{H}_\alpha$.

Formulae (IV.35,36) give us

$$\begin{aligned} j_1(r, -+) = {} & \frac{e}{m}\frac{1}{V}\sum_{p,q}(2\boldsymbol{p}-\boldsymbol{q})e^{i(q\cdot r)}W_q(p)\Theta_q(p) \\ & -\frac{e^2}{m}\left[\sum_q A(\boldsymbol{q})e^{i(q\cdot r)}\right]\left\{\varrho_0+\sum_{p,q'}\chi_{q'}(p)\vartheta_{q'}(p)e^{i(q'\cdot r)}\right\},\end{aligned} \tag{IV.37}$$

$$j_{2,\gamma}(r, -+) = \gamma\frac{e}{2m}\frac{1}{V}\sum_{p,q}e^{i(q\cdot r)}q_\gamma\Theta_q(p), \quad \gamma = -1, 0, 1, \tag{IV.38}$$

$$j_{2,1}(r, ++--) = \frac{e}{2m}\frac{1}{V}\sum_{p,q}e^{i(q\cdot r)}W_q(p)q_0\tilde{\vartheta}_q(p), \tag{IV.39}$$

$$j_{2,-1}(r, ++--) = -\frac{e}{2m}\frac{1}{V}\sum_{p,q}e^{i(q\cdot r)}W_q(p)q_0\tilde{\vartheta}^*_{-q}(-p), \tag{IV.40}$$

$$\begin{aligned} j_{2,0}(r, ++--) = {} & -\frac{e}{4m}\frac{1}{V}\sum_{p,q}e^{i(q\cdot r)}W_q(p)[-q_1\tilde{\vartheta}^*_{-q}(-p) \\ & +q_{-1}\tilde{\vartheta}_q(p)],\end{aligned} \tag{IV.41}$$

$$
\begin{aligned}
\Theta_q(p) &= C_{-+}(p_1, p_2) - C^*_{-+}(-p_2, -p_1), \\
\vartheta_q(p) &= C_{-+}(p_1, p_2) + C^*_{-+}(-p_2, -p_1), \\
\tilde{\vartheta}_q(p) &= C_{--}(p_1, p_2) + C^*_{++}(-p_2, -p_1), \\
-\tilde{\vartheta}_{-q}(-p) &= C_{++}(p_1, p_2) + C^*_{--}(-p_2, -p_1).
\end{aligned} \tag{IV.42}
$$

We now will consider, as was done by Bogoliubov (1959), only $\boldsymbol{j}_1$ and the perturbation given by $\delta\hat{H}_A(-+)$. Equations (IV.14) can now be written

$$
\begin{aligned}
L_q(\Theta_q) &= \frac{e}{m}(2\boldsymbol{p}-\boldsymbol{q})\boldsymbol{A}(\boldsymbol{q})\,W_q(p), \\
M_q(\vartheta_q) &= 0.
\end{aligned} \tag{IV.43}
$$

From (IV.43) we have $\vartheta_q = 0$ and we can write for the Fourier components

$$
\boldsymbol{j}_1(q) = \frac{1}{2V}\sum_p e(2\boldsymbol{p}-\boldsymbol{q})\,W_q(p)\,\Theta_q(p) - e^2 A(q)\frac{2}{mV}\sum_p v^2(p). \tag{IV.44}
$$

We denote by $T_\alpha(p, q)$ the solution of the equation

$$
L_q(T_\alpha) = -\frac{2p_\alpha - q_\alpha}{m}W_q(p), \qquad \alpha = 1, 2, 3. \tag{IV.45}
$$

As L_q is a linear and homogeneous operator, we have from (IV.43,45)

$$
\Theta_q(p) = e\sum_\alpha T_\alpha(p, q)A_\alpha(q). \tag{IV.46}
$$

For the current we have the expression

$$
j^\alpha(q) = -\frac{e\varrho_0}{m}\sum_\beta \left((\delta_{\alpha\beta} - S_{\alpha\beta}(q)\right) A_\beta(q) \tag{IV.47}
$$

where ϱ_0 is defined by (IV.31) and

$$
S_{\alpha\beta}(q) = -\frac{1}{V}\sum_p \frac{2p_\alpha - q_\alpha}{2\varrho_0}W_q(p)\,T_\beta(p, q). \tag{IV.48}
$$

Using (IV.45) we can write

$$
S_{\alpha\beta}(q) = \frac{1}{V}\frac{m}{\varrho_0}\sum_p L_q\left(T_\alpha(p, q)\right)T_\beta. \tag{IV.49}
$$

From (IV.49) we see that

$$L_q[S_{\alpha\beta}(q)-S_{\beta\alpha}(q)] \equiv 0$$

and $S_{\alpha\beta}$ is symmetric

$$S_{\alpha\beta} = S_{\beta\alpha}. \tag{IV.50}$$

The $S_{\alpha\beta}$ play the role of phenomenological coefficients in the Onsager equations and (IV.50) are a kind of Onsager relation.

The identity (III.52) can be rewritten as

$$L_q(\chi_q) = \frac{2(\boldsymbol{p}\cdot\boldsymbol{q})-q^2}{2m} W_q(p).$$

So from (IV.48) we have

$$\sum_\alpha q_\alpha S_{\alpha\beta}(q) = \frac{m}{\varrho_0 V}\sum_p L_q(\chi_q)T_\beta = \frac{1}{V}\frac{m}{\varrho_0}\sum_p \chi_q L_q(T_\beta)$$

$$= \frac{1}{V\varrho_0}\sum_p (2p_\beta - q_\beta)W_q(p)\chi_q(p) =$$

$$= \frac{1}{V\varrho_0}\sum_p (2p_\beta-q_\beta)\{u^2(p)v^2(p-q)-v^2(p)u^2(p-q)\}$$

$$= \frac{1}{V\varrho_0}\sum_p (2p_\beta-q_\beta)v^2(p-q)-\frac{1}{V\varrho_0}\sum_p(2p_\beta-q_\beta)v^2(p)$$

$$= q_\beta \frac{2}{V\varrho_0}\sum_p v^2(p). \tag{IV.51}$$

We have used here the fact that L_q is an hermitian operator. From (IV.51,31) it follows that $S_{\alpha\beta}$ fulfills the relation which is equivalent to Buckingham's sum rule (IV.20)

$$\sum_\alpha q_\alpha S_{\alpha\beta}(q) = q_\beta. \tag{IV.52}$$

Consequently, we have gauge invariance for the current. Thus in the expression for the current we have the projection operator $\frac{q_\mu q_\nu}{q^2} - \delta_{\mu\nu}$ (see (IV.25)) which gives current conservation (for the static case) $(\boldsymbol{q}\cdot \boldsymbol{j}(q)) = 0$.

Because of the projection operator, $\boldsymbol{j}(q)$ depends on the transverse part of the vector potential which we denote by $A_\perp$

$$j^\alpha(q) = \frac{e^2\varrho_0}{m}\sum_\beta \bigl(S_{\alpha\beta}(q)-\delta_{\alpha\beta}\bigr)A_{\perp\beta},$$

$$A_{\perp\beta} = A_\beta - \frac{(\boldsymbol{q}\cdot\boldsymbol{A})}{q^2}q_\beta, \quad (\boldsymbol{q}\cdot\boldsymbol{A}_\perp) = 0. \tag{IV.53}$$

We can now write (IV.43) in the form

$$L_q(\Theta_q) = 2\frac{e}{m}(\boldsymbol{p}_\perp \cdot \boldsymbol{A}_\perp) W_q(p) \tag{IV.54}$$

where $\boldsymbol{p}_\perp$ is perpendicular to $\boldsymbol{q}$. We choose the z-axis, in momentum space, parallel to $\boldsymbol{A}_\perp$, and the x-axis parallel to $\boldsymbol{q}$; the more general case is considered by Czerwonko (1966b). Hence we have

$$\Theta_q(p) = eA_{\perp z}(q) T_z(q, p). \tag{IV.55}$$

For $T_z(q, p)$ we have the equation

$$L_q(T_z) = \frac{2}{m} p_z W_q(p). \tag{IV.56}$$

The function $\Theta_q(p)$ describes the response of the system to an external field. For constant A the component $A(0)$ is different from zero. Since in this case the response must be equal to zero, we look for a solution of (IV.56) in the form

$$T_z(p, q) = q\tau_1(p) + q^2\tau_2(p) + + . \tag{IV.57}$$

After substitution of (IV.55) into (IV.44) we can write

$$\boldsymbol{j}(q) = \frac{e^2 \varrho_0}{m}\{S(q) - \boldsymbol{e}_z\} A(q), \tag{IV.58}$$

where $|\boldsymbol{e}_z| = 1$ and

$$S(q) = -\sum_p \frac{2\boldsymbol{p} - \boldsymbol{q}}{2\varrho_0} T_z(p, q) W_q(p). \tag{IV.59}$$

Since T_z is first order in q as $q \to 0$, S will go to zero as q^2. Therefore, for small q we have

$$\boldsymbol{j}(q) = -\frac{e^2 \varrho_0}{m} \boldsymbol{A}_\perp(q), \tag{IV.60}$$

i.e. we have obtained the Meissner effect.

The method of Bogoliubov used in this section is based on the generalized $\{u, v\}$ transformation so as to preserve the gauge invariance of the theory. A method very similar to this one is given by Blatt (1960).

Bardeen (1957) called attention to the fact that the collective oscillations play an essential role in preserving the gauge invariance of the theory. Such papers as those by Rickayzen (1959) and Anderson (1958) are based on this idea.

An alternative way to obtain gauge-invariant results is to use the Green functions approach (Nambu, 1960; Gor'kov, 1958; Kadanoff and Martin, 1961; Ambegaokar and Kadanoff, 1961).

Different methods of treating the problem of gauge invariance are discussed in detail by Blatt (1964) and Rickayzen (1965).

PART II

QUANTUM FLUIDS

Introduction

WE SHALL consider now a second, extremely interesting phenomenon observed at low temperatures: superfluidity. Whereas a period of 46 years elapsed between the discovery of superconductivity in 1911 and the creation of its theory in 1957, the theoretical explanation of superfluidity, elaborated from 1941 to 1944, closely followed this phenomenon's discovery in 1938.

Superfluidity is observed only in ^{4}He. After the initial liquefaction of helium (boiling temperature 4.22°K) by Kamerlingh Onnes in 1908, further cooling was attempted, and in the vicinity of 2°K a discontinuity in the temperature dependence of many parameters was observed, initially by Kamerlingh Onnes (1911b). Onnes' first results, obtained by measuring the temperature dependence of the density, were confirmed by his later experiments (Kamerlingh Onnes and Boks, 1924). In 1927, Keesom and Wolfke (1928a,b) observed a similar discontinuity by measuring the temperature dependence of the dielectric constant. To explain these phenomena, they suggested the existence of two allotropic modifications of helium which they called "helium I" and "helium II". This terminology was universally accepted after Keesom and Clusius (1932a,b) showed a singularity in the specific heat curve. The curve which represents the temperature dependence of the specific heat has a shape resembling the Greek letter lambda with a jump at 2.19°K. In the transitions the authors did not observe a latent heat which is characteristic of first-order phase transitions. The temperature at which the transition takes place was called "the λ point".

Liquid helium can be solidified at temperatures near 0°K under a pressure of 25 atm. Therefore, the diagram in the p–Θ plane has, for helium, the shape given in Fig. 11.

We see that there is no triple point between the solid, liquid and gaseous states.

W. H. and A. P. Keesom (1936) discovered that helium II has a very large thermal conductivity. Its maximum, at 1.9°K, is 810 cal/cm sec deg, i.e. 10^3 times larger than the thermal con-

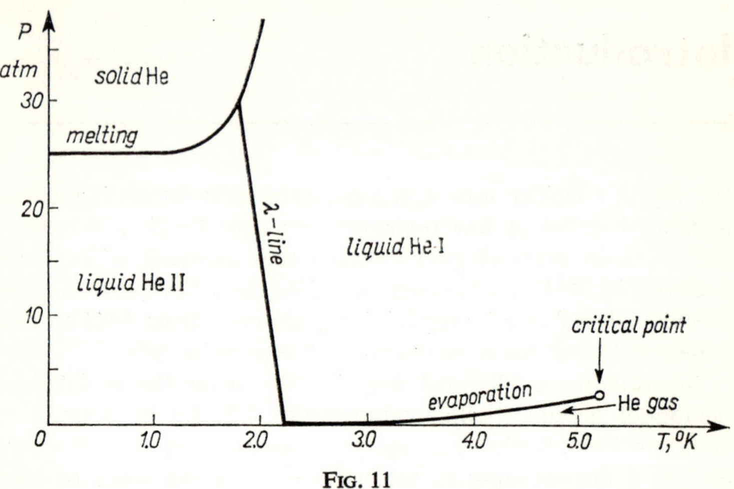

FIG. 11

ductivity of pure copper. So large a conductivity was called thermal superconductivity. Allen, Peierls and Uddin (1937) observed that the heat current is not proportional to the temperature gradient. This experiment indicated that the mechanism of the heat conductivity in helium II is different from that in other fluids.

However, He II's most characteristic feature was not the thermal superconductivity but rather the vanishingly small viscosity, discovered by Kapitza (1938). Kapitza measured the viscosity of helium by means of the capillary flow method. The flow velocity is inversely proportional to the viscosity coefficient η. For He I, the capillary filled in a few minutes; however, after the temperature was lowered below the λ point, He II filled the capillary in a few seconds. The first estimates gave $\eta_{II} < 1/1500\eta_I$. Further considerations of experimental results led to the conclusion that one can take $\eta_{II} = 0$. The He II property of flowing through a capillary without viscosity is called superfluidity.

These results of Kapitza seemed to be in contradiction with results also obtained in 1938 by Keesom and Mac Wood (1938). These authors measured the viscosity of helium with the rotating disc method and found that, below the λ point, it does not differ markedly from the viscosity of He I (which is about ten times less than that of water).

In the same year Allen and Jones (1938) discovered the thermomechanical (fountain) effect. A small vessel with a fine capillary at the bottom was immersed in a reservoir of helium. Heat was supplied to the vessel electrically. With the electric heater switched off, the levels of helium in the vessel and the reservoir were equal. After switching on the heater, the helium level in the vessel rose slightly above that in the reservoir. The observed level difference was about twenty cm. Thus a temperature difference produces a pressure difference between the reservoir and the immersed vessel. In the second variation of this experiment, a glass capillary packed with emery powder and with one end widened was employed. If the powder was heated by radiation, a liquid fountain emerged from the capillary. The height of the fountain reached thirty centimeters.

The inverse of the thermomechanical effect was observed by Daunt and Mendelssohn (1939). A small Dewar with a hole at the bottom end was filled with emery powder and equipped with an electrical resistance thermometer. If the level of He II in the Dewar and the reservoir is the same, one observes no temperature difference. If, however, a pressure difference is produced by withdrawing the Dewar, the helium flows out of it, and one observes an increased temperature in the Dewar. When one lowers the Dewar into the bath, helium flows into it and a temperature decrease is observed. Experiments revealed these and many other interesting and surprising properties of He II, all requiring theoretical explanation. A first attempt was made by London (1938); he advanced the hypothesis that the transition of liquid helium I to helium II might be caused by the condensation mechanism of the degenerate Bose–Einstein gas. The volume of helium is three times as large as van der Waals forces would predict, thus its density is very small. From this point of view, it is rather more similar to a gas than a solid, whence the possibility of treating it as a gas. One must stress, however, that helium II in no way resembles an ideal gas. Below the critical temperature Θ_c corresponding to the λ point, the kinetic energy becomes very small and the mean de Broglie wavelength is of the order of the mean molecular distance. For a finite fraction of molecules which "condense" into the lowest quantum state with zero momentum, the de Broglie wavelength becomes infinite.

While in the many-body system the de Broglie wavelength of particles is comparable to molecular spacing, it is practically impossible to speak about individual particles. Such a system can be treated rather as a big macromolecule than an ideal gas.

According to the London ideal Bose-gas model of superfluid helium, below the defined critical temperature, one fraction of atoms is condensed in the lowest energy level and the other occupies the excited states similar to those appearing in the Bloch theory of metals. On the basis of the London's assumptions, Tisza (1938 a, b; see also 1940, 1967) formulated his macroscopic theory of liquid helium, the so-called "two-fluid" theory. He found that the condensed atoms do not participate in the dissipation of momentum. Therefore, he attributed the viscosity to the excited helium atoms. By means of this simple "two-fluid" model Tisza was able to explain both the contradictory results of the viscosity measurements obtained by Kapitza (1938) and, independently, by Keesom and Mac Wood (1938), and the mechanism of the fountain effect.

The two-fluid model was confirmed by Kapitza's (1941 a, b) experiments. Those experiments, which will be described later, were performed at the time when Landau (1941) formulated his new theory of superfluidity.

Although the new theory is also called the two-fluid theory, it is based on the assumption of the possibility of the existence of two independent movements in He II. Those movements take place practically without momentum transfer from one to the other. It was discovered later that an interaction arises only near the so-called critical velocity which will be defined later. From the point of view of Landau's theory, Tisza's assumption about the existence of two fluids (one of them consisting of the excited helium atoms and the second of the atoms condensed in the lowest energy state) is wrong.

The movements in helium II are connected with effective masses or densities. One of the movements has the properties of those in the normal liquid, the second of those in the superfluid. Therefore, it is convenient to speak about normal and superfluid components of He II. The normal component which behaves like a real fluid is connected with all thermal excitations of the liquid whose viscosity is due to mutual collisions of phonons and rotons.

These assumptions explained the results of the above experiments. It is immediately clear that the two different methods of measurement of the viscosity must lead to different results. In the rotating disc method the viscosity of the normal component (normal "fluid") causes a damping of the torsional oscillations. In the capillary flow method we observe the second, superfluid, component.

The thermomechanical effect can be explained as follows. The ratios of the densities of the normal and of the superfluid component to the whole density are functions of temperature. In the extreme cases, at $\Theta = 0°$ and above $\Theta = \Theta_c$, we have either the superfluid or the normal component, respectively. Hence, heating the helium in the small vessel leads to an increase of the normal component. However, the normal component cannot penetrate through the capillary into the reservoir as easily as the superfluid component can penetrate from the reservoir to the vessel. This effect resembles osmosis. Here the capillary plays the role of a semi-permeable membrane. The observed pressure difference is similar to osmotic pressure. In the mechanocaloric effect, due to the pressure difference, the superfluid component flows through the emery powder. This motion, however, is not connected with heat transfer. Therefore, the original amount of heat continues to reside in the diminished mass of He II, leading to a temperature increase in the vessel.

One of the most important problems investigated by Kapitza (1941a,b) was the mechanism of thermal conductivity in He II. His experiments show that, if heat is applied to the bulb (a Dewar vessel connected by a capillary with the reservoir containing He II), there is a steady outflow of liquid from the vessel. As the vessel never empties, it follows that a flow into the bulb occurs. The stream going out of the vessel carries heat and has viscosity, on the other hand, the opposite stream does not transfer heat and has no viscosity. The first stream can be identified with the normal, and the latter with the superfluid component. From this and further experiments one concludes that the normal component can interpenetrate the superfluid component. At the present time, this is considered to be the true mechanism of thermal superconductivity, discovered by W. H. and A. P. Keesom (1936).

We have mentioned that the ratio of the density of the normal component to the total density is a function of temperature. A measurement of this ratio, and therefore the demonstration of the existence of two fluids, was given by Andronikashvili (1946). A series of closely spaced thin discs, suspended in a cryostat with He II, can perform torsional oscillations. The oscillating discs pull along only the normal fluid. Therefore, the moment of inertia (discs plus the liquid trapped between and moving with the discs) is a function of temperature. This can be demonstrated experimentally.

Tisza in 1938, on the basis of his theory, and Landau, independently, both predicted a very interesting thermal property of He II, namely the possibility of a wave propagation process, distinct from pressure waves associated with ordinary sound. Tisza argued that this new process can lead to a levelling out of temperature differences, just as pressure differences are leveled out in ordinary sound waves. Landau called these new waves "second sound". They are also called thermal or entropy waves, a more apt name, since, as is stressed by London (1954), microphones do not react to them. After the prediction of second sound by Landau in 1941, Shalnikov and Sokolov attempted to generate first sound by oscillating piezo-quartz, but without success. Lifshitz (1944a,b) considered theoretically various methods of exciting second sound. From his analysis it was clear that if second sound is excited "mechanically", the ratio of radiation intensities of second and first sound, which is equal to $I_2/I_1 \cong \alpha^2 \Theta c_2^3/c_1$, is very small. Here, α is the thermal expansion coefficient, and $c_{1,2}$ are the velocities of first and second sound, respectively. For $\Theta \sim 2°\text{K}$, for example, $I_2/I_1 \sim 2\times 10^{-6}$. If the source of second sound is a periodically varying temperature, we have $I_2/I_1 \sim c_v/\Theta \alpha^2 c_1 c_2$ (where c_v is the specific heat). For $\Theta \sim 2°\text{K}$ we have $I_2/I_1 \sim 5\times 10^3$. In 1944 Peshkov produced and observed standing thermal waves in He II (Peshkov, 1944, 1946a,b, 1948, 1949). Periodic fluctuations of the temperature were created by a plane electric heater (frequency of electric current 100–10,000 Hz). He found that the velocity of second sound is practically independent of frequency. Peshkov found that below 2.2°K the velocity of second sound has (at 1.65°K) a maximum of 20.3 m/sec, and (at 1.1°K) a minimum of 18.4 m/sec. The velocity rises again to a value of about 150 m/sec if the temperature is lowered (Atkins and Osborne, 1950). A similar dependence was observed by Peshkov (1960) who notes, however, that "under existing experimental conditions there is no point in speaking about second sound at temperatures below 0.5°K". The velocity of first sound increases below 2.2°K from 220 to 250 m/sec.

In first sound waves, periodic pressure changes are so fast that they occur adiabatically without oscillations of the temperature. The normal and superfluid components move in the same direction and with the same speed.

In second sound waves, we have periodic temperature changes. Since the thermal expansion of helium is very small, one does not observe pressure oscillations. The normal and superfluid currents move in opposite directions, in such a manner that the total current is zero.

Tisza's theory had its greatest success in the prediction of second sound. However, according to this theory, the velocity of second sound tends to zero as the temperature decreases. Landau's theory which also predicted second sound (Landau, 1941) leads to $c_1/c_2 = \sqrt{3}$ for $\Theta = 0$. Experiments (Atkins and Osborne, 1950) resolved this discrepancy by verifying Landau's theory.

Landau treated liquid helium more as a solid than as an ideal Bose gas. In his theory, an essential role is played by the excitations of a weakly excited many-particle system. These excitations are quantized, the quanta being quasi-particles or elementary excitations. Landau's approach resembles, in a certain sense, Debye's where the spectrum of the weakly excited solid is described by the spectrum of quasi-particles, i.e. phonons.

In the first version Landau (1941) considered a spectrum of elementary excitations consisting of two superimposed branches. The first branch gave the energy of excitations as a linear function of momentum and described sound quanta, phonons. The second, with an energy gap, represented the energy as a quadratic function of momentum and described excitations named "rotons".

A quantitative comparison of experimental data concerning the velocity of second sound (Peshkov, 1946) with theoretical calculations based on the two-branch spectrum showed a noticeable discrepancy. This and the theoretical results of Bogoliubov (1947b) induced Landau (1947) to attribute the form shown in Fig. 12 ($\varepsilon(p)$ (°K), p(Å^{-1})) to the energy spectrum of He II.

We see that a phonon energy dependence on momentum in the form

$$\varepsilon(p) = cp \tag{1}$$

(where c is the velocity of sound) is insufficient to explain the whole spectrum of Fig. 12. In the vicinity of the momentum $p_0 \sim 1/a$ (a is the mean interatomic distance), $\varepsilon(p)$ can be expanded in a series in $p-p_0$ and has the form

$$\varepsilon(p) = \Delta + \frac{(p-p_0)^2}{2\mu} \tag{2}$$

where (Henshaw and Woods, 1961) $\Delta = 8.6°\text{K}$, $\mu = 0.16\ m_{\text{He}}$, $p_0 = 1.9 \times 10\ \text{cm}^{-1}\ \hbar$ (m_{He} is the mass of the atom ^{4}He).

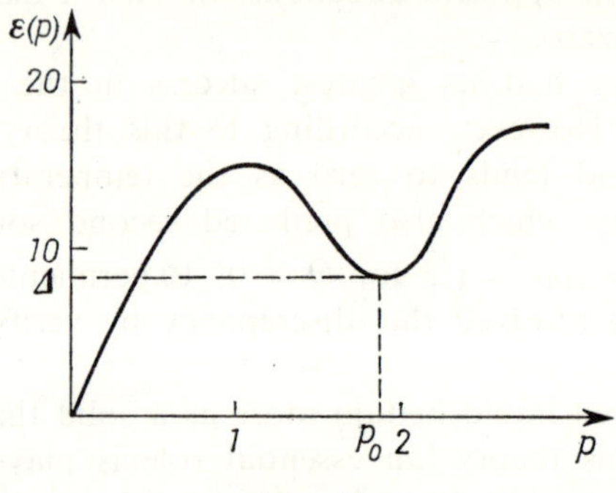

Fig. 12

Now $\varepsilon(p)$ given by (2) is treated as the spectrum of rotons. Landau stressed that in view of the new spectrum it is impossible to regard rotons and phonons as qualitatively different types of elementary excitations. By means of the total (phonon and roton) spectrum Landau obtained all the thermodynamic functions for He II. These are in good agreement with experiment. In calculations, phonons are treated by means of Bose–Einstein statistics and rotons by means of Boltzmann statistics. Owing to the large gap Δ in the energy term (2), the Bose–Einstein distribution function can be replaced by the Boltzmann distribution function.

The phonons describe the potential motion of the liquid. According to Landau, the rotons may describe certain rotational or vortex motions which distinguish a fluid from a solid. In order to explain the nature of rotons, Lee and Mohling (1959) examined experimental data of the total cross-section for the inelastic scattering of cold neutrons in helium II. The authors concluded that the angular momentum of rotons is equal to zero, i.e. that they had no rotational character.

Chester (1963) considered roton excitation as a ^{4}He atom with a modified mass. For $p \approx p_0$ there are essentially no local density fluctuations, characteristic of phonon excitations, and what remains is a single particle propagation through the liquid. This picture of the roton is consistent with the results of Miller, Pines and Nozières (1962) and with those in the early paper of Bogoliubov (1947), and seems to be the most adequate at the moment.

The parameters of the Landau spectrum (Δ, p_0, μ) can be determined by neutron-scattering experiments: single-energy neutrons scattered elastically by rotons or phonons. By measuring the initial and final momenta of the neutrons, as well as the angle of scattering, one can reproduce the entire spectrum of elementary excitations (see, for example, Palevsky, Otnes and Larsson, 1959; Yarnell *et al.*, 1959; Henshaw and Woods, 1961; Feynman, 1954). This spectrum actually has the form shown in Fig. 12).

The energy spectrum of elementary excitations was expressed in 1953 by Feynman (1954; see also Pitaevskii, 1954) through the formula

$$\varepsilon(p) = \frac{p^2}{2mS(p)} \tag{3}$$

where $S(p)$ is (in X-ray analysis) the so-called "structure factor". The factor $S(p)$ describes not only the interaction of the liquid with γ-quanta (X-rays) but also the interaction with particles at 0°K in general. It cannot be found theoretically but can be determined from experimental data (for example, from the measurement of scattered X-rays, see Beaumont and Reekie, 1955). On the other hand, some general properties of $\varepsilon(p)$ give us some information about the momentum dependence of S. In order to get an idea of the spectrum of elementary excitations in Bose systems, Bogoliubov (1947b) considered a simple model of a "nearly perfect" Bose gas. He obtained in the limit of small p, $\varepsilon(p) = cp$, and in the limit of large p, $\varepsilon(p) = p^2/2m$. For these cases, we have $S \approx p/2mc$ and $S \approx 1$, respectively.

For $p \sim 1/a$ $(p = p_0)$ the structure factor must have a maximum (or be a monotonic function of p). If S has a maximum, $\varepsilon(p)$ will be of the form assumed by Landau.

In his paper, Landau (1941) gives not only the thermodynamic but also the full system of hydrodynamic equations of He II (without dissipative terms). From the linearized equations Landau has obtained the velocities of the two sounds in He II. The system of hydrodynamic equations of He II, obtained on the basis of the conservation laws of momentum and entropy, and Galileo's principle of relativity, was given by Khalatnikov (1952c). Hydrodynamic equations with dissipative terms were obtained by Khalatnikov (1952a) and used to consider the attenuation of first and second sound (Khalatnikov, 1952b). The derivation of the temperature dependence of the kinetic coefficients, based on the phonon–roton model, were given in papers by Khalatnikov (1952b), Landau and Khalatnikov (1949) and Khalatnikov (1950).

In his paper, Landau introduced the notion of a critical velocity v_c. Below this velocity (at $\Theta = 0°$) no phonons or rotons are produced and we have a frictionless motion, without energy dissipation. Landau's formula for the critical velocity is

$$v_c = \min_{(p)} \frac{\varepsilon(p)}{p}. \tag{4}$$

This minimum is realized at the point of the curve $\varepsilon(p)$ where $\varepsilon/p = \mathrm{d}\varepsilon/\mathrm{d}p$. From Fig. 12 one can determine this point and find that the corresponding velocity is $v \sim 60$ m/sec. There is, however, a great discrepancy with experimental data, since the largest observed critical velocities are ~ 70 cm/sec. In 1955, Feynman (1955) called attention to the fact that during the flow of He II through a capillary, in the superfluid component vortex lines are formed, which possess a self-energy. They can be treated as new elementary excitations. If we insert the energy of these elementary excitations into formula (4), we obtain $v_c \sim 100$ cm/sec, in much better agreement with experiment than offered by the phonon–roton spectrum. If we simplify the formula obtained by Feynman, we find the following "uncertainty relation"

$$mv_c d \sim \hbar$$

for the critical velocity of He II in the capillary (where d is a diameter of the capillary).

The vortex lines are also responsible for the fact observed by Osborne (1950) that helium II rotates like a normal liquid. The scattering of rotons and phonons on vortex lines leads to the momentum transfer between the two movements.

Thus far, we have discussed the properties of only one, though the most interesting quantum fluid, He II.† It is a fluid formed by the atoms of ^{4}He. As these atoms obey Bose–Einstein statistics, we call He II a Bose quantum fluid. Moreover, we believe that a system of atoms of the isotope ^{6}He (which is very unstable) must form a superfluid.

Many-particle systems (such as liquid ^{3}He, the electron gas in metals at low temperatures, and the systems of nucleons in atomic nuclei) formed by particles obeying Fermi–Dirac statistics are called Fermi fluids.

In liquid ^{3}He we do not observe superfluidity. The superfluidity of the electron gas we call superconductivity. A system of nucleons is treated as an ordinary quantum fluid, as well as a superfluid.

We call ^{3}He and ^{4}He quantum fluids because they cannot be solidified (at normal pressure) at temperatures approaching absolute zero. This is due to the quantum properties of many-body systems which differ from the classical ones as they approach absolute zero. More precisely, according to classical theory, at absolute zero the kinetic energy of the particles must be equal to zero and the potential energy must have a minimum. The minimum of the potential energy leads to an ordered structure (crystal lattice) which occurs in the solid state. At temperatures near absolute zero, the atoms can perform small oscillations about their equilibrium positions. We can consider these oscillations classically or by quantum-mechanical methods. The latter are connected with so-called zero-point oscillations and the zero-point energy, an idea completely alien to classical physics. One may verify the existence of zero-point oscillations by examining X-ray scattering in crystals. When experimental data are compared with predictions offered by both of these theories, the quantum theory connected with zero-point oscillations clearly provides a sounder interpretation than the classical theory. Also, as we have stressed in Part I, the effective attraction of electrons in superconductors is a consequence of their interaction with the zero-point oscillations of the crystal lattice.

Examination shows that the zero-point energy is too large to make solidification of He energetically favored (London, 1954).

† More details about the development of experimental and theoretical investigations of superfluidity of ^{4}He are discussed by Galasiewicz (1969 b).

This is a consequence of the small mass and weak van der Waals forces of helium. On the other hand, similar considerations of the zero-point energy for H_2 yield the possibility of solidification. Although the mass of H_2 is half the mass of He, the van der Waals forces between hydrogen molecules are about twelve times larger than between helium atoms.

The theory of infinite systems of interacting Fermi particles with spin 1/2, called Fermi fluids, was given by Landau (1956, 1957). This is, in fact, the theory of liquid ^{3}He.

The theory for finite systems of interacting Fermi particles and its application to the theory of nuclei is described in detail in Migdal's monograph (1965).

The most important assumption in the Landau theory is that the single-particle energy spectrum has the same character for a Fermi liquid as for an ideal Fermi gas. As the particles of a liquid interact, the energy spectrum is not connected with the particles but with the Fermi type quasi-particles. These quasi-particles which are "superpositions" of particles have definite momenta. The distribution function in momentum space is different for interacting and non-interacting particles. But in both cases, in the ground state, the Fermi momentum p_F at which the distribution function has a jump is the same. This momentum is defined by the particle density, independent of the interaction (Luttinger and Ward, 1960; Pitaevskii, 1959), through the formula

$$\varrho = \frac{p_F^3}{3\pi^2}. \tag{5}$$

From this ground state we obtain the excited states by occupation of the levels for $p > p_F$. In this case we obtain "holes" with $p < p_F$ and "particles" with $p > p_F$. "Holes" and "particles" are the quasi-particles which describe a weakly excited system of Fermi particles. In an ideal Fermi gas these excitations do not interact, whereas in a Fermi liquid they do. Through the specific type of interaction (pairing correlations) we obtain for a Fermi liquid an energy spectrum with an energy gap (Larkin and Migdal, 1963) and in this case we have superconductivity (electrons in metals) or superfluidity (nucleons in nuclei). Instead of a jump we have a smearing-out of the distribution function in the vicinity of p_F.

After the discovery of superconductivity, the possibility of attraction between elementary excitations in ^{3}He was examined

(Brueckner and Gammel, 1958; Pitaevskii, 1959; Brueckner *et al.*, 1960; Emery and Sessler, 1960; Gor'kov and Pitaevskii, 1960; Anderson and Morel, 1961; Morel and Nozières, 1962; Cohen and Abrahams, 1966). It was found that for elementary excitations with sufficiently large angular momentum there exists an effective attraction between elementary excitations. For excitations with $l = 2$ (d-pairing) one finds a critical temperature 10^{-3}–10^{-4} °K. Experimental investigations of ^{3}He, down to 10^{-3} °K, have not confirmed the existence of the transition (Peshkov, 1964; Abel *et al.*, 1965).

The interaction between quasi-particles makes them unstable. The possibility of the decay of a quasi-particle with momentum p is proportional to $(p-p_F)^2$. This probability is small for momenta $p \approx p_F$. Therefore, quasi-particles with an energy $\sim(p-p_F)$ are stable for $p \sim p_F$ and in this region of momenta one may speak of elementary excitations (Abrikosov, Gor'kov and Dzyaloshinskii, 1962).

In the case of interacting Fermi systems, no small parameter enters the theory and hence one cannot use perturbation methods. Therefore, we ought to introduce certain universal constants into the theory, as in the case of dispersion relations, where the masses of particles and the constants characterizing the interaction are taken from experiment. In the case of infinite Fermi systems, in order to determine the spectrum of elementary excitations, it is sufficient to introduce one constant, the effective mass m^* of a quasi-particle. (For finite systems, besides m^* one must introduce parameters characterizing the potential well.) This constant enters into the formula giving the temperature dependence of the specific heat,

$$c_v = \tfrac{1}{3} m^* p_F \Theta. \tag{6}$$

Having found p_F by measuring the density of liquid ^{3}He ($p_F \approx n \cdot 0.8 \times 10^8$ cm^{-1}), we find $m^* = 2.4\, m_{^3\mathrm{He}}$ from measuring the specific heat.

The spectrum of collective oscillations is described by the expansion coefficients (of the Legendre polynomials of the quasi-particle scattering amplitude for momenta near p_F and angle zero). These coefficients are also parameters introduced into the theory. For example, the experimental data for the velocity of sound in ^{3}He is $c = 183$ m/sec, and gives for the "zero" coefficient

$l_0 = 7.6$ and for the first coefficient $l_1 = 3.0$. For this sound characterized by a frequency ω we have the condition $\omega\tau \ll 1$, where τ is the collision time. This condition is fulfilled for a very small collision time, i.e. for frequent collisions. Therefore it describes the collision-dominated region. Frequent collisions lead to a state of local thermodynamic equilibrium. It can be described by local quantities found from the hydrodynamic equations. Under these circumstances a perturbation in the form of organized density oscillations, i.e. ordinary sound, can be propagated in liquid. When the temperature Θ is lowered, the mean free path l of quasi-particles increases proportionally to $1/\Theta^2$. The collision time increases, and collisions are infrequent. It is a collision-free region in which we cannot have thermodynamic equilibrium and ordinary sound cannot propagate. In his theory of ^{3}He, Landau considered the case $\omega\tau \gg 1$, corresponding to the extreme case of a collision-free region. He found that for this region disturbances can propagate in ^{3}He, but these are not ordinary sound waves. To observe such disturbances, called "zero sound" by Landau, one needs to excite the liquid at high frequencies, so that the period of oscillations $\sim 1/\omega$ is much smaller then τ. "Zero sound" was observed experimentally for the first time by Abel, Anderson and Wheatley (1966), in full agreement with the predictions of Landau's theory.

In 1963, the Landau equations for a Bose superfluid were derived by Bogoliubov (1963) from the microscopic Hamiltonian. The spectrum of elementary excitations, in the hydrodynamic region, was obtained by calculating the single-particle Green functions. In addition, the hydrodynamic equations for ordinary fluids were derived from the microscopic Hamiltonian. Galasiewicz (1963b) considered the equations of the Bose superfluid by taking into account the viscosity of the normal component. In that paper, damping phenomena, i.e. the attenuation of first sound, were also considered by means of the single-particle Green functions.

Kadanoff and Martin (1963) studied the properties of the correlation functions by using the hydrodynamic equations for an ordinary fluid. The basic idea of this method is similar to that used by Bogoliubov (1963). Consider a system of many particles described by means of Green or correlation functions. When a small perturbation is introduced, which leads to a very slow variation in space and time of all physical quantities, the system may be described by the linearized hydrodynamic equations. In this special case the two alternative approaches: (i) the Green or correlation function description, and (ii) the "hydrodynamic" description, are equivalent. This allows us to find expressions for the Green and correlation functions in the hydrodynamic region. The description in terms of correlation functions leads naturally to expressions for the transport coefficients which are Kubo-type formulae. For superfluid helium the methods derived by Kadanoff and Martin (1963) were used by Hohenberg and Martin (1965). Here, we shall use the description in terms of Green functions proposed by Bogoliubov (1963).

CHAPTER V
Basic identities and relations

§ 1 Time derivatives of some "local quantities" for Bose systems†

Consider a system of identical Bose particles, with pairing forces. The Hamiltonian of the system in the second quantization representation has the form

$$\hat{H}[\eta, U, A] = \hat{H} + \hat{H}_t[\eta, U, A],$$

$$\hat{H} = \frac{1}{2m}\int (\nabla\psi^+(t, r)\cdot\nabla\psi(t, r))\,d^3\boldsymbol{r} - \lambda\int \psi^+(t, r)\psi(t, r)\,d^3\boldsymbol{r}$$
$$+\tfrac{1}{2}\int V(r-r')\psi^+(t, r)\psi^+(t, r')\psi(t, r')\psi(t, r)\,d^3\boldsymbol{r}\,d^3\boldsymbol{r}',$$

$$\hat{H}_t[\eta, U, A] = \int\{\eta(t, r)\psi^+(t, r) + \eta^*(t, r)\psi(t, r)\}\,d^3\boldsymbol{r}$$
$$+\int U(t, r)\hat{\varrho}(t, r)\,d^3\boldsymbol{r} - \frac{1}{m}\int(\hat{\boldsymbol{j}}(t, r)\cdot\boldsymbol{A}(t, r))\,d^3\boldsymbol{r}$$
$$-\frac{1}{2m}\int\hat{\varrho}(t, r)A^2(t, r)\,d^3\boldsymbol{r},$$

$$\hat{\varrho}(t, r) = \psi^+(t, r)\psi(t, r), \quad \hat{\boldsymbol{j}}(t, r) = \hat{\boldsymbol{j}}^0(t, r) - \hat{\varrho}(t, r)\boldsymbol{A}(t, r),$$
$$\hat{\boldsymbol{j}}^0(t, r) = \tfrac{i}{2}\left[(\nabla\psi^+(t, r))\psi(t, r) - \psi^+(t, r)(\nabla\psi(t, r))\right] \qquad \text{(V.1)}$$

where λ is a constant, and $\psi^+(t, r)$, $\psi(t, r)$ are Bose operators in the Heisenberg representation for the Hamiltonian $\hat{H}_t[\eta, U, A]$. (We write r instead of $\boldsymbol{r}$ in the argument of the functional dependence, $\hbar = 1$).

In the method proposed by Bogoliubov, in addition to the usual

† The content of this section is based on work by Bogoliubov (1963, Section 2).

terms in the Hamiltonian, there are additional terms corresponding to "sources of particles" $\eta(t, r)$, $\eta^*(t, r)$, to the external scalar field $U(t, r)$ and the external vector field $A(t, r)$ introduced by Krasnikov (1966, 1967 b). All these functions are in general given time-dependent c-number functions. The introduction of the external independent perturbations enables us to calculate the variations of hydrodynamical expectation values with respect to them and, in consequence, to obtain the formulae for the Green functions.

In order to obtain the hydrodynamic equations we must derive a number of identities giving expressions for the time derivatives of the following "local quantities":

$$\begin{aligned}
\varrho(t, r) &= \langle\hat{\varrho}(t, r)\rangle, \qquad \Phi(t, r) = \langle\psi(t, r)\rangle,\\
j_\alpha(t, r) &= \langle\hat{j}^0_\alpha(t, r)\rangle - \langle\hat{\varrho}(t, r)\rangle A_\alpha(t, r), \quad (\alpha = 1, 2, 3),\\
\varrho(t, r)\varepsilon(t, r) &= \frac{-1}{4m}\langle(\nabla^2\psi^+(t, r)\psi(t, r) + \psi^+(t, r)(\nabla^2\psi((t, r))\rangle\\
&\quad + \frac{1}{2}\int V(r-r')\langle\psi^+(t, r)\psi^+(t, r')\psi(t, r')\psi(t, r)\rangle\, d^3\boldsymbol{r}
\end{aligned} \tag{V.2}$$

where ϱ, $\boldsymbol{j}$, $\varrho\varepsilon$ denote the mean density, the mean current and the mean energy, respectively (ε is the mean energy per particle). Further $\langle\ldots\rangle$ denotes the non-equilibrium expectation values of the field operators (defined in § 3, formula (V.25)). As was shown by Penrose and Onsager (1956), Bogoliubov (1961), the expectation value of the Bose-field operator $\langle\psi(t, r)\rangle$ in the presence of the condensate (the superfluid case) is non-zero.

In order to find the time derivative of (V.2) we use the equations of motion for the Hamiltonian (V.1). We have

$$\begin{aligned}
i\frac{\partial\psi(t, r)}{\partial t} &= [\psi(t, r), H(\eta, U, A)]\\
&= -\lambda\psi(t, r) - \frac{\nabla^2}{2m}\psi(t, r) + \int V(r-r')\psi^+(t, r')\psi(t, r')\,dr'\psi(t, r)\\
&\quad + U(t, r)\psi(t, r) + \eta(t, r) + \frac{i}{m}\big((\nabla\psi(t, r))\cdot A(t, r)\big)\\
&\quad + \frac{i}{2m}\psi(t, r)\big(\nabla\cdot A(t, r)\big) + \frac{1}{2m}\psi A^2(t, r).
\end{aligned} \tag{V.3}$$

After differentiating $\varrho(t, r)$ and using (V.3) we have

$$\frac{\partial \varrho(t,r)}{\partial t} = \left\langle \frac{\partial \psi^+(t,r)}{\partial t}\psi(t,r) + \psi^+(t,r)\frac{\partial \psi(t,r)}{\partial t} \right\rangle$$

$$= \frac{i}{2m}\langle -(\nabla^2\psi^+(t,r))\psi(t,r) + \psi^+(t,r)(\nabla^2\psi(t,r))\rangle$$

$$+ \int V(r-r') < \psi^+(t,r)\psi^+(t,r')\psi(t,r')\psi(t,r)$$

$$-\psi^+(t,r)\psi^+(t,r')\psi(t,r')\psi(t,r) > \mathrm{d}^3\mathbf{r}'$$

$$+ i\eta^*(t,r)\Phi(t,r) - i\eta(t,r)\Phi^*(t,r).$$

If the operators $\psi(t,r)$, $\psi^+(t,r)$ are in the Heisenberg representation (with respect to the Hamiltonian $\hat{H}[\eta, U, A]$), the density matrix does not depend on time (see § 3). Using

$$(\nabla\langle(\nabla\psi^+(t,r))\psi(t,r) - \psi^+(t,r)(\nabla\psi(t,r))\rangle)$$

$$= \langle(\nabla^2\psi^+(t,r))\psi(t,r) - \psi^+(t,r)(\nabla^2\psi(t,r))\rangle$$

we obtain the continuity equation

$$m\frac{\partial \varrho(t,r)}{\partial t} + \sum_\alpha \frac{\partial j_\alpha(t,r)}{\partial r_\alpha}$$

$$= im[\eta^*(t,r)\Phi(t,r) - \eta(t,r)\Phi^*(t,r)]. \quad \text{(V.4)}$$

We see that for a system without pairing interactions, the right-hand side of (V.4) is zero.

For the current density we obtain

$$\frac{\partial j_\alpha}{\partial t} = \frac{1}{4m}\frac{\partial}{\partial r_\alpha}\nabla^2\varrho - \frac{1}{2m}\sum_\beta \frac{\partial}{\partial r_\beta}\left\langle \frac{\partial \psi^+}{\partial r_\alpha}\frac{\partial \psi}{\partial r_\beta} + \frac{\partial \psi^+}{\partial r_\beta}\frac{\partial \psi}{\partial r_\alpha} \right\rangle$$

$$-\int \frac{\partial V(r-r')}{\partial r_\alpha}\langle\psi^+(t,r)\psi^+(t,r')\psi(t,r')\psi(t,r)\rangle \mathrm{d}\mathbf{r}'$$

$$+\varrho\frac{\partial}{\partial r_\alpha}(\lambda - U) + \frac{1}{2}\left(\frac{\partial \Phi^*}{\partial r_\alpha}\eta + \frac{\partial \Phi}{\partial r_\alpha}\eta^* - \Phi^*\frac{\partial \eta}{\partial r_\alpha} - \Phi\frac{\partial \eta^*}{\partial r_\alpha}\right)$$

(in order to simplify the calculations we put $A = 0$). When no misunderstanding is possible, we omit the arguments t, r, of the functions. In the following, the description of equilibrium and non-equilibrium processes is simplified by introducing $D_t(r, r-r')$,

$$\langle\psi^+(t,r)\psi^+(t,r')\psi(t,r')\psi(t,r)\rangle \equiv D_t(r, r'-r)$$

$$= D_t(r', r-r') = \tfrac{1}{2}\{D_t(r', r-r') + D_t(r, r'-r)\}.$$

After introducing $\boldsymbol{R} = \boldsymbol{r} - \boldsymbol{r}'$ we finally obtain

$$\frac{\partial j_\alpha}{\partial t} = \frac{1}{4m}\frac{\partial}{\partial r_\alpha}\nabla^2\varrho - \frac{1}{2m}\sum_\beta \frac{\partial}{\partial r_\beta}\left\langle \frac{\partial\psi^+}{\partial r_\alpha}\frac{\partial\psi}{\partial r_\beta} + \frac{\partial\psi^+}{\partial r_\beta}\frac{\partial\psi}{\partial r_\alpha}\right\rangle$$

$$-\frac{1}{2}\int \frac{\partial V(R)}{\partial R_\alpha}\{D_t(r, -R)+D_t(r-R, R)\}\,\mathrm{d}^3\boldsymbol{R} \qquad \text{(V.5)}$$

$$+\varrho\frac{\partial}{\partial r_\alpha}(\lambda - U) + \frac{1}{2}\left(\frac{\partial\Phi^*}{\partial r_\alpha}\eta + \frac{\partial\Phi}{\partial r_\alpha}\eta^* - \Phi^*\frac{\partial\eta}{\partial r_\alpha} - \Phi\frac{\partial\eta^*}{\partial r_\alpha}\right)$$

For $\varrho\varepsilon$ we have

$$\frac{\partial(\varrho\varepsilon)}{\partial t} = -\frac{1}{4m}\left\langle\left(\nabla^2\frac{\partial\psi^+}{\partial t}\right)\psi + (\nabla^2\psi^+)\frac{\partial\psi}{\partial t} + \frac{\partial\psi^+}{\partial t}\nabla^2\psi + \psi^+\left(\nabla^2\frac{\partial\psi}{\partial t}\right)\right\rangle$$

$$+\tfrac{1}{2}\int V(r-r')\frac{\partial}{\partial t}\langle\psi^+(t,r)\psi^+(t,r')\psi(t,r')\psi(t,r)\rangle\,\mathrm{d}^3\boldsymbol{r}'.$$

After using the commutation relations we can write the last term as

$$\frac{\partial}{\partial t}\langle\psi^+(t,r)\psi^+(t,r')\psi(t,r')\psi(t,r)\rangle = \frac{i}{2m}\langle\psi^+(t,r)\psi^+(t,r')$$

$$\times\left(\nabla^2\psi(t,r')\right)\psi(t,r) + \psi^+(t,r)\psi^+(t,r')\psi(t,r')\left((\nabla^2\psi(t,r)\right)$$

$$-\left(\nabla^2\psi^+(t,r')\right)\psi^+(t,r)\psi(t,r')\psi(t,r) - \psi^+(t,r)\left(\nabla^2\psi^+(t,r')\right)$$

$$\times\,\psi(t,r')\psi(t,r)\rangle + i\eta^*(t,r)\langle\psi^+(t,r')\psi(t,r')\psi(t,r)\rangle$$

$$+i\eta^*(t,r')\langle\psi^+(t,r)\psi(t,r')\psi(t,r)\rangle - i\eta(t,r)\langle\psi^+(t,r)\psi^+(t,r')$$

$$\times\,\psi(t,r')\rangle - i\eta(t,r')\langle\psi^+(t,r)\psi^+(t,r')\psi(t,r\rangle.$$

Thus, owing to the last identity and to (V.3), we have

$$\frac{\partial(\varrho\varepsilon)}{\partial t} = \frac{i}{8m^2}\langle(\nabla^4\psi^+)\psi - (\nabla^2\psi^+)(\nabla^2\psi) + (\nabla^2\psi^+)(\nabla^2\psi) - \psi^+(\nabla^4\psi)\rangle$$

$$+\frac{i}{4m}\Big\langle -\left(\nabla^2\int V(r-r')\psi^+(t,r)\psi^+(t,r')\psi(t,r')\,\mathrm{d}^3\boldsymbol{r}'\right)\psi(t,r)$$

$$+\int V(r-r')\left[(\nabla^2\psi^+(t,r))\psi^+(t,r')\psi(t,r')\psi(t,r)\right.$$

$$\left.-\psi^+(t,r)\psi^+(t,r')\psi(t,r')\left(\nabla^2\psi(t,r)\right)\right]\mathrm{d}^3\boldsymbol{r}'$$

$$+\psi^+(t,r)\left(\nabla^2\int V(r-r')\psi^+(t,r')\psi(t,r')\psi(t,r)\,\mathrm{d}^3\boldsymbol{r}'\right)\Big\rangle$$

$$+\frac{i}{4m}\int V(r-r')\langle\psi^+(t,r)\psi^+(t,r')\left(\nabla^2\psi(t,r')\right)\psi(t,r)$$

$$+\psi^+(t,r)\psi^+(t,r')\psi(t,r')(\nabla^2\psi(t,r))-(\nabla^2\psi^+(t,r))\psi^+(t,r')$$
$$\times\psi(t,r')\psi(t,r)-\psi^+(t,r)(\nabla^2\psi^+(t,r'))\psi(t,r')\psi(t,r)\rangle\,d^3\boldsymbol{r}'$$
$$+\frac{i}{2}\int V(r-r')\{\eta^*(t,r')\langle\psi^+(t,r)\psi(t,r')\psi(t,r)\rangle$$
$$+\eta^*(t,r)\langle\psi^+(t,r')\psi(t,r')\psi(t,r)\rangle-\eta(t,r)\langle\psi^+(t,r)$$
$$\times\psi^+(t,r')\psi(t,r')\rangle-\eta(t,r')\langle\psi^+(t,r)\psi^+(t,r')\psi(t,r)\rangle\}\,d^3\boldsymbol{r}'$$
$$-\frac{i}{4m}\langle(\nabla^2\eta^*)\psi+\eta^*\nabla^2\psi-(\nabla^2\psi^+)\eta-\psi^+(\nabla^2\eta)\rangle$$
$$-\frac{i}{2m}\sum_\alpha\left[\frac{\partial}{\partial r_\alpha}(U-\lambda)\left\langle\frac{\partial\psi^+}{\partial r_\alpha}\psi-\psi^+\frac{\partial\psi}{\partial r_\alpha}\right\rangle\right].\tag{V.6}$$

In order to simplify (V.6) we use the formula

$$\sum_\beta\frac{\partial}{\partial r_\beta}\left\{\left(\frac{\partial}{\partial r_\beta}\nabla^2\psi^+\right)\psi-(\nabla^2\psi^+)\frac{\partial\psi}{\partial r_\beta}-\psi^+\frac{\partial}{\partial r_\beta}\nabla^2\psi+\frac{\partial\psi^+}{\partial r_\beta}\nabla^2\psi\right\}$$
$$=(\nabla^4\psi^+)\psi-\psi^+(\nabla^4\psi)$$
$$=\sum_\beta\frac{\partial^2}{\partial r_\beta^2}\{(\nabla^2\psi^+)\psi-\psi^+(\nabla^2\psi)\}$$
$$-2\sum_\beta\frac{\partial}{\partial r_\beta}\left\{(\nabla^2\psi^+)\frac{\partial\psi}{\partial r_\beta}-\frac{\partial\psi^+}{\partial r_\beta}(\nabla^2\psi)\right\}$$

and transform the terms with potential V. Finally, we have (V.6) in the form

$$\frac{\partial(\varrho\varepsilon)}{\partial t}=\frac{i}{8m^2}\nabla^2\langle(\nabla^2\psi^+)\psi-\psi^+(\nabla^2\psi)\rangle+\frac{i}{4m}\sum_\beta\frac{\partial}{\partial r_\beta}$$
$$\times\left\{-\frac{1}{m}\left\langle(\nabla^2\psi^+)\frac{\partial\psi}{\partial r_\beta}-\frac{\partial\psi^+}{\partial r_\beta}(\nabla^2\psi)\right\rangle+\int V(r-r')\left\langle\psi^+(t,r)\psi^+(t,r')\right.\right.$$
$$\left.\left.\times\psi(t,r')\frac{\partial\psi(t,r)}{\partial r_\beta}-\frac{\partial\psi^+(t,r)}{\partial r_\beta}\psi^+(t,r')\psi(t,r')\psi(t,r)\right\rangle d^3\boldsymbol{r}'\right\}$$
$$+\frac{i}{4m}\sum_\beta\int\frac{\partial V(r-r')}{\partial r_\beta}\left\langle\psi^+(t,r)\psi^+(t,r')\psi(t,r')\frac{\partial\psi(t,r)}{\partial r_\beta}\right.$$

$$-\frac{\partial\psi^+(t,r)}{\partial r_\beta}\psi^+(t,r')\psi(t,r')\psi(t,r)+\psi^+(t,r)\left[\psi^+(t,r')\frac{\partial\psi(t,r')}{\partial r'_\beta}\right.$$

$$\left.-\frac{\partial\psi^+(t,r')}{\partial r'_\beta}\psi(t,r')\right]\psi(t,r)\Big\rangle\,\mathrm{d}^3\boldsymbol{r}'-\frac{1}{m}\sum_\beta j_\beta\frac{\partial}{\partial r_\beta}(U-\lambda)$$

$$+\frac{i}{4m}(\eta\nabla^2\Phi^*-\eta^*\nabla^2\Phi+\Phi^*\nabla^2\eta-\Phi\nabla^2\eta^*)+$$

$$+\frac{i}{2}\int V(r-r')\,\{\eta^*(t,r)\langle\psi^+(t,r')\psi(t,r')\psi(t,r)\rangle+\eta^*(t,r')$$

$$\times\langle\psi^+(t,r)\psi(t,r')\psi(t,r)\rangle-\eta(t,r')\langle\psi^+(t,r)\psi^+(t,r')\psi(t,r)\rangle$$

$$-\eta(t,r)\langle\psi^+(t,r)\psi^+(t,r')\psi(t,r')\rangle\}\,\mathrm{d}^3\boldsymbol{r}'.$$

We define the new function

$$G_t^{(\alpha)}(r,r'-r)=\frac{i}{4m}\Big\langle\psi^+(t,r')\Big((\psi^+(t,r)\frac{\partial\psi(t,r)}{\partial r_\alpha}-\frac{\partial\psi^+(t,r)}{\partial r_\alpha}\psi(t,r)\Big)\psi(t,r')\Big\rangle \tag{V.7}$$

and put $\boldsymbol{R}=\boldsymbol{r}'-\boldsymbol{r}$. Finally,

$$\frac{\partial(\varrho\varepsilon)}{\partial t}=\frac{i}{8m^2}\nabla^2\langle(\nabla^2\psi^+)\psi-\psi^+(\nabla^2\psi)\rangle+\sum_\beta\frac{\partial}{\partial r_\beta}\left\{\frac{i}{4m^2}\Big\langle\frac{\partial\psi^+}{\partial r_\beta}\nabla^2\psi\right.$$

$$\left.-(\nabla^2\psi^+)\frac{\partial\psi}{\partial r_\beta}\Big\rangle+\int V(R)G_t^{(\beta)}(r,R)\,\mathrm{d}^3\boldsymbol{R}\right\}$$

$$+\sum_\beta\int\frac{\partial V(R)}{\partial R_\beta}\{G_t^{(\beta)}(r,-R)+G_t^{(\beta)}(r-R,R)\}\,\mathrm{d}^3\boldsymbol{R}$$

$$-\frac{1}{m}\sum_\beta j_\beta\frac{\partial}{\partial r_\beta}(U-\lambda)+\frac{i}{4m}(\eta\nabla^2\Phi^*-\eta^*\nabla^2\Phi+\Phi^*\nabla^2\eta-\Phi\nabla^2\eta^*)$$

$$+\frac{i}{2}\int V(r-r')\,[\eta^*(t,r)\langle\psi^+(t,r')\psi(t,r')\psi(t,r)\rangle$$

$$+\eta^*(t,r')\langle\psi^+(t,r)\,\psi(t,r')\psi(t,r)\rangle$$

$$-\eta(t,r')\langle\psi^+(t,r)\psi^+(t,r')\psi(t,r)\rangle$$

$$-\eta(t,r)\langle\psi^+(t,r)\psi^+(t,r')\psi(t,r')\rangle]\,\mathrm{d}^3\boldsymbol{r}'. \tag{V.8}$$

The equation for $\langle\psi\rangle$ gives the generalized Schrödinger equation for the "condensate wave function" $\Phi(t, r)$:

$$i\frac{\partial\Phi(t,r)}{\partial t} = -\lambda\Phi(t,r) - \frac{\nabla^2}{2m}\Phi(t,r) \tag{V.9}$$
$$+\int V(r-r')\langle\psi^+(t,r')\psi(t,r')\psi(t,r)\rangle \mathrm{d}^3r' + U(t,r)\Phi(t,r) + \eta(t,r).$$

Such an equation, for the equilibrium case, was considered by Gross (1961), Pitaevskii (1961), and Weller (1964).

§ 2 Time derivatives of some "local quantities" for Fermi systems

Consider now a system of identical Fermi particles, with pairing forces (Galasiewicz, 1966a). The Hamiltonian of the system in the second quantization representation has the form

$$\hat{H}[\tilde{\eta}, U, A] = \hat{H} + \hat{H}_t[\tilde{\eta}, U, A],$$
$$\hat{H} = \frac{1}{2m}\int \nabla\psi^+(t,x)\nabla\psi(t,x)\mathrm{d}x$$
$$-\lambda\int \psi^+(t,x)\psi(t,x)\mathrm{d}x + \frac{1}{2}\iint V(r-r')$$
$$\times\psi^+(t,x)\psi^+(t,x')\psi(t,x')\psi(t,x)\mathrm{d}x\mathrm{d}x', \tag{V.10}$$
$$\hat{H}_t[\tilde{\eta}, U, A] = \iint [\tilde{\eta}(t|x,x')\psi^+(t,x)\psi^+(t,x')$$
$$+\tilde{\eta}^*(t|x,x')\psi(t,x')\psi(t,x)]\mathrm{d}x\mathrm{d}x'$$
$$+\int U(t,r)\hat{\varrho}(t,x)\mathrm{d}x - \frac{1}{m}\int (\hat{j}(t,x)\cdot A(t,r))\mathrm{d}x$$

where λ is a constant, $x = (\boldsymbol{r}, s)$, $\boldsymbol{r}$ denotes the space variables, s the spin index, and $\psi(t, x)$, $\psi^+(t, x)$ are the Fermi operators in the Heisenberg representation with the usual commutation relations. The Fermi operators are expressed in terms of the creation operators a^+_{ks} and annihilation operators a_{ks}:

$$\psi(t,x) = \frac{1}{\sqrt{V}}\sum_{(k)} a_{ks}(t)e^{i(k\cdot r)}, \quad \psi^+(t,x) = \frac{1}{\sqrt{V}}\sum_{(k)} a^+_{ks}(t)e^{-i(k\cdot r)}, \tag{V.11}$$

Here, $\int dx$ denotes integration over the space variables and summation over s, $\hbar = 1$, $e = 1$, $c = 1$.

Now for the case of Fermi particles an additional term containing the "sources of pairs" $\tilde{\eta}$, $\tilde{\eta}^*$ is introduced into the Hamiltonian. These sources are given c-number functions depending on x, t.

In order to obtain the hydrodynamical equations we calculate the time derivatives of the following "local quantities":

$$\varrho(t, r) = \sum_{s} \langle \psi^+(t, x)\psi(t, x)\rangle,$$

$$\Phi(t|x_1, x_2) = \langle \psi(t, x_1)\psi(t, x_2)\rangle,$$

$$j_\alpha(t, r) = \frac{1}{2} i \sum_{s} \left\langle \frac{\partial \psi^+(t, x)}{\partial r_\alpha} \psi(t, x) - \psi^+(t, x) \frac{\partial \psi(t, x)}{\partial r_\alpha} \right\rangle$$

$$-\varrho(t, r)A_\alpha(t, r) = j^0_\alpha(t, r) - \varrho(t, r)A_\alpha(t, r)$$

$$(\alpha = 1, 2, 3), \quad \text{(V.12)}$$

$$\varrho(t, r)\varepsilon(t, r) = -\frac{1}{4m} \sum_{s} \langle (\nabla^2 \psi^+(t, x))\psi(t, x)$$

$$+\psi^+(t, x)\nabla^2\psi(t, x)\rangle$$

$$+\frac{1}{2}\int V(r-r') \sum_{s, s'} \langle \psi^+(t, x)\psi^+(t, x')\psi(t, x')$$

$$\times \psi(t, x)\rangle \, d^3\boldsymbol{r}$$

where ϱ is the mean density of particles, $\boldsymbol{j}$ is the mean current density and ε the mean energy per particle. As shown by Bogoliubov (1961), for systems with a condensate, $\langle \psi(t, x_1)\psi(t, x_2)\rangle$ is different from zero. In order to obtain the time derivatives of interest, we shall use the equations of motion which follow from the Hamiltonian (V.10). They have the form

$$i \frac{\partial \psi(t, x)}{\partial t} = -\lambda\psi(t, x) - \frac{\nabla_r^2}{2m}\psi(t, x) + \int V(r-r')\psi^+(t, x')$$

$$\times \psi(t, x')\,dx'\psi(t, x) + 2\int \tilde{\eta}(t|x, x')\,dx'$$

$$+U(t, r)\psi(t, x) + \frac{i}{m}(A(t, r)\cdot\nabla\psi(t, x)) \quad \text{(V.13)}$$

$$+\frac{i}{2m}\psi(t,x)(\nabla\cdot A(t,r))+\frac{1}{2m}\psi(t,x)A^2(t,r).$$

After calculations similar to those for Bose systems ($A = 0$) we have

$$m\frac{\partial\varrho(t,r)}{\partial t}+\nabla j(t,r)=S_\varrho[\tilde{\eta},\tilde{\eta}^*], \tag{V.14}$$

$$\frac{\partial j_\alpha(t,r)}{\partial t}=\frac{1}{4m}\frac{\partial}{\partial r_\alpha}\nabla^2\varrho(t,r)-\frac{1}{2m}\sum_\beta\frac{\partial}{\partial r_\beta}\sum_s\left\langle\frac{\partial\psi^+}{\partial r_\alpha}\frac{\partial\psi}{\partial r_\beta}+\frac{\partial\psi^+}{\partial r_\beta}\frac{\partial\psi}{\partial r_\alpha}\right\rangle+\varrho(t,r)\frac{\partial}{\partial r_\alpha}(\lambda-U(t,r))-\frac{1}{2}\int\frac{\partial V(R)}{\partial R_\alpha}$$
$$\times\{D_t(r,-R)+D_t(r-R,R)\}\,d^3\boldsymbol{R}+S_j^{(\alpha)}[\tilde{\eta},\tilde{\eta}^*], \tag{V.15}$$

$$\frac{\partial\varrho(t,r)\varepsilon(t,r)}{\partial t}=\frac{1}{8m^2}\nabla^2\sum_s\langle(\nabla^2\psi^+)\psi-\psi^+(\nabla^2\psi)\rangle$$
$$+\sum_\beta\frac{\partial}{\partial r_\beta}\left\{\frac{i}{4m^2}\sum_s\left\langle\frac{\partial\psi^+}{\partial r_\beta}\nabla^2\psi-(\nabla^2\psi^+)\frac{\partial\psi}{\partial r_\beta}\right\rangle\right.$$
$$\left.+\int V(R)G_t^{(\beta)}(r,R)\,d^3\boldsymbol{R}\right\}+\sum_\beta\int\frac{\partial V}{\partial R_\beta}\{G_t^{(\beta)}(r,-R)$$
$$+G_t^{(\beta)}(r-R,R)\}\,d^3\boldsymbol{R}-\frac{1}{m}\big(j(t,r)\cdot\nabla(U(t,r)-\lambda)\big)$$
$$+S_{\varrho\varepsilon}[\tilde{\eta},\tilde{\eta}^*]. \tag{V.16}$$

In (V.15) and (V.16) the following notations are introduced:

$$S_\varrho[\tilde{\eta},\tilde{\eta}^*]=i2m\sum_{s,s'}\int[\tilde{\eta}^*(t|x,x')\Phi(t|x,x')-\tilde{\eta}(t|x,x')\Phi^*(t|x,x')]\,d^3\boldsymbol{r}' \tag{V.14a}$$

$$\sum_{s,s'}\langle\psi^+(t,x)\psi^+(t,x')\psi(t,x')\psi(t,x)\rangle\equiv D_t(r,r'-r)$$
$$=D_t(r',r-r')=\tfrac{1}{2}\{D_t(r',r-r')+D_t(r,r'-r)\}, \tag{V.15a}$$

$$\sum_{s,s'}\frac{i}{4m}\left\langle\psi^+(t,x')\left(\psi^+(t,x)\frac{\partial\psi(t,x)}{\partial r_\alpha}-\frac{\partial\psi^+(t,x)}{\partial r_\alpha}\psi(t,x)\right)\psi(t,x')\right\rangle$$
$$\equiv G_j^{(\alpha)}(r,r'-r),\qquad \boldsymbol{R}=\boldsymbol{r}-\boldsymbol{r}', \tag{V.16a}$$

$$S_j^{(\alpha)}[\eta, \eta^*] \equiv \sum_{s, s'} \int \left[\tilde{\eta}^*(t|x, x') \frac{\partial}{\partial r_\alpha} \Phi(t|x', x) \right.$$

$$\left. - \Phi(t|x', x) \frac{\partial \tilde{\eta}^*(t|x, x')}{\partial r_\alpha} \right] \mathrm{d}^3\boldsymbol{r}' + \sum_{s, s'} \int \left[\tilde{\eta}(t|x, x') \frac{\partial}{\partial r_\alpha} \Phi^*(t|x', x) \right.$$

$$\left. - \Phi^*(t|x', x) \frac{\partial \tilde{\eta}(t|x, x')}{\partial r_\alpha} \right] \mathrm{d}^3\boldsymbol{r}', \qquad \text{(V.17)}$$

$$S_{\varrho\varepsilon}[\tilde{\eta}, \tilde{\eta}^*] \equiv i \sum_{s, s'} \int V(r-r') [\tilde{\eta}^*(t|x, x') \Phi(t|x', x)$$

$$- \tilde{\eta}(t|x, x') \Phi^*(t|x', x)] \mathrm{d}^3\boldsymbol{r}' + \frac{i}{2m} \sum_{s, s'} \int [\tilde{\eta}(t|x, x') \nabla_{\boldsymbol{r}}^2 \Phi^*(t|x', x)$$

$$+ \Phi^*(t|x', x) \nabla_{\boldsymbol{r}}^2 \tilde{\eta}(t|x, x')] \mathrm{d}^3\boldsymbol{r}' - \frac{i}{2m} \sum_{s, s'} \int [\tilde{\eta}^*(t|x, x') \nabla_{\boldsymbol{r}}^2 \Phi(t|x', x)$$

$$+ \Phi(t|x', x) \nabla_{\boldsymbol{r}}^2 \tilde{\eta}^*(t|x, x')] \mathrm{d}^3\boldsymbol{r}' + i \sum_{s, s'} \iint V(r-r') \tilde{\eta}^*(t|x, x'')$$

$$\times \sum_{s'} \langle \psi^+(t, x') \psi(t, x') \psi(t, x'') \psi(t, x) \rangle \mathrm{d}^3\boldsymbol{r}' \mathrm{d}^3\boldsymbol{r}''$$

$$+ i \sum_{s', s''} \iint V(r-r') \tilde{\eta}^*(t|x', x'') \sum_{s} \langle \psi^+(t, x) \psi(t, x) \psi(t, x'')$$

$$\times \psi(t, x') \rangle \mathrm{d}^3\boldsymbol{r}' \mathrm{d}^3\boldsymbol{r}'' - i \sum_{s, s''} \iint V(r-r') \tilde{\eta}(t|x, x'')$$

$$\times \sum_{s'} \langle \psi^+(t, x') \psi(t, x') \psi(t, x'') \psi(t, x) \rangle^* \mathrm{d}^3\boldsymbol{r}' \mathrm{d}^3\boldsymbol{r}''$$

$$- i \sum_{s', s''} \iint V(r-r') \tilde{\eta}(t|x', x'') \sum_{s} \langle \psi^+(t, x) \psi(t, x)$$

$$\times \psi(t, x'') \psi(t, x') \rangle^* \mathrm{d}^3\boldsymbol{r}' \mathrm{d}^3\boldsymbol{r}''. \qquad \text{(V.18)}$$

Since

$$\Phi(t|x_1, x_2) = -\Phi(t|x_2, x_1), \qquad \tilde{\eta}(t|x, x') = -\tilde{\eta}(t, x', x),$$

we can put

$$\Phi(t|x_1, x_2) \equiv \varepsilon(s_1) \Delta(s_1+s_2) \tilde{\Phi}(t|r_1, r_2),$$
$$\tilde{\eta}(t|x, x') \equiv \varepsilon(s) \Delta(s+s') \tilde{\eta}(t|r, r'), \qquad \text{(V.19)}$$

$$\Delta(s) = \begin{cases} 1, & s = 0 \\ 0, & s \neq 0 \end{cases}$$

where $\Delta(s)$ denotes the discrete δ function and $\varepsilon(s) = \text{sgn}\, s$. Moreover, if write

$$\sum_{s'} \langle \psi^+(t, x')\psi(t, x')\psi(t, x_1)\psi(t, x_2)\rangle$$
$$\equiv Y_t(r', r'-r_1, r'-r_2)\varepsilon(s_1)\Delta(s_1+s_2), \quad \text{(V.20)}$$

the equation for $\tilde{\Phi}$ has the form

$$i\frac{\partial\tilde{\Phi}(t|r_1, r_2)}{\partial t} = \left[-\frac{1}{2m}(\nabla_1^2+\nabla_2^2)-2\lambda+V(r_2-r_1)\right]\tilde{\Phi}(t|r_1, r_2)$$
$$+\int d^3\boldsymbol{r}'[V(r_1-r')+V(r_2-r')]Y_t(r', r'-r_1, r'-r_2)+2\tilde{\eta}(t|r_1, r_2)$$
$$+\int [\tilde{\eta}(t|r_1, r')\tilde{F}(t|r', r_2)+\tilde{\eta}(t|r_2, r')\tilde{F}(t|r', r_1)]d^3\boldsymbol{r}' \quad \text{(V.21)}$$

where $\tilde{F}$ is defined by

$$\langle\psi^+(t, x_1)\psi(t, x_2)\rangle \equiv F(t|x_1, x_2) = F^*(t|x_2, x_1)$$
$$\equiv \Delta(s_1-s_2)\tilde{F}(t|r_1, r_2),$$
$$F(t|x, x) = \varrho(t, x).$$

Equations (V.14–16) and (V.21) are the basis from which the hydrodynamical equations for the Fermi superfluid are obtained. Further we shall assume that $|\tilde{\Phi}(t|r_1, r_2)|$ and $|\tilde{\eta}(t|r_1, r_2)|$ vanish if $|\boldsymbol{R}| \equiv |\boldsymbol{r}_1-\boldsymbol{r}_2| > R_0$, where R_0 is the radius of a pair in the condensate. We shall find the order of magnitude of R_0 from the theory of superconductivity.

§ 3 The relations between the retarded Green functions and the variations of average values

The relations between the retarded Green functions and the variations of average values (see, for instance, Zubarev, 1961; Galasiewicz, 1963b; Tyablikov, 1965) are obtained by considering the response of a quantum-mechanical system to a perturbation which is introduced adiabatically. The perturbation changes the average values of the dynamical variables.

The two time retarded thermodynamic Green functions

$$\langle\langle \hat{A}(t); \hat{B}(\tau)\rangle\rangle^r$$

are defined by

$$\langle\langle \hat{A}(t); \hat{B}(\tau)\rangle\rangle^r \equiv G_r(t-\tau) \equiv -i\Theta(t-\tau)\langle[\hat{A}(t), \hat{B}(\tau)]\rangle_{eq}, \quad \text{(V.22)}$$

$$\Theta(t) = \begin{cases} 1, & t > 0 \\ 0, & t < 0 \end{cases}$$

where $\langle\ldots\rangle_{eq}$ denotes grand canonical averaging

$$\langle\ \rangle_{eq} = \frac{\mathrm{Tr}(e^{-\frac{\hat{H}}{\Theta}}\ldots)}{\mathrm{Tr}(e^{-\frac{\hat{H}}{\Theta}})}, \quad \Theta = kT, \quad k = 1. \quad \text{(V.23)}$$

$\hat{A}(t)$, $\hat{B}(t)$ are operators in the Heisenberg representation with the Hamiltonian $\hat{H}$.

The Fourier components of (V.22) have the form

$$\langle\langle \hat{A}; \hat{B}\rangle\rangle^r_E = \frac{1}{2\pi}\int_{-\infty}^{+\infty} \langle\langle \hat{A}(t); \hat{B}(\tau)\rangle\rangle^r e^{iE(t-\tau)}\mathrm{d}(t-\tau). \quad \text{(V.24)}$$

Let us consider an operator $\hat{A}$ not depending explicitly on time, and a statistical operator $D(t)$ in the Schrödinger representation. Then the average value of $\hat{A}$ is equal to

$$\langle \hat{A}_t\rangle \equiv \mathrm{Tr}\{D(t)A\}. \quad \text{(V.25)}$$

The operator $D(t)$ is found from the equation (Schrödinger representation)

$$i\frac{\mathrm{d}D(t)}{\mathrm{d}t} = [\hat{H}+\delta\hat{H}_t, D(t)] \quad \text{(V.26)}$$

where $\hat{H}$ is the Hamiltonian (explicitly independent of time) of the system considered, in thermodynamical equilibrium; $\delta\hat{H}_t$ is a small perturbation, adiabatically introduced, which satisfies

$$\delta\hat{H}_{t=-\infty} = 0. \quad \text{(V.27)}$$

The initial condition for the statistical operator (at $t = -\infty$) has the form

$$D(-\infty) = D_0 = \frac{e^{-\frac{\Theta}{\hat{H}}}}{\mathrm{Tr}(e^{-\frac{\hat{H}}{\Theta}})}. \quad \text{(V.28)}$$

This means that at $t = -\infty$ the system is in thermodynamic equilibrium. Assuming that the system is perturbed slightly from equilibrium, we look for $D(t)$ in the form

$$D(t) = D_0 + \delta D(t). \quad \text{(V.29)}$$

From (V.28) we see that

$$\delta D(-\infty) = 0.$$

We put (V.29) in (V.26) and neglect the product $\delta D \delta \hat{H}_t$. To solve (V.26), we introduce an auxiliary operator

$$\delta D_1 = e^{i\hat{H}t} \delta D e^{-i\hat{H}t} \tag{V.30}$$

which obeys the equation

$$i \frac{\mathrm{d}}{\mathrm{d}t} \delta D_1 = e^{i\hat{H}t} [\delta \hat{H}_t, D_0] e^{-i\hat{H}t} \tag{V.31}$$

with the initial condition

$$\delta D_1(-\infty) = 0. \tag{V.32}$$

The integration of (V.31) and application of the condition (V.32) leads to

$$\delta D(t) = \frac{1}{i} \int_{-\infty}^{t} e^{i\hat{H}(\tau - t)} [\delta \hat{H}_\tau, D_0] e^{-i\hat{H}(\tau - t)} \mathrm{d}\tau. \tag{V.33}$$

We put (V.33) in (V.29) and then (V.29) in (V.25), and use the fact that under the trace sign the operators can be cyclically displaced, to obtain

$$\langle \hat{A}_t \rangle = \mathrm{Tr}\{D_0 \hat{A}\} - i \int_{-\infty}^{t} \frac{\mathrm{Tr}\{e^{-\frac{\hat{H}}{\Theta}} [\hat{A}(t), \delta \hat{H}_\tau(\tau)]\}}{\mathrm{Tr}\{e^{-\frac{\hat{H}}{\Theta}}\}} \mathrm{d}\tau$$

$$= \langle \hat{A} \rangle_{\mathrm{eq}} - i \int_{-\infty}^{+\infty} \Theta(t-\tau) \langle [\hat{A}(t), \delta \hat{H}_\tau(\tau)] \rangle_{\mathrm{eq}} \mathrm{d}\tau. \tag{V.34}$$

Here $\hat{A}(t)$, $\delta \hat{H}(\tau)$ are operators in the Heisenberg representation (as regards the Hamiltonian $\hat{H}$).

Let the variation of the Hamiltonian depend periodically on time,

$$\delta \hat{H}_\tau = \sum_{\Omega} e^{-i\Omega\tau + \varepsilon\tau} \hat{V}_\Omega, \quad (\varepsilon > 0, \varepsilon \to 0) \tag{V.35}$$

where $\hat{V}_\Omega$ is an operator which does not depend explicitly on time. For the variation $\delta\langle \hat{A} \rangle \equiv \langle \hat{A} \rangle - \langle \hat{A} \rangle_{\mathrm{eq}}$ we have

$$\delta \langle \hat{A}_t \rangle = -i \sum_{\Omega} \int_{-\infty}^{+\infty} \Theta(t-\tau) \langle [\hat{A}(t), \hat{V}_\Omega(\tau)] \rangle_{\mathrm{eq}} e^{-i(\Omega + i\varepsilon)\tau} \mathrm{d}\tau$$

$$= \sum_{\Omega} \int_{-\infty}^{+\infty} \langle\langle \hat{A}(t); \hat{V}_\Omega(\tau) \rangle\rangle^r e^{-i(\Omega + i\varepsilon)\tau} \mathrm{d}\tau. \tag{V.36}$$

Recalling (V.24) we obtain

$$\delta\langle\hat{A}_t\rangle = 2\pi \sum_{\Omega} e^{-i\Omega t+\varepsilon t}\langle\langle\hat{A}; \hat{V}_{\Omega}\rangle\rangle^r_{E=\Omega+i\varepsilon}. \tag{V.37}$$

In the special case of one frequency, $\Omega = \pm\omega$, we have

$$\delta\langle\hat{A}_t\rangle = 2\pi e^{-i\omega t+\varepsilon t}\langle\langle\hat{A}, \hat{V}_{\omega}\rangle\rangle^r_{E=\omega+i\varepsilon}+2\pi e^{i\omega t+\varepsilon t}\langle\langle\hat{A}; \hat{V}_{\omega}\rangle\rangle^r_{E=-\omega+i\varepsilon}. \tag{V.38}$$

Thus we have obtained the formulae (V.37), (V.38) which connect the variation of the average value with the Green functions. If we were able to obtain the variation of the average value (and, consequently, the left-hand side of (V.38)) then the Fourier transforms of the retarded Green functions could also be obtained. The linearized hydrodynamic equations enable us to calculate the variations $\delta\varrho$, $\delta\boldsymbol{j}$ and $\nabla\delta\langle\psi\rangle$. This gives us the possibility of calculating, in the hydrodynamic approximation, Green functions of the form: $\langle\langle\psi; \psi\rangle\rangle, \langle\langle\psi; \tilde{\varrho}\rangle\rangle, \langle\langle\tilde{\varrho}; \tilde{\varrho}\rangle\rangle, \langle\langle\psi; \hat{\boldsymbol{j}}\rangle\rangle, \langle\langle\tilde{\varrho}; \hat{\boldsymbol{j}}\rangle\rangle, \langle\langle\hat{\boldsymbol{j}}; \hat{\boldsymbol{j}}\rangle\rangle$. If a product of two, three, ... operators ψ, ψ^+ appears in the expression of Green functions, we call them two-, three-, ... leg Green functions, respectively. In Chapters VI–VIII we will derive the hydrodynamical equations without the assumption that H_t^1 is small. For this purpose let us consider the mean values when the time-dependent part of the Hamiltonian cannot be treated as a small perturbation. The Hamiltonian has now the form

$$\hat{H}_\eta = \hat{H}+\hat{H}^1[\eta(t)] \tag{V.39}$$

where η is a c-number.

In the Schrödinger representation the statistical operator $D_S(t)$ is found from the equation

$$i\frac{\mathrm{d}D_S(t)}{\mathrm{d}t} = [\hat{H}_\eta, D_S(t)]_-. \tag{V.40}$$

We introduce an operator $D_I(t)$ ("interaction" representation as far as D is concerned)

$$D_I(t) = e^{i\hat{H}t}D_S(t)e^{-i\hat{H}t}. \tag{V.41}$$

We have the equation

$$i\frac{\mathrm{d}D_I(t)}{\mathrm{d}t} = [\hat{H}_I^1[\eta(t)], D_I(t)]_- \tag{V.42}$$

with the following initial condition at $t = -\infty$:

$$D_I(-\infty) = D_S(-\infty) = D_0 = \frac{e^{-\frac{\hat{H}}{\Theta}}}{\mathrm{Tr}\,(e^{-\frac{\hat{H}}{\Theta}})} \tag{V.43}$$

The integration of (V.42) with the aid of (V.43) gives

$$D_I(t) = D_0 + \frac{1}{i}\int_{-\infty}^{t} [\hat{H}_I^1[\eta(\tau)], D_I(\tau)]\,\mathrm{d}\tau$$

$$= \sum_{n/0}^{\infty} \frac{(-i)^n}{n!} T \underbrace{\int_{-\infty}^{t}\int_{-\infty}^{t} \dots \int_{-\infty}^{t}}_{n} [\hat{H}_I^1(\tau_1)[\dots,[\hat{H}_I^1(\tau_n), D_0]\dots]$$

$$\times\, \mathrm{d}\tau_1 \dots \mathrm{d}\tau_n = SD_0S^{-1} \tag{V.44}$$

where

$$S = Te^{-i\int_{-\infty}^{t} \hat{H}_I^1(\tau)\,\mathrm{d}\tau}. \tag{V.45}$$

The letter T denotes here a time-ordered product. From (V.44) and (V.41) we have

$$D_S(t) = e^{-i\hat{H}t} SD_0S^{-1}e^{i\hat{H}t}. \tag{V.46}$$

The mean value of the operator $\hat{A}_S$, which is not explicitly time dependent in the Schrödinger representation, has the form

$$\langle \hat{A}_t \rangle \equiv \mathrm{Tr}\{D_S(t)\hat{A}_S\} = \mathrm{Tr}\{e^{-i\hat{H}t}SD_0S^{-1}e^{i\hat{H}t}\hat{A}\} = \mathrm{Tr}\{D_0\hat{A}_{H_\eta}(t)\}. \tag{V.47}$$

In (V.47)

$$\hat{A}_{H_\eta}(t) = S^{-1}\hat{A}_I(t)S \tag{V.48}$$

is the operator $\hat{A}$ in the Heisenberg representation, described by the equation

$$\frac{\mathrm{d}\hat{A}_{H_\eta}(t)}{\mathrm{d}t} = [\hat{A}_{H_\eta}(t), \hat{H}_\eta(t)]. \tag{V.49}$$

The non-equilibrium mean value (V.47) can be treated either as the mean value of the "equilibrium" operator $\hat{A}_S$ with the non-equilibrium statistical operator $D_S(t)$, or as the mean value of the "non-equilibrium" operator $\hat{A}_{H_\eta}(t)$ with the equilibrium statistical operator D_0. If $\hat{H}^1[\eta(t)]$ is a small perturbation ($\hat{H}^1[\eta(t)] = \delta\hat{H}_t$), the second term in the expansion (V.44) gives the same $\delta D(t)$ as eqn. (V.33).

Notice, that for $\hat{H}^1 \sim \hat{V}\eta(t)$, where $\hat{V}$ is an operator which does not depend explicitly on time, $\hat{H}_I^1$ has the form $\hat{V}_I(t)\eta(t)$. From the point of view of the non-equilibrium operator $\hat{H}_\eta$, $\hat{V}_I$ is an operator in the interaction representation, while from the point of view of the equilibrium Hamiltonian $\hat{H}$, $\hat{V}_I$ is an operator in the Heisenberg representation. In (V.36) we express the change of the non-equilibrium mean value by Green functions in which all quantities are taken at thermodynamic equilibrium. Therefore, we view the operators under the Green function sign as being in the Heisenberg representation.

§ 4 Gauge transformations and Green functions

The aim of the present section is to derive some connexions between the Green functions of superfluid and normal Bose systems. These connections are obtained by applying to the Bose operators ψ, ψ^+ time independent and time dependent gauge transformations (Galasiewicz, 1967a). Part of the results is compared with those which follow from the "hydrodynamic approach" to the Green functions (Chapters VI, VII).

Consider a superfluid Bose system with Hamiltonian

$$\hat{H} = \frac{1}{2m}\int (\nabla\psi^+(t, r)\cdot\nabla\psi(t, r))\mathrm{d}^3\boldsymbol{r} - \lambda\int \psi^+(t, r)\psi(t, r)\mathrm{d}^3\boldsymbol{r}$$
$$+\frac{1}{2}\int V(r-r')\psi^+(t, r)\psi^+(t, r')\psi(t, r')\psi(t, r)\mathrm{d}\boldsymbol{r}^3\mathrm{d}^3\boldsymbol{r}'. \tag{V.50}$$

If the equilibrium of the system is perturbed by adiabatically introducing, for instance, the "sources of particles", the Hamiltonian has the form

$$\hat{H}_\eta = \hat{H} + \delta\hat{H}_t \tag{V.51}$$

where

$$\delta\hat{H}_t = \int [\delta\eta(t, r)\psi^+(t, r) + \delta\eta^*(t, r)\psi(t, r)]\mathrm{d}^3\boldsymbol{r}. \tag{V.52}$$

In order to introduce the sources adiabatically, we put

$$\delta\eta = e^{-i\omega t+\varepsilon t+i(k\cdot r)}\delta\eta_k + e^{i\omega t+\varepsilon t-\ (k\cdot r)}\delta\eta_{-k}, \tag{V.53}$$
$$\varepsilon > 0,\ \varepsilon \to 0.$$

By means of the Hamiltonian (V.51) one can obtain the hydrodynamic equations, with sources, for the Bose superfluid, describing the deviation of the system from thermodynamic equilibrium. This deviation is caused by the introduction of the infinitesimal sources $\delta\eta$ and $\delta\eta^*$.

The continuity equation for the Fourier components has the form (VII.83)

$$-mE\,\delta\varrho(k)+(\boldsymbol{k}\cdot\delta\boldsymbol{j}(k)) = m\Delta\eta, \tag{V.54}$$

where

$$\Delta\eta = \sqrt{\varrho_0}(\delta\eta^*_{-k}-\delta\eta_k), \tag{V.55}$$

ϱ_0 is the density of the condensate in equilibrium and ϱ and $\boldsymbol{j}$ are defined by (V.1). After introducing the term (V.52) adiabatically, the change in the mean values is given by (V.38) where

$$\delta\langle\hat{A}\rangle = 2\pi e^{-i\omega t+\varepsilon t}\langle\langle A; V_{+\omega}\rangle\rangle^r_{E=\omega+i\varepsilon}+2\pi e^{i\omega t+\varepsilon t}\langle\langle A; V_{-\omega}\rangle\rangle^r_{E=-\omega+i\varepsilon}. \tag{V.56}$$

For the Fourier components of the variations $\delta\varrho$ and $\delta\boldsymbol{j}$ we obtain

$$\delta\varrho(k) = \frac{2\pi}{\sqrt{V\varrho_0}}\sum_q \langle\langle a^+_{q-\frac{k}{2}}a_{q+\frac{k}{2}}; a_{-k}\rangle\rangle^r_E\Delta\eta = \frac{2\pi}{\sqrt{\varrho_0}}\langle\langle\hat{\varrho}_k; a_{-k}\rangle\rangle^r_E\Delta\eta,$$

$$\delta\boldsymbol{j}(k) = \frac{2\pi}{\sqrt{V\varrho_0}}\sum_q \boldsymbol{q}\langle\langle a^+_{q-\frac{k}{2}}a_{q+\frac{k}{2}}; a_{-k}\rangle\rangle^r_E\Delta\eta = \frac{2\pi}{\sqrt{\varrho_0}}\langle\langle\boldsymbol{j}_k; a_{-k}\rangle\rangle^r_E\Delta\eta,$$

$$\hat{\varrho}_k = \frac{1}{\sqrt{V}}\sum_q a^+_{q-\frac{k}{2}}a_{q+\frac{k}{2}},$$

$$\boldsymbol{j} = \frac{1}{\sqrt{V}}\sum_q \boldsymbol{q}a^+_{q-\frac{k}{2}}a_{q+\frac{k}{2}}. \tag{V.57}$$

We have made use of the formula which follows from the hydrodynamic equations

$$\langle\langle a^+_{q-\frac{k}{2}}a_{q+\frac{k}{2}}; a^+_k\rangle\rangle^r_E = -\langle\langle a^+_{q-\frac{k}{2}}a_{q+\frac{k}{2}}; a_{-k}\rangle\rangle^r_E \tag{V.58}$$

(all variations are proportional to $\Delta\eta$).

Moreover, in our considerations we shall use the formula giving (in the complex plane) the connexion between the retarded and advanced Green functions (Tyablikov, 1965)

$$\langle\langle\hat{A}(k); \hat{B}(-k)\rangle\rangle_{E=\omega+i\varepsilon} = \langle\langle\hat{B}(-k); \hat{A}(k)\rangle\rangle_{-E=\omega+i\varepsilon}. \tag{V.59}$$

From (V.22) and the definition of advanced Green functions

$$\langle\langle \hat{A}(t); \hat{B}(\tau)\rangle\rangle^a = i\Theta(\tau-t)\langle \hat{A}(t)\hat{B}(\tau)-\hat{B}(\tau)\hat{A}(t)\rangle$$

we have

$$\langle\langle \hat{A}(t); \hat{B}(\tau)\rangle\rangle^r = \langle\langle \hat{B}(\tau); \hat{A}(t)\rangle\rangle^a.$$

We are interested in the expressions, containing integration over space variables, which according to the last relation between the Green functions can be written, after Fourier transformation, in the form

$$\begin{aligned}\int_{-\infty}^{+\infty} d\tau \int d^3\boldsymbol{r}' \langle\langle \hat{A}(t, \boldsymbol{r}); \hat{B}(\tau, \boldsymbol{r}')\rangle\rangle^r \delta\eta(\tau, \boldsymbol{r}')\\ = 2\pi e^{-i(\omega+i\varepsilon)t+i(\boldsymbol{k}\cdot\boldsymbol{r})}\langle\langle \hat{A}(k); \hat{B}(-k)\rangle\rangle^r_{E=\omega+i\varepsilon}\delta\eta_k\\ +2\pi e^{-i(-\omega+i\varepsilon)t-i(\boldsymbol{k}\cdot\boldsymbol{r})}\langle\langle \hat{A}(-k); \hat{B}(k)\rangle\rangle^r_{E=-\omega+i\varepsilon}\delta\eta_{-k}\\ = 2\pi e^{-i(\omega+i\varepsilon)t+i(\boldsymbol{k}\cdot\boldsymbol{r})}\langle\langle \hat{B}(-k); \hat{A}(k)\rangle\rangle^a_{-E=\omega+i\varepsilon}\delta\eta_k\\ +2\pi e^{-i(-\omega+i\varepsilon)t-i(\boldsymbol{k}\cdot\boldsymbol{r})}\langle\langle \hat{B}(k); \hat{A}(-k)\rangle\rangle^a_{-E=-\omega+i\varepsilon}\delta\eta_{-k}.\end{aligned}$$

A comparison of the corresponding coefficients of plane waves gives formula (V.59).

After applying (V.57), the continuity equation can be written in the form

$$2\pi[\boldsymbol{k}\langle\langle \boldsymbol{j}_k; a_{-k}\rangle\rangle^r_{E=\omega+i\varepsilon}-mE\langle\langle \hat{\varrho}_k; a_{-k}\rangle\rangle^r_{E=\omega+i\varepsilon}] = m\sqrt{\varrho_0}. \quad \text{(V.60)}$$

We shall use now the procedure proposed by Bogoliubov (1961) for the case of thermodynamic equilibrium ($E = 0$). To $\psi(0, \boldsymbol{r})$, $\psi^+(0, \boldsymbol{r})$ (in the Schrödinger representation) we apply the infinitesimal gauge transformation

$$\begin{aligned}\psi(0, \boldsymbol{r}) \to e^{-i\delta\chi(\boldsymbol{r})}\psi(0, \boldsymbol{r}) = \psi(0, \boldsymbol{r})-i\psi(0, \boldsymbol{r})\delta\chi(\boldsymbol{r}),\\ \psi^+(0, \boldsymbol{r}) \to e^{+i\delta\chi(\boldsymbol{r})}\psi^+(0, \boldsymbol{r}) = \psi^+(0, \boldsymbol{r})+i\psi^+(0, \boldsymbol{r})\delta\chi(\boldsymbol{r})\end{aligned} \quad \text{(V.61)}$$

where $\delta\chi$ is real and has the form

$$\delta\chi(\boldsymbol{r}) = e^{i(\boldsymbol{k}\cdot\boldsymbol{r})}\delta\eta_k+e^{-i(\boldsymbol{k}\cdot\boldsymbol{r})}\delta\eta_{-k}. \quad \text{(V.62)}$$

Transformation (V.61) leads to the following variation of the Hamiltonian (V.50)

$$\begin{aligned}\delta\hat{H} = -\frac{1}{m}\int \boldsymbol{j}(0, \boldsymbol{r})\nabla\delta\chi(\boldsymbol{r})d^3\boldsymbol{r} = \delta\hat{H}_k+\delta\hat{H}_{-k},\\ \delta\hat{H}_k = -\frac{1}{m}\int e^{i(\boldsymbol{k}\cdot\boldsymbol{r})}\boldsymbol{k}\boldsymbol{j}(0, \boldsymbol{r})d^3\boldsymbol{r}\,\delta\eta_k.\end{aligned} \quad \text{(V.63)}$$

The variation of the Hamiltonian, described by formula (V.63), is connected with the variation of the mean value $\langle\psi\rangle_H \to \langle\psi\rangle_{H+\delta H}$ which follows from (V.61). Therefore, we treat this variation as in the hydrodynamic approach, that is, as a consequence of the variation (V.63) of the Hamiltonian

$$\delta\langle\psi\rangle \equiv \langle\psi\rangle_{H+\delta H} - \langle\psi\rangle_H = -i\frac{2\pi}{m}\{e^{i(\boldsymbol{k}\cdot\boldsymbol{r})}\boldsymbol{k}\langle\langle a_k;\hat{\boldsymbol{j}}_{-k}\rangle\rangle_0^r\delta\eta_k - e^{-i(\boldsymbol{k}\cdot\boldsymbol{r})}(\boldsymbol{k}\langle\langle a_{-k};\hat{\boldsymbol{j}}_k\rangle\rangle_0^r)\,\delta\eta_{-k}\}. \quad \text{(V.64)}$$

On the other hand, $\langle\psi\rangle_{H+\delta H}$ can be calculated directly by changing $\langle\ldots\rangle_{H+\delta H}$ to $\langle\ldots\rangle_H$ and performing the transformation which is the inverse of the transformation (V.61) (Bogoliubov, 1961). We obtain

$$\langle\psi\rangle_{H+\delta H} = \langle\psi\rangle_H + i\langle\psi\rangle_H\,\delta\chi. \quad \text{(V.65)}$$

Hence,

$$\delta\langle\psi\rangle = i\langle\psi\rangle_H\,\delta\chi = i\sqrt{\varrho_0}(e^{i(\boldsymbol{k}\cdot\boldsymbol{r})}\delta\eta_k + e^{-i(\boldsymbol{k}\cdot\boldsymbol{r})}\delta\eta_{-k}). \quad \text{(V.66)}$$

From (V.66) and (V.64) we have

$$-2\pi\,(\boldsymbol{k}\langle\langle a_{-k};\hat{\boldsymbol{j}}_k\rangle\rangle_0^r) = m\sqrt{\varrho_0}. \quad \text{(V.67)}$$

Using (V.59) we get (V.67) in the form

$$2\pi\,(\boldsymbol{k}\langle\langle\hat{\boldsymbol{j}}_k;\,a_{-k}\rangle\rangle_0^r) = m\sqrt{\varrho_0} \quad \text{(V.68)}$$

$(\boldsymbol{k} \to -\boldsymbol{k})$. Thus we have obtained the continuity equations (V.54,60) for the case $E = 0$. We see that the Green function in (V.68) has a $1/k$ type singularity.

Consider now the general case for $E \neq 0$. Using the Schrödinger equation we will find what the variation of the Hamiltonian $\delta\hat{H}_t$ is under a time dependent gauge transformation. This transformation can be written in the form

$$U(t) = e^{i\hat{f}(t)}. \quad \text{(V.69)}$$

Therefore, from

$$i\frac{\partial}{\partial t}|n\rangle = \hat{H}|n\rangle \quad \text{(V.70)}$$

we change to

$$i\frac{\partial}{\partial t}|n'\rangle = \hat{H}'|n'\rangle,\ |n'\rangle = U|n\rangle, \qquad \hat{H}' = U'\hat{H}U^{-1} - \frac{\partial\hat{f}}{\partial t}. \quad \text{(V.71)}$$

The transformation represented by (V.69) must be an infinitesimal transformation of the form given by (V.61). We therefore have

$$U\psi U^{-1} = \psi + i[\hat{f}, \psi] = \psi - i\psi\,\delta\chi(t, r), \tag{V.72}$$

and hence

$$\hat{f}(t) = \int \psi^{+}(0, r)\psi(0, r)\,\delta\chi(t, r)\,\mathrm{d}^3\boldsymbol{r}. \tag{V.73}$$

So, instead of (V.64), we now have

$$\delta\hat{H}_t = -\frac{1}{m}\int\left\{\hat{j}(0, r)\nabla\,\delta\chi(t, r) + m\varrho(0, r)\frac{\partial\,\delta\chi(t, r)}{\partial t}\right\}\mathrm{d}^3\boldsymbol{r}. \tag{V.74}$$

If we take $\delta\chi$ as a real quantity of the form

$$\delta\chi(t, r) = e^{-i\omega t + \varepsilon t + i(k\cdot r)}\delta\eta_k + e^{i\omega t + \varepsilon t - i(k\cdot r)}\delta\eta_{-k}, \tag{V.75}$$

$$\varepsilon > 0, \quad \varepsilon \to 0,$$

then

$$\delta\hat{H}_t = e^{-i\omega t + \varepsilon t}V_{+\omega} + e^{i\omega t + \varepsilon t}V_{-\omega}, \tag{V.76}$$

where

$$V_{\pm\omega} = -\frac{i}{m}\int e^{\pm i(k\cdot r)}[\pm k\hat{j}(0, r) - (\pm\omega + i\varepsilon)\hat{\varrho}(0, r)]\,\delta\eta_{\pm k}, \tag{V.77}$$

$$\varepsilon > 0, \quad \varepsilon \to 0.$$

The variation $\delta\hat{H}_t$, which is adiabatically introduced and given by (V.76), leads to a variation of the mean value $\langle\psi\rangle$ which can be found with the help of (V.38)

$$\begin{aligned}\delta\langle\psi\rangle = -i\frac{2\pi}{m}\,\{&e^{-i\omega t + \varepsilon t + i(k\cdot r)}\,[(k\langle\langle a_k; \hat{j}_{-k}\rangle\rangle^r_{E=\omega+i\varepsilon})\\ &-m(\omega + i\varepsilon)\langle\langle a_k; \hat{\varrho}_{-k}\rangle\rangle^r_{E=\omega+i\varepsilon}]\,\delta\eta_k - e^{i\omega t + \varepsilon t - i(k\cdot r)}\\ &\times[-k\langle\langle a_{-k}; \hat{j}\rangle\rangle^r_{E=+\omega+i\varepsilon} + m(\omega - i\varepsilon)\langle\langle a_{-k}; \hat{\varrho}_k\rangle\rangle^r_{E=-\omega+i\varepsilon}]\,\delta\eta_{-k}\},\end{aligned} \tag{V.78}$$

$$\varepsilon > 0, \; \varepsilon \to 0.$$

On the other hand,

$$\delta\langle\psi\rangle = i\sqrt{\varrho_0}\,[e^{-i\omega t + \varepsilon t + i(k\cdot r)}\delta\eta_k + e^{i\omega t + \varepsilon t - i(k\cdot r)}\delta\eta_{-k}]. \tag{V.79}$$

Using (V.79), (V.78) and (V.59) we obtain the continuity equations (V.54,60) for the case $E \neq 0$. Formula (V.60) is a particular form of one of Ward's identities which express the continuity equation in the Green functions formalism.

From formula (V.74) for $\delta\hat{H}_t$ it is clear that a gauge transformation leads to terms in the Hamiltonian (V.50) involving the scalar potential and the longitudinal vector potential. This fact allows us to obtain some relations between the Green functions

$$\langle\langle\hat{\varrho}_k;\hat{\varrho}_{-k}\rangle\rangle^r_E,\quad \langle\langle\hat{\boldsymbol{j}}_k;\hat{\varrho}_{-k}\rangle\rangle^r_E,\quad \langle\langle\hat{\boldsymbol{j}}_k;\hat{\boldsymbol{j}}_{-k}\rangle\rangle^r_E.$$

From the formula (V.73) for $\hat{f}$ it follows that

$$\begin{aligned}\delta\langle\hat{\varrho}\rangle &= \langle\hat{\varrho}\rangle_{H+\delta H}-\langle\hat{\varrho}\rangle_H = 0,\\ \delta\langle\hat{\boldsymbol{j}}\rangle &= \langle\hat{\boldsymbol{j}}\rangle_{H+\delta H}-\langle\hat{\boldsymbol{j}}\rangle_H = \varrho\nabla\delta\chi.\end{aligned} \tag{V.80}$$

Now, from (V.38,76,77,80) we obtain

$$k\langle\langle\hat{\varrho}_k;(\hat{\boldsymbol{j}}_{-k})_{||}\rangle\rangle^r_E-mE\langle\langle\hat{\varrho}_k;\hat{\varrho}_{-k}\rangle\rangle^r_E = 0 \tag{V.81}$$

and

$$-mE\langle\langle\hat{\boldsymbol{j}}_k;\varrho_{-k}\rangle\rangle^r_E+k\langle\langle\hat{\boldsymbol{j}}_k;(\hat{\boldsymbol{j}}_{-k})_{||}\rangle\rangle^r_E = -\varrho\frac{m}{2\pi}\boldsymbol{k}. \tag{V.82}$$

As a consequence of (V.81,82) we have

$$k^2\langle\langle(\hat{\boldsymbol{j}}_k)_{||};(\hat{\boldsymbol{j}}_{-k})_{||}\rangle\rangle^r_E-m^2E^2\langle\langle\hat{\varrho}_k;\hat{\varrho}_{-k}\rangle\rangle^r_E = -\varrho\frac{m}{2\pi}\boldsymbol{k}^2 \tag{V.83}$$

where $(\hat{\boldsymbol{j}}_k)_{||}$ denotes the component of $\hat{\boldsymbol{j}}_k$ parallel to $\boldsymbol{k}$. Formulae (V.81–83) should also result from adopting the hydrodynamic approach to the Green functions, after adding to the Hamiltonian (V.50) the terms involving the scalar and vector potential. Identities (V.81–83) are valid for both, normal fluids and superfluids. They are proved by Petru (1968, 1969a).

CHAPTER VI

Ordinary Bose and Fermi fluids

§ 1 Hydrodynamic equations for ordinary Bose and Fermi fluids†

All calculations will be performed for a Bose fluid. In the case of a Fermi fluid we have only an additional summation over spin indices. Since for ordinary fluids $\langle\psi\rangle = 0$ and $\langle\psi\psi\rangle = 0$, the functions η, η^* disappear from eqns. (V.4,5,8,14), and the functions $\tilde{\eta}$, $\tilde{\eta}^*$ disappear from eqns. (V.14–16). Therefore, we can set these functions equal to zero in the Hamiltonians (V.1,10). In order to simplify the calculations, we also take $A = 0$.

Consider first the state of thermodynamic equilibrium, which is characterized by five parameters: two thermodynamic parameters and the components of the drift velocity $\boldsymbol{v}$ of the system. For the thermodynamic parameters we choose the particle density ϱ and the temperature Θ. The expectation values determined by these parameters are denoted by $\langle\ldots\rangle_{\varrho,\Theta,v}$. By means of the transformations of the field operators given by

$$\psi \rightarrow \psi e^{im(\boldsymbol{v}\cdot\boldsymbol{r})},$$

we can express the expectation values $\langle\ldots\rangle_{\varrho,\Theta,v}$ in terms of the expectation values $\langle\ldots\rangle_{\varrho,\Theta,0}$ for thermodynamic equilibrium and the fluid at rest.

For example, for $\boldsymbol{j}$ and ε defined by (V.2) we have

$$\boldsymbol{j} = \langle\hat{\boldsymbol{j}}^0\rangle_{\varrho,\Theta,v} = \langle\hat{\boldsymbol{j}}^0\rangle_{\varrho,\Theta,0} + m\varrho\boldsymbol{v} = m\varrho\boldsymbol{v},$$

$$\varrho\varepsilon = -\frac{1}{4m}\langle(\nabla^2\psi^+)\psi+\psi^+(\nabla^2\psi)\rangle_{\varrho,\Theta,0} + \tfrac{1}{2}\int V(r-r')\langle\psi^+\psi^+\psi\psi\rangle\, d^3\boldsymbol{r}' + \varrho\frac{mv^2}{2}. \qquad \text{(VI.1)}$$

† This section is based on work by Bogoliubov (1963).

Hence

$$\varepsilon = E(\varrho, \Theta) + \frac{mv^2}{2}. \tag{VI.2}$$

The expectation value $\langle \hat{j} \rangle_{\varrho, \Theta, 0}$ vanishes since for $\boldsymbol{v} = 0$ we have invariance with respect to the transformation $\boldsymbol{r} \to -\boldsymbol{r}$. The mean energy, per particle, for thermodynamic equilibrium and the fluid at rest, is $E(\varrho, \Theta)$.

Below we shall deal with expectation values of the type

$$\begin{aligned} \mathfrak{A} &= \langle (D_1\psi^+(t, r))\,(D\psi(t, r))\rangle_{\varrho, \Theta, v} = \mathfrak{A}(\varrho, \Theta, v), \\ \mathfrak{B} &= \langle (D_1\psi^+(t, r))\,(D_2\psi^+(t, r'))\,(D_3\psi(t, r'))\,(D_4\psi(t, r))\rangle_{\varrho, \Theta, v} \\ &= \mathfrak{B}(\varrho, \Theta, v|r-r') \end{aligned} \tag{VI.3}$$

where D_ν are linear forms of constants and spatial differentiation operators with respect to the space variables (see, for example, the expressions (V.7) for D and $G^{(\alpha)}$). In the case of thermodynamic equilibrium there is no time dependence and, because of space homogeneity, only a dependence on coordinate differences appears.

Consider now the hydrodynamic non-equilibrium processes for which non-equilibrium mean values of the type

$$\begin{aligned} \mathfrak{A}(t, r) &= \langle (D_1\psi^+(t, r))\,(D_2\psi(t, r))\rangle, \\ \mathfrak{B}(t, r, r-r') &= \langle (D_1\psi^+(t, r))\,(D_2\psi^+(t, r'))\,(D_3\psi(t, r'))\,(D_4\psi(t, r))\rangle, \end{aligned} \tag{VI.4}$$

and the external fields $U(t, r)$ vary slowly under the time and space translations

$$t \to t + t_0, \quad r \to r + r_0, \quad |t_0| \approx T, \quad |r_0| \approx l$$

where T is the largest of the relaxation times and l the mean free path. In the vicinity of each point $(t, \boldsymbol{r})$ there are only small deviations from the equilibrium described by the local variables $\varrho(t, r)$, $\Theta(t, r)$, $v(t, r)$. (We assume that deviations from thermodynamic equilibrium vanish in such a way that it is possible to determine the order of magnitude of T and l.)

We assume that the quantities (VI.3), where Θ, ϱ, v are constant in space and time, differ now very slightly from the local equilibrium quantities

$$\begin{aligned} &\mathfrak{A}\big(\varrho(t, r), \quad \Theta(t, r), \quad v(t, r)\big), \\ &\mathfrak{B}\big(\varrho(t, r), \quad \Theta(t, r), \quad v(t, r)|R\big). \end{aligned} \tag{VI.5}$$

The smaller the gradients of ϱ, Θ, $v^{(\alpha)}$, and U, the smaller is the difference between (VI.4) and (VI.5). The local quantities $\Theta(t, r)$, $\boldsymbol{v}(t, r)$ are defined by (VI.1,2) and they are thus expressed in terms of $\boldsymbol{j}(t, r)$ and $\varepsilon(t, r)$. We assume then that for suitably small gradients of ϱ, Θ, $v^{(\alpha)}$, U, we have, in each small region, a local quasi-equilibrium.

We note that the same situation usually occurs in the derivation of the hydrodynamic equations, for ordinary fluids, from the Boltzmann equation, if we look for a solution differing very little from equilibrium and expandable in a power series of gradients (Chapman and Cowling, 1952; Bogoliubov, 1946).

Indeed, we will consider the gradients $\dfrac{\partial}{\partial t}$, $\dfrac{\partial}{\partial \boldsymbol{r}}$ as of "the first order of smallness", while gradients like $\dfrac{\partial^2}{\partial r_\alpha \partial r_\beta}$ are of "the second order of smallness", etc. In order to formulate this assumption in a form suitable for calculations, a small parameter μ is introduced (Bogoliubov, 1946, 1963). This parameter characterizes the "smallness" of the gradients. With the help of this parameter we write (VI.4) in the form

$$\mathfrak{A}(t, r) = \tilde{\mathfrak{A}}(\mu t, \mu r; \mu) = \tilde{\mathfrak{A}}(\tau, \xi; \mu),$$

$$\mathfrak{B}(t, r) = \tilde{\mathfrak{B}}(\mu t, \mu r, r - r'; \mu) = \tilde{\mathfrak{B}}(\tau, \xi, R; \mu),$$

$$U(t, r) = \tilde{U}(\mu t, \mu r) = U(r, \xi),$$

$$r = \mu t, \qquad = \mu \boldsymbol{r}, \qquad \boldsymbol{R} = \boldsymbol{r} - \boldsymbol{r}',$$

and further

$$\tilde{\mathfrak{A}}(r, \xi; \mu) = \tilde{\mathfrak{A}}^{(0)}(r, \xi) + \mu \tilde{\mathfrak{A}}^{(1)}(r, \xi) + \mu^2 \tilde{\mathfrak{A}}^{(2)}(r, \xi) + \ldots +,$$

$$\tilde{\mathfrak{B}}(r, \xi; R; \mu) = \tilde{\mathfrak{B}}^{(0)}(r, \xi, R) + \mu \tilde{\mathfrak{B}}^{(1)}(r, \xi, R) + \mu^2 \tilde{\mathfrak{B}}^{(2)}(r, \xi, R) + \ldots +, \tag{VI.6}$$

$$\tilde{\mathfrak{A}}^{(0)}(r, \xi) = \tilde{\mathfrak{A}}(\tilde{\varrho}(r, \xi), \tilde{\Theta}(r, \xi), \tilde{v}(r, \xi)),$$

$$\tilde{\mathfrak{B}}^{(0)}(r, \xi, R) = \tilde{\mathfrak{B}}(\tilde{\varrho}(r, \xi), \tilde{\Theta}(r, \xi), \tilde{v}(r, \xi) | R)$$

where $\tilde{\mathfrak{A}}^{(0)}$, $\tilde{\mathfrak{B}}^{(0)}$ are obtained from (VI.4) by changing there the non-equilibrium mean value $\langle \ldots \rangle$ into the corresponding mean value $\langle \ldots \rangle_{\varrho, \Theta, v}$ (VI.5). The $\mathfrak{A}^{(1)}$ and $\mathfrak{B}^{(1)}$ must be linear forms of the gradients of ϱ, Θ, $v^{(\alpha)}$, etc. They describe the deviations from local equilibrium.

In particular, we can write

$$\varrho(t, r) = \tilde{\varrho}(\tau, \xi), \quad \boldsymbol{j}(t, r) = \tilde{\boldsymbol{j}}(\tau, \xi), \quad \varepsilon(t, r) = \tilde{\varepsilon}(\tau, \xi)$$

where $\tilde{\Theta}$ and $\tilde{v}$ are defined by the equations

$$\boldsymbol{v} = \frac{1}{m\tilde{\varrho}}\tilde{\boldsymbol{j}},$$
$$\tilde{\varepsilon} = E(\tilde{\varrho}, \tilde{\Theta}) + \frac{m\tilde{v}^2}{2}. \tag{VI.7}$$

Using a small parameter μ we now go over to the derivation of the hydrodynamic equations without and with viscous terms. Noting that D_t and G_t are of the type of $\mathfrak{B}$ we obtain from (V.4,5,8)

$$m\frac{\partial\varrho(\tau, \xi)}{\partial\tau} + \sum_\beta \frac{\partial j_\beta(\tau, \xi)}{\partial\xi_\beta} = 0, \tag{VI.8}$$

$$\frac{\partial j_\alpha}{\partial\tau} = \sum_\beta \frac{\partial\mathscr{T}_{\alpha\beta}}{\partial\xi_\beta} - \varrho\frac{\partial}{\partial\xi_\alpha}(U(\tau, \xi) - \lambda), \tag{VI.9}$$

$$\frac{\partial(\varrho\varepsilon)}{\partial\tau} = \sum_\beta \frac{\partial\mathscr{I}_\beta}{\partial\xi_\beta} - \frac{1}{m}\sum_\beta j_\beta \frac{\partial}{\partial\xi_\beta}(U - \lambda) \tag{VI.10}$$

where

$$\mathscr{T}_{\alpha\beta} = \mathscr{T}^{(0)}_{\alpha\beta} + \mu\mathscr{T}^{(1)}_{\alpha\beta},$$

$$\mathscr{T}^{(0)}_{\alpha\beta} = -\frac{1}{2m}\left\langle\frac{\partial\psi^+}{\partial r_\alpha}\frac{\partial\psi}{\partial r_\beta} + \frac{\partial\psi^+}{\partial r_\beta}\frac{\partial\psi}{\partial r_\alpha}\right\rangle^{(0)}$$
$$+\tfrac{1}{2}\int\frac{\partial V(R)}{\partial R_\alpha}R_\beta \boldsymbol{D}^{(0)}_\tau(\xi, R)\,\mathrm{d}^3\boldsymbol{R},$$

$$\mu\mathscr{T}^{(1)}_{\alpha\beta} = -\frac{\mu}{2m}\left\langle\frac{\partial\psi^+}{\partial r_\alpha}\frac{\partial\psi}{\partial r_\beta} + \frac{\partial\psi^+}{\partial r_\beta}\frac{\partial\psi}{\partial r_\alpha}\right\rangle^{(1)}$$
$$+\mu\tfrac{1}{2}\int\frac{\partial V(R)}{\partial R_\alpha}R_\beta D^{(1)}_\tau(\xi, R)\,\mathrm{d}^3\boldsymbol{R}, \tag{VI.11}$$

$$\mathscr{I}_\beta = \mathscr{I}^{(0)}_\beta + \mu\mathscr{I}^{(1)}_\beta,$$

$$\mathscr{I}^{(0)}_\beta = \left[\frac{i}{4m^2}\left\langle\frac{\partial\psi^+}{\partial r_\beta}(\nabla^2\psi) - (\nabla^2\psi^+)\frac{\partial\psi}{\partial r_\beta}\right\rangle^{(0)} + \int V(R)\overset{(0)}{G}{}^{(\beta)}(\xi, R)\,\mathrm{d}^3\boldsymbol{R}\right.$$
$$\left.-\sum_\alpha\int\frac{\partial V(R)}{\partial R_\alpha}R_\beta \overset{(0)}{G}{}^{(\alpha)}_\tau(\xi, R)\,\mathrm{d}^3\boldsymbol{R}\right],$$

$$\mu \mathscr{I}_\beta^{(1)} = \mu \Bigg[\frac{\partial}{\partial \xi_\beta} \frac{i}{8m^2} \langle (\nabla^2 \psi^+)\psi - \psi^+(\nabla^2\psi) \rangle + \frac{i}{4m^2} \left\langle \frac{\partial \psi}{\partial r_\beta} (\nabla^2 \psi) - (\nabla^2 \psi) \frac{\partial \psi}{\partial r_\beta} \right\rangle^{(1)} + \int V(R) \overset{(1)}{G}{}_\tau^{(\beta)}(\xi, R) \mathrm{d}^3\boldsymbol{R}$$
$$- \sum_\beta \int \frac{\partial V(R)}{\partial R_\alpha} R_\beta \overset{(1)}{G}{}_\tau^{(\alpha)}(\xi, R)\mathrm{d}^3\boldsymbol{R}$$
$$+ \frac{1}{2} \sum_{\alpha, \gamma} \frac{\partial}{\partial \xi_\gamma} \int \frac{\partial V(R)}{\partial R_\alpha} R_\beta R_\gamma \overset{(0)}{G}{}_\tau^{(\alpha)}(\xi, R) \mathrm{d}^3\boldsymbol{R} \Bigg]. \tag{VI.12}$$

Consider first the case of an "ideal" fluid without terms of order μ (without linear forms in the gradients of ϱ, Θ, $v^{(\alpha)}$). We have

$$\mathscr{T}_{\alpha\beta}^{(0)} = \mathscr{T}_{\alpha\beta}(\tilde{\varrho}, \tilde{\Theta}, \tilde{v}), \qquad \mathscr{I}_\alpha^{(0)} = \mathscr{I}_\alpha(\tilde{\varrho}, \tilde{\Theta}, \tilde{v})$$

where these expressions are obtained if in (VI.11,12) the non-equilibrium expectation values $\langle \ldots \rangle$ are replaced by the suitable equilibrium expectation values $\langle \ldots \rangle_{\varrho, \Theta, v}$.

Consider now the expression for $\mathscr{T}_{\alpha\beta}^{(0)}$ (putting $U = 0$). By means of the transformation $\psi \to \psi e^{im(\boldsymbol{v}\cdot\boldsymbol{r})}$ we express the expectation value $\langle \ldots \rangle_{\varrho, \Theta, v}$ in terms of $\langle \ldots \rangle_{\varrho, \Theta, 0}$. From (VI.11) we see that only the first term is affected by this transformation. We have

$$\left\langle \frac{\partial \psi^+}{\partial r_\alpha} \frac{\partial \psi}{\partial r_\beta} + \frac{\partial \psi^+}{\partial r_\beta} \frac{\partial \psi}{\partial r_\alpha} \right\rangle_{\varrho, \Theta, v} = \left\langle \frac{\partial \psi^+}{\partial r_\alpha} \frac{\partial \psi}{\partial r_\beta} + \frac{\partial \psi^+}{\partial r_\beta} \frac{\partial \psi}{\partial r_\alpha} \right\rangle_{\varrho, \Theta, 0} + 2\varrho m^2 v_\alpha v_\beta .$$

At equilibrium with $\boldsymbol{v} = 0$ the terms with one derivative vanish owing to invariance under the transformation $\boldsymbol{r} \to -\boldsymbol{r}$. For $\boldsymbol{v} = 0$, $\mathscr{T}_{\alpha\beta}$ must be an isotropic tensor of the form

$$\mathscr{T}_{\alpha\beta}(\varrho, \Theta, 0) = -\delta_{\alpha\beta} P(\varrho, \Theta).$$

Therefore, for $\boldsymbol{v} \neq 0$ the stress tensor is

$$\mathscr{T}_{\alpha\beta}(\varrho, \Theta, v) = -m\varrho v_\alpha v_\beta - \delta_{\alpha\beta} P(\varrho, \Theta) \tag{VI.13}$$

and for any $\alpha = 1, 2, 3$ we have

$$-P(\varrho, \Theta) = -\frac{1}{m} \left\langle \frac{\partial \psi^+}{\partial r_\alpha} \frac{\partial \psi}{\partial r_\alpha} \right\rangle_{\varrho, \Theta, 0} + \frac{1}{2} \int \frac{\partial V(R)}{\partial R_\alpha} R_\alpha D(R|\varrho, \Theta) \mathrm{d}^3\boldsymbol{R}, \tag{VI.14}$$

$$D(r - r'|\varrho, \Theta) = \langle \psi^+(r)\psi^+(r')\psi(r')\psi(r) \rangle_{\varrho, \Theta, 0}.$$

Let us denote the free energy per particle by $F(\varrho, \Theta)$. If N is the number of particles, the total free energy is

$$NF(\varrho, \Theta) = -\Theta \ln \mathrm{Tr}\{e^{-\frac{\hat{H}}{\Theta}}\}.$$

We shall demonstrate that P, introduced in (VI.14), is equal to

$$P(\varrho, \Theta) = -\frac{\partial NF\left(\frac{V}{N}, \Theta\right)}{\partial V} = \varrho^2 \frac{\partial F(\varrho, \Theta)}{\partial \varrho} \qquad \text{(VI.15)}$$

which is the definition of the pressure. In order to prove (VI.15) let us perform a linear dilatation of the volume V in one direction only, for instance along the α-axis. This gives $\boldsymbol{r} \to \boldsymbol{r}'$, $r'_\beta = r_\beta (\beta \neq \alpha)$, $r'_\alpha = L r_\alpha$, $\mathrm{d}\boldsymbol{r} \to L \mathrm{d}\boldsymbol{r}$, $V \to LV$. We also perform a canonical transformation of the field operators (Martin and Schwinger, 1959)

$$\psi \to L^{-1/2} \psi.$$

In the new coordinate system we have

$$\hat{H}_L = \frac{1}{2m} \int \nabla_L \psi^+(r) \nabla_L \psi(r) \mathrm{d}^3\boldsymbol{r}$$

$$+ \frac{1}{2} \iint V_L(r-r') \psi^+(r) \psi^+(r') \psi(r') \psi(r) \mathrm{d}^3\boldsymbol{r}\, \mathrm{d}^3\boldsymbol{r}'$$

where

$$\nabla_L = \left(\frac{1}{L}\frac{\partial}{\partial r_\alpha}, \frac{\partial}{\partial r_\beta}, \frac{\partial}{\partial r_\gamma}\right), \qquad \beta, \gamma \neq \alpha$$

and $V_L(R)$ is obtained from $V(R)$ after the transformations

$$R'_\alpha = L R_\alpha, \quad R'_\beta = R_\beta (\beta \neq \alpha),$$

$$\left(\frac{\partial V_L(R')}{\partial L}\right)_{L=1} = \left(\frac{\partial V}{\partial R'_\alpha}\frac{\partial R'_\alpha}{\partial L}\right)_{L=1} = \frac{\partial V}{\partial R_\alpha} R_\alpha.$$

We have now

$$P(\varrho, \Theta) = \left(\frac{\partial N F_L}{\partial V L}\right)_{L=1} = -\frac{\Theta}{V}\left[\frac{\partial}{\partial L} \ln \mathrm{Tr}\{e^{-\frac{\hat{H}L}{\Theta}}\}\right]_{L=1}$$

$$= \frac{1}{V}\left\langle\left(\frac{\partial \hat{H}_L}{\partial L}\right)_{L=1}\right\rangle_{\varrho, \Theta, 0} = -\frac{1}{m}\left\langle\frac{\partial \psi^+}{\partial r_\alpha}\frac{\partial \psi}{\partial r_\alpha}\right\rangle_{\varrho, \Theta, 0}$$

$$+ \frac{1}{2}\int \frac{\partial V(R)}{\partial R_\alpha} R_\alpha D(R|\varrho, \Theta) \mathrm{d}^3\boldsymbol{R}, \qquad \text{(VI.16)}$$

that is, formula (VI.14). The expectation values in (VI.6) do not depend on $\boldsymbol{r}$, therefore the integration over $\boldsymbol{r}$ is equivalent to multiplication by V.

After substituting (VI.13) into (VI.9) we have

$$\frac{\partial \tilde{j}_\alpha}{\partial \tau} = -\sum_\beta \frac{\partial}{\partial \xi_\beta}(m\tilde{\varrho}\tilde{v}_\alpha \tilde{v}_\beta + \delta_{\alpha\beta} P(\tilde{\varrho}, \tilde{\Theta})) - \tilde{\varrho}\frac{\partial U}{\partial \xi_\alpha}. \qquad \text{(VI.17)}$$

Now we wish to transform equation (VI.10) for energy in the same way as eqn. (VI.9).

Notice that

$$G^{(\alpha)}(r-r'|\varrho, \Theta, v) = \frac{i}{4m}\langle \psi^+(r')[\psi^+(r) imv_\alpha \psi(r)$$

$$+\psi^+(r) imv_\alpha \psi(r)]\psi(r')\rangle_{\varrho,\Theta,0} = -\frac{v_\alpha}{2} D(r-r'|\varrho, \Theta).$$

The first term in the expression (VI.12) for the energy current $\mathscr{I}_\beta^{(0)}$ can be written

$$-\frac{i}{4m}\Big\langle \sum_\beta \Big(\frac{\partial^2 \psi^+}{\partial r_\beta^2}\frac{\partial \psi}{\partial r_\alpha} - \frac{\partial \psi^+}{\partial r_\alpha}\frac{\partial^2 \psi}{\partial r_\beta^2}\Big)\Big\rangle_{\varrho,\Theta,v}$$

$$= -\frac{m}{2}v_\alpha \sum_\beta v_\beta^2 \langle \psi^+\psi\rangle_{\varrho,\Theta,0} - \frac{1}{2m}\sum_\beta v_\beta \Big\langle \frac{\partial \psi^+}{\partial r_\alpha}\frac{\partial \psi}{\partial r_\beta}$$

$$+\frac{\partial \psi^+}{\partial r_\beta}\frac{\partial \psi}{\partial r_\alpha}\Big\rangle_{\varrho,\Theta,0} + \frac{1}{4m}v_\alpha \sum_\beta \frac{\partial}{\partial r_\beta}\Big\langle \frac{\partial \psi^+}{\partial r_\beta}\psi + \psi^+\frac{\partial \psi}{\partial r_\beta}\Big\rangle_{\varrho,\Theta,0}$$

$$-\frac{1}{2m}v_\alpha \sum_\beta \Big\langle \frac{\partial \psi}{\partial r_\beta}\frac{\partial \psi}{\partial r_\beta}\Big\rangle_{\varrho,\Theta,0}$$

$$= -\varrho v_\alpha \frac{mv^2}{2} - \frac{v_\alpha}{m}\Big\langle \frac{\partial \psi^+}{\partial r_\alpha}\frac{\partial \psi}{\partial r_\alpha}\Big\rangle_{\varrho,\Theta,0} - \frac{v_\alpha}{2m}\sum_\beta \Big\langle \frac{\partial \psi^+}{\partial r_\beta}\frac{\partial \psi}{\partial r_\beta}\Big\rangle_{\varrho,\Theta,0}.$$

Therefore, we have

$$\mathscr{I}_\alpha(\varrho, \Theta, v) = -\varrho v_\alpha \frac{mv^2}{2}$$

$$-\frac{v_\alpha}{2m}\sum_\beta \Big\langle \frac{\partial \psi^+}{\partial r_\beta}\frac{\partial \psi}{\partial r_\beta}\Big\rangle_{\varrho,\Theta,0} - \frac{v_\alpha}{2}\int V(R) D(R|\varrho, \Theta)\,\mathrm{d}^3\boldsymbol{R}$$

$$-\frac{v_\alpha}{m}\Big\langle \frac{\partial \psi^+}{\partial r_\alpha}\frac{\partial \psi}{\partial r_\alpha}\Big\rangle_{\varrho,\Theta,0} + \frac{v_\alpha}{2}\int \frac{\partial V(R)}{\partial R_\alpha} R_\alpha D(R|\varrho, \Theta)\,\mathrm{d}^3\boldsymbol{R}.$$

On the other hand,

$$\frac{1}{2m}\sum_{\beta}\left\langle\frac{\partial\psi^{+}}{\partial r_{\beta}}\frac{\partial\psi}{\partial r_{\beta}}\right\rangle_{\varrho,\Theta,0}+\frac{1}{2}\int V(R)D(R|\varrho,\Theta)\mathrm{d}^{3}\boldsymbol{R}$$
$$=\varrho E(\varrho,\Theta)=\varrho\left\{F(\varrho,\Theta)-\Theta\frac{\partial F(\varrho,\Theta)}{\partial\Theta}\right\}$$

(see (V.2)). After taking (VI.14) into account, we have for the energy density

$$\frac{\partial(\tilde{\varrho}\tilde{\varepsilon})}{\partial\tau}=-\sum_{\alpha}\frac{\partial}{\partial\xi_{\alpha}}\tilde{v}_{\alpha}\left\{\tilde{\varrho}\left(E(\tilde{\varrho},\tilde{\Theta})+\frac{m\tilde{v}^{2}}{2}\right)\right.$$
$$\left.+P(\tilde{\varrho},\tilde{\Theta})\right\}-\frac{1}{m}\sum_{\alpha}\tilde{j}_{\alpha}\frac{\partial U}{\partial\xi_{\alpha}} \qquad \text{(VI.18)}$$

$$=-\sum_{\alpha}\frac{\partial}{\partial\xi_{\alpha}}\tilde{v}_{\alpha}\tilde{\varrho}\left(E(\tilde{\varrho},\tilde{\Theta})+\frac{m\tilde{v}^{2}}{2}\right)+\sum_{\alpha,\beta}\frac{\partial}{\partial r_{\beta}}(\tilde{v}_{\alpha}\mathscr{T}_{\alpha\beta}(\tilde{\varrho},\tilde{\Theta},0)),$$

$$v_{\alpha}\left\{\varrho\left(E(\tilde{\varrho},\tilde{\Theta})+\frac{mv^{2}}{2}\right)+P(\tilde{\varrho},\tilde{\Theta})\right\}=\mathscr{I}_{\alpha},$$

$$\mathscr{T}_{\alpha\beta}(\varrho,\Theta,0)=-\delta_{\alpha\beta}P(\varrho,\theta).$$

We introduce the entropy, per particle, by means of the formula

$$S(\varrho,\Theta)=-\frac{\partial F(\varrho,\Theta)}{\partial\Theta}. \qquad \text{(VI.19)}$$

Moreover, we will use the formulae

$$E(\varrho,\Theta)=F(\varrho,\Theta)-\Theta\frac{\partial F(\varrho\Theta)}{\partial\Theta},\qquad P(\varrho,\Theta)=\varrho^{2}\frac{\partial F(\varrho,\Theta)}{\partial\varrho},$$

$$\Lambda(\varrho,\Theta)=F(\varrho,\Theta)+\varrho\frac{\partial F(\varrho,\Theta)}{\partial\varrho},\qquad \boldsymbol{j}=m\varrho\boldsymbol{v} \qquad \text{(VI.20)}$$

where Λ is the chemical potential per particle.

Now instead of eqn. (VI.18) for the energy density we wish to obtain and use the equation for the entropy density.

Using (VI.2,19,20) we obtain for the left-hand side of (VI.18)

$$\frac{\partial(\tilde{\varrho}\tilde{\varepsilon})}{\partial\tau}=\frac{\partial}{\partial\tau}\left[\tilde{\varrho}\left(F-\tilde{\Theta}\frac{\partial F}{\partial\tilde{\Theta}}\right)+\frac{m\tilde{\varrho}\tilde{v}^{2}}{2}\right]$$
$$=\Lambda\frac{\partial\tilde{\varrho}}{\partial\tau}+\tilde{\Theta}\frac{\partial}{\partial\tau}(\varrho S)+\frac{m\tilde{v}^{2}}{2}\frac{\partial\tilde{\varrho}}{\partial\tau}+\tilde{\boldsymbol{j}}\frac{\partial\tilde{\boldsymbol{v}}}{\partial\tau}.$$

From the continuity equation (V.4) we obtain

$$\left(\tilde{j}\cdot\frac{\partial\tilde{v}}{\partial\tau}\right)=\left(\tilde{v}\cdot\frac{\partial\tilde{j}}{\partial\tau}\right)-m\tilde{v}^2\frac{\partial\tilde{\varrho}}{\partial\tau}.$$

Equation (VI.17) gives

$$\left(\tilde{v}\cdot\frac{\partial\tilde{j}}{\partial\tau}\right)=-\tilde{v}^2(\nabla\cdot\tilde{j})-\frac{1}{2}(j\cdot\nabla)\tilde{v}^2-(\tilde{v}\cdot\nabla P)-\tilde{\varrho}(\tilde{v}\cdot\nabla U).$$

We finally have

$$\frac{\partial(\tilde{\varrho}\tilde{\varepsilon})}{\partial\tau}=-\frac{1}{m}\Lambda(\nabla\cdot\tilde{j})-\frac{1}{2}(\nabla\cdot(\tilde{v}^2\tilde{j}))-(\tilde{v}\cdot\nabla P)+\tilde{\Theta}\frac{\partial}{\partial\tau}(\varrho S)$$
$$-\tilde{\varrho}(\tilde{v}\cdot\nabla U). \qquad \text{(VI.21)}$$

Calculating the right-hand side of (VI.18) we obtain

$$\frac{\partial(\tilde{\varrho}\tilde{\varepsilon})}{\partial\tau}=-(\nabla\cdot(\tilde{\varrho}\tilde{v}E))-\frac{1}{2}(\nabla\cdot(\tilde{v}^2\tilde{j}))-(\nabla\cdot(P\tilde{v}))-\frac{1}{m}(\tilde{j}\cdot\nabla U),$$

$$(\nabla\cdot(\tilde{\varrho}\,\tilde{v}\,E))=(\nabla\cdot(\tilde{\varrho}\,\tilde{v}\,F))-\left(\nabla\cdot\left(\tilde{\varrho}\tilde{v}\Theta\frac{\partial F}{\partial\Theta}\right)\right)$$
$$=\Lambda\frac{1}{m}(\nabla\cdot\tilde{j})-P(\nabla\cdot\tilde{v})+\Theta(\nabla\cdot(\tilde{\varrho}S\tilde{v})). \qquad \text{(VI.22)}$$

From (VI.21,22) we finally have

$$\frac{\partial}{\partial\tau}(\tilde{\varrho}S)+(\nabla\cdot(\tilde{\varrho}S\tilde{v}))=0. \qquad \text{(VI.23)}$$

We transform the equations (VI.8,17,18,23) from the auxiliary variables τ, $\boldsymbol{\xi}$, to the original variables t, $\boldsymbol{r}$. These equations are the hydrodynamic equations describing an ordinary liquid without dissipative terms

$$m\frac{\partial\varrho(t,r)}{\partial t}+(\nabla\cdot j(t,r))=0, \qquad \text{(VI.24)}$$

$$\frac{\partial j_\alpha}{\partial t}=-\sum_\beta\frac{\partial}{\partial r_\beta}[m\varrho v_\alpha v_\beta+\delta_{\alpha\beta}P]-\varrho\frac{\partial U}{\partial r_\alpha}, \qquad \text{(VI.25)}$$

$$\frac{\partial}{\partial t}(\varrho S)+\nabla(\varrho S\boldsymbol{v})=0. \qquad \text{(VI.26)}$$

Now we wish to obtain the hydrodynamic equations with viscous terms (Galasiewicz, 1966b; Krasnikov, 1966).

After considering viscosity, we have

$$\mathcal{T}_{\alpha\beta} = \mathcal{T}^{(0)}_{\alpha\beta} + \mathcal{T}^{(1)}_{\alpha\beta}, \quad \mathcal{I}_{\beta} = \mathcal{I}^{(0)}_{\beta} + \overset{1}{\mathcal{I}}{}^{(1)}_{\beta} + \overset{2}{\mathcal{I}}{}^{(1)}_{\beta}, \tag{VI.27}$$

where $\mathcal{T}^{(1)}_{\alpha\beta}$, $\overset{1}{\mathcal{I}}{}^{(1)}_{\beta}$ must be the most general linear forms of the space derivatives of v_{α}, ϱ, Θ. Therefore, we put

$$\begin{gathered}\mathcal{T}^{(1)}_{\alpha\beta} = \delta_{\alpha\beta}\,\zeta\,\mathrm{div}\,\boldsymbol{v} + \eta\left(\frac{\partial v_{\alpha}}{\partial r_{\beta}} + \frac{\partial v_{\beta}}{\partial r_{\alpha}} - \frac{2}{3}\,\delta_{\alpha\beta}\,\mathrm{div}\,\boldsymbol{v}\right),\\ \overset{1}{\mathcal{I}}{}^{(1)}_{\beta} = D\frac{\partial\varrho}{\partial r_{\beta}} + \varkappa\frac{\partial\Theta}{\partial r_{\beta}}, \quad \overset{2}{\mathcal{I}}{}^{(1)}_{\beta} = \sum_{\alpha} v_{\alpha}\mathcal{T}_{\alpha\beta}(\varrho, \Theta, v)\end{gathered} \tag{VI.28}$$

where ζ, η are the viscosity and bulk viscosity coefficients, D is the diffusion coefficient and $\varkappa$ the coefficient of thermal conductivity.

The form of $\overset{2}{\mathcal{I}}{}^{(1)}_{\beta}$ follows from the appearance of the factor $\sum_{\alpha} v_{\alpha}\mathcal{T}_{\alpha\beta}(\varrho, \Theta, 0)$ in (VI.18). The elements of the stress tensor $\mathcal{T}^{(0)}_{\alpha\beta}(\varrho, \Theta, 0)$ are expressed in terms of the mean values $\langle\ \rangle^{(0)}_{\varrho,\Theta,0}$. However, the quantity $\mathcal{I}^{(1)}_{\beta}$ is evaluated in terms of the mean values $\langle\ \rangle^{(1)}_{\varrho,\Theta,v}$. Therefore, in the expression for $\mathcal{I}_{\beta}$ the term $\sum_{\alpha} v_{\alpha}\mathcal{T}^{(1)}_{\alpha\beta}(\varrho, \Theta, v)$ should appear.

Let us introduce a small parameter μ which characterizes the slowness of the time variations and the smallness of the deviation from local isotropy. From (VI.17) we see that for the normal fluid the potential U is of the same order in μ as v_{α} and consequently, of the same order as the parameters describing liquid. (For a superfluid the sources η, η^* are quantities of the order of the derivatives of the parameters and therefore, terms linear in η, η^* must appear in $\mathcal{T}^{(1)}_{\alpha\beta}$). In considering viscosity we must substitute $\mathcal{I}_{\beta} = \mathcal{I}^{(0)}_{\beta} + \mathcal{I}^{(1)}_{\beta}$ in (VI.13). Equation (VI.26) takes the form

$$\frac{\partial(\varrho S)}{\partial t} + \sum_{\beta}\frac{\partial}{\partial r_{\beta}}(\varrho\, S\, v_{\beta}) = \frac{1}{\Theta}\left[D\nabla^2\varrho + \varkappa\nabla^2\Theta + \sum_{\alpha\beta}\frac{\partial}{\partial r_{\beta}} v_{\alpha}\mathcal{T}^{(1)}_{\alpha\beta}\right]. \tag{VI.29}$$

Consider the possibility of the existence of thermodynamic equilibrium with $U = U(r)$. Since in thermodynamic equilibrium $\Theta = \text{const.}$, $v = 0$ and all quantities are time-independent, eqns.

(VI.24,25,29) yield

$$\frac{\partial P}{\partial \varrho}\frac{\partial \varrho}{\partial r_\alpha}+\varrho\frac{\partial \tilde{U}}{\partial r_\alpha}=0, \quad D\nabla^2\varrho=0. \tag{VI.30}$$

Hence, $D=0$, and we see that a normal fluid is characterized by three transport coefficients.

Thus the hydrodynamic equations with viscous terms have the form

$$\frac{\partial \varrho}{\partial t}+\sum_\alpha \frac{\partial(\varrho v_\alpha)}{\partial r_\alpha}=0,$$

$$m\frac{\partial(\varrho v_\alpha)}{\partial t}=\sum_\beta \frac{\partial}{\partial r_\beta}\Big\{\delta_{\alpha\beta}[-P(\varrho,\Theta)+\zeta\,\mathrm{div}\,\boldsymbol{v}]$$

$$-m\varrho v_\alpha v_\beta+\eta\left(\frac{\partial v_\alpha}{\partial r_\beta}+\frac{\partial v_\beta}{\partial r_\alpha}-\frac{2}{3}\,\delta_{\alpha\beta}\,\mathrm{div}\,\boldsymbol{v}\right)\Big\}-\varrho\frac{\partial U}{\partial r_\alpha}, \tag{VI.31}$$

$$\frac{\partial(\varrho S)}{\partial t}+\sum_\beta \frac{\partial}{\partial r_\beta}(\varrho S v_\beta)=\frac{\varkappa}{\Theta}\nabla^2\Theta+\frac{1}{\Theta}R$$

where the dissipation function R has the form (we omit summation over α, β and use formula (VII.40))

$$R=\frac{\partial v_\alpha}{\partial r_\beta}\mathscr{T}^{(1)}_{\alpha\beta}=\frac{\eta}{2}\left(\frac{\partial v_\alpha}{\partial r_\beta}+\frac{\partial v_\beta}{\partial r_\alpha}-\frac{2}{3}\,\delta_{\alpha\beta}\nabla\boldsymbol{v}\right)^2+\zeta(\nabla\boldsymbol{v})^2.$$

From Chapter V, § 2 it is clear that the same hydrodynamic equations can be obtained for Fermi systems with the Hamiltonian (V.10) for the local quantities defined by (V.12).

§ 2 Linearized hydrodynamic equations and the retarded Green functions

Consider, by means of eqns. (VI.31), an infinitesimal deviation from thermodynamic equilibrium. The deviation is induced by an infinitesimal scalar potential, and according to Krasnikov (1966), by the vector potential $A(t, r)$.

We write

$$\varrho=\varrho^0+\delta\varrho(t,r), \quad \Theta=\Theta^0+\delta\Theta(t,r), \quad A(t,r)=\delta A(t,r),$$
$$\boldsymbol{v}=\delta\boldsymbol{v}(t,r), \quad (\boldsymbol{v}^0=0), \quad S=S^0+\delta S(t,r),$$
$$U(t,r)=\delta U(t,r).$$

We neglect the terms proportional to $\delta v_\alpha \delta v_\beta$, $\delta\varrho\, \delta v_\alpha$, $\delta\varrho \dfrac{\delta U}{\partial r_\alpha}$, $\delta v\, \delta A$ and obtain, considering now the terms involving δA (Krasnikov, 1966),

$$m\frac{\partial \delta\varrho}{\partial t}+\varrho\sum_\beta \frac{\partial \delta v_\beta}{\partial r_\beta}=0,$$

$$\begin{aligned} m\frac{\partial v_\alpha}{\partial t} &= -\left(\frac{\partial P}{\partial \varrho}\right)_\Theta \frac{\partial \delta\varrho}{\partial r_\alpha}-\left(\frac{\partial P}{\partial \Theta}\right)_\varrho \frac{\partial \delta\Theta}{\partial r_\alpha}+\left(\zeta+\frac{1}{3}\eta\right)\frac{\partial}{\partial r_\alpha}\operatorname{div}\boldsymbol{v} \\ &\quad +\eta\nabla^2 v_\alpha-\varrho\frac{\partial \delta U}{\partial r_\alpha}-\varrho\frac{\partial A^\alpha}{\partial t}, \end{aligned} \tag{VI.32}$$

$$\varrho\frac{\partial \delta S}{\partial t}+S\frac{\partial \delta\varrho}{\partial t}+\varrho S\sum_\beta \frac{\partial \delta v_\beta}{\partial r_\beta}=\frac{\varkappa}{\Theta}\nabla^2\delta\Theta$$

(the upper index of ϱ, Θ, S is omitted).

Hence, we obtain linearized hydrodynamic equations, that is, equations in the acoustic approximation.

In the Hamiltonian (V.1) we now have

$$\hat{H}_t^1 \equiv \delta\hat{H}_t=\int \hat{\varrho}(0,r')\delta U(t,r')\,d^3r'-\frac{1}{m}\int (\hat{j}^0(0,r')\cdot\delta A(t,r'))\,d^3\boldsymbol{r}'. \tag{VI.33}$$

In order to introduce (VI.33) adiabatically, it is necessary to write

$$\begin{aligned} \delta U(t,r) &= e^{-i\omega t+\varepsilon t+i(k\cdot r)}\delta U(k)+e^{i\omega t+\varepsilon t-i(k\cdot r)}\delta U(-k), \\ \delta A(t,r) &= e^{-i\omega t+\varepsilon t+i(k\cdot r)}\delta A(k)+e^{i\omega t+\varepsilon t-i(k\cdot r)}\delta A(-k). \end{aligned} \tag{VI.34}$$

Then

$$\delta\hat{H}_\tau = e^{-i\omega\tau+\varepsilon\tau}V_\omega+e^{i\omega\tau+\varepsilon\tau}V_{-\omega} \tag{VI.35}$$

where

$$\begin{aligned} V_{\pm\omega} &= V^U_{\pm\omega}+V^A_{\pm\omega}, \\ V^U_\pm &= \int e^{\pm i(k\cdot r)}\hat{\varrho}(0,r')\,d^3\boldsymbol{r}'\,\delta U(\pm k), \\ V^A_\pm &= -\frac{1}{m}\int e^{\pm i(k\cdot r)}(\hat{j}(0,r')\cdot\delta A(\pm k))\,d^3\boldsymbol{r}'. \end{aligned} \tag{VI.36}$$

The adiabatically introduced variation of the Hamiltonian induces variations of the mean values given by (V.38).

From (VI.36) and (V.38) we have, for Bose systems,

$$\delta\varrho(t,r)=2\pi e^{-i\omega t+\varepsilon t}\int e^{i(k\cdot r')}\Big[\langle\langle\hat{\varrho}(r);\hat{\varrho}(r')\rangle\rangle^r_{E=\omega+i\varepsilon}\delta U(k)$$

$$-\frac{1}{m}(\langle\langle\hat{\varrho}(r);\hat{j}^0(r')\rangle\rangle^r_{E=\omega+i\varepsilon}\cdot\delta A(k))\Big]\,d^3r' \tag{VI.37}$$

$$+2\pi e^{i\omega t+\varepsilon t}\int e^{-i(k\cdot r')}\Big[\langle\langle\hat{\varrho}(r);\,\hat{\varrho}(r')\rangle\rangle^r_{E=-\omega+i\varepsilon}\delta U(-k)$$

$$-\frac{1}{m}(\langle\langle\hat{\varrho}(r);\hat{j}^0(r')\rangle\rangle^r_{E=-\omega+i\varepsilon}\cdot\delta A(-k))\Big]\,d^3r'.$$

We introduce the Fourier components

$$\delta\varrho(t,r)=e^{-i\omega t+\varepsilon t+i(k\cdot r)}\delta\varrho(k)+e^{i\omega t+\varepsilon t-i(k\cdot r)}\delta\varrho(k). \tag{VI.38}$$

After writing (VI.37) in the momentum representation we have

$$\delta\varrho(t,r)=2\pi e^{-i\omega t+\varepsilon t+i(k\cdot r)}\Big[\langle\langle\hat{\varrho}_{kj};\,\hat{\varrho}_{-k}\rangle\rangle^r_{E=\omega+i\varepsilon}\delta U(k)$$

$$-\frac{1}{m}(\langle\langle\hat{\varrho}_k;\hat{j}^0_{-k}\rangle\rangle^r_{E=\omega+i\varepsilon}\cdot\delta A(k))\Big]$$

$$+2\pi e^{i\omega t+\varepsilon t-i(k\cdot r)}\Big[\langle\langle\hat{\varrho}_k;\,\hat{\varrho}_{-k}\rangle\rangle^r_{E=-\omega+i\varepsilon}\delta U(k)$$

$$-\frac{1}{m}(\langle\langle\hat{\varrho}_k;\hat{j}^0_{-k}\rangle\rangle^r_{E=-\omega+i\varepsilon}\cdot\delta A(-k))\Big],$$

$$\hat{\varrho}_k=\frac{1}{\sqrt{V}}\sum_q a^+_{q-\frac{k}{2}}a_{q+\frac{k}{2}};\qquad \hat{j}^0_k=\frac{1}{\sqrt{V}}\sum_q q a^+_{q-\frac{k}{2}}a_{q+\frac{k}{2}}. \tag{VI.39}$$

From (VI.38,39) we obtain

$$\delta\varrho(k)=2\pi\langle\langle\hat{\varrho}_k;\,\hat{\varrho}_{-k}\rangle\rangle^r_{E=\omega+i\varepsilon}\delta U(k)-\frac{2\pi}{m}(\langle\langle\hat{\varrho}_k;\hat{j}^0_{-k}\rangle\rangle^r_{E=\omega+i\varepsilon}\cdot\delta A(k)). \tag{VI.40}$$

Similarly,

$$\delta j^0(k)=2\pi\langle\langle\hat{j}^0_k;\,\hat{\varrho}_{-k}\rangle\rangle^r_{E=\omega+i\varepsilon}\delta U(k)-\frac{2\pi}{m}\langle\langle\hat{j}^0_k;\,(\hat{j}^0_{-k}\rangle\rangle^r_{E=\omega+i\varepsilon}\delta A(k)), \tag{VI.41}$$

$$\delta j(k)=\delta j^0(k)-\varrho\delta A(k)=m\varrho\delta v(k)$$

where j^0 and j are defined by (V.1). These formulae connect the hydrodynamic quantities which can be found from eqns. (VI.32), with Green functions. It must be emphasized that since the hydrodynamic equations describe a many-body system only for "slow" changes of hydrodynamic quantities, these connexions have an asymptotic character and are valid only for $k \ll 1/l$, $E \ll 1/T$ (where l is the mean free path and T is the relaxation time).

From (VI.40,41) it follows that

$$\frac{\delta\varrho(k)}{\delta U(k)} = 2\pi\langle\langle\hat{\varrho}_k;\hat{\varrho}_{-k}\rangle\rangle_E, \qquad \frac{\delta\varrho(k)}{\delta A(k)^{\alpha}_{||}} = -\frac{2\pi}{m}\langle\langle\hat{\varrho}_k;\hat{j}^{0\alpha}_{-k||}\rangle\rangle_E,$$

$$\frac{\delta j(k)^{0\alpha}_{\perp}}{\delta A(k)^{\beta}_{\perp}} = -\frac{2\pi}{m}\langle\langle\hat{j}^{0\alpha}_{k\perp};\hat{j}^{0\beta}_{-k\perp}\rangle\rangle_E, \quad \frac{\delta j(k)^{0\alpha}_{||}}{\delta A(k)^{\beta}_{||}} = -\frac{2\pi}{m}\langle\langle\hat{j}^{0\alpha}_{k\perp};\hat{j}^{0\beta}_{-k\perp}\rangle\rangle_E,$$

$$\delta A(k) = \delta A(k)_{||} + \delta A(k)_{\perp}, \qquad \frac{\delta j(k)^{0\alpha}}{\delta U(k)} = 2\pi\langle\langle\hat{j}^0_k;\hat{\varrho}_{-k}\rangle\rangle_E$$

$$\delta A(k)_{||}\|\boldsymbol{k}, \quad \delta A(k)_{\perp}\perp\boldsymbol{k}. \tag{VI.42}$$

For a Fermi system

$$\delta\varrho(k) = 2\pi\langle\langle\hat{\varrho}_k;\hat{\varrho}_{-k}\rangle\rangle_E\,\delta U(k) - \frac{2\pi}{m}\left(\langle\langle\hat{\varrho}_k;\hat{j}^0_{-k}\rangle\rangle_E\cdot\delta\boldsymbol{A}(k)\right), \tag{VI.43}$$

$$\hat{\varrho}_k = \frac{1}{\sqrt{V}}\sum_{p,s} a^+_{p-\frac{k}{2},s}\,a_{p+\frac{k}{2},s}, \qquad \hat{j}^0_k = \frac{1}{\sqrt{V}}\sum_{p,s} \boldsymbol{p}a^+_{p-\frac{k}{2},s}\,a_{p+\frac{k}{2},s}$$

and the derivatives $\frac{\delta\varrho}{\delta U}, \frac{\delta j^{\alpha}}{\delta U}$ have the same form as for Bose systems.

Now, we endeavour to calculate $\delta\varrho$ and $(\boldsymbol{k}\cdot\delta\boldsymbol{v}(k))$ from eqns. (VI.32), for the case of an ideal fluid. After conversion to Fourier components (as in (VI.38)) we have

$$\varrho\,\delta Y - E\,\delta\varrho(k) = 0,$$

$$m\varrho E\,\delta Y - k^2\left(\frac{\partial P}{\partial\varrho}\right)_{\Theta}\delta\varrho(k) - k^2\left(\frac{\partial P}{\partial\Theta}\right)_{\varrho}\delta\Theta(k) = \varrho k^2\delta U(k) - \varrho E(\boldsymbol{k}\cdot\delta\boldsymbol{A}(k)),$$

$$-\varrho S\,\delta Y + E\left[\varrho\left(\frac{\partial S}{\partial\varrho}\right)_{\Theta} + S\right]\delta\varrho(k) + E\varrho\left(\frac{\partial S}{\partial\Theta}\right)_{\varrho}\delta\Theta(k) = 0,$$

$$m\delta v^{\alpha}_{\perp} = -\delta A^{\alpha}(k)_{\perp}, \tag{VI.44}$$

$$\delta Y \equiv (\boldsymbol{k}\cdot\delta\boldsymbol{v}(k)) = (\boldsymbol{k}\cdot\delta\boldsymbol{v}_{||}(k)), \quad \delta\boldsymbol{v} = \delta\boldsymbol{v}_{||} + \delta\boldsymbol{v}_{\perp}.$$

We denote the determinant of this system of equations by $D(E)$, and by D_{ϱ} the determinant which is obtained from $D(E)$ when we change the second column to the right-hand side of (VI.39).

We obtain

$$D(E) = m\varrho^2 \left(\frac{\partial S}{\partial \Theta}\right)_\varrho E[E^2 - c^2k^2], \quad c^2 = \frac{1}{m}\left(\frac{\partial P}{\partial \varrho}\right)_S, \quad \text{(VI.45)}$$

$$D_\varrho = \varrho^2 \left(\frac{\partial S}{\partial \Theta}\right)_\varrho E\varrho[k^2 \delta U(k) - E(\mathbf{k}\cdot\delta \mathbf{A}(k))],$$

$$D_Y = \varrho^2 \left(\frac{\partial S}{\partial \Theta}\right)_\varrho E^2[k^2 \delta U(k) - E(\mathbf{k}\cdot\delta \mathbf{A}(k))]. \quad \text{(VI.46)}$$

Hence

$$\frac{\delta\varrho(k)}{\delta A(k)_{||}} = -\frac{\varrho E k}{m(E^2 - c^2k^2)} = -\frac{1}{m}\,\frac{\delta j_{k||}}{\delta U(k)},$$

$$\frac{\delta\varrho(k)}{\delta U(k)} = \frac{\varrho k^2}{m[E^2 - c^2k^2]}, \quad \frac{\delta j_{k||}}{\delta A_{k||}} = -\frac{E\varrho^2}{E^2 - c^2k^2} \quad \text{(VI.47)}$$

where c is the sound velocity and ck the energy of sound quanta, phonons.

By considering the viscosity we have, instead of (VI.44),

$$\varrho\,\delta Y - E\,\delta\varrho(k) = 0,$$

$$\left[Em\varrho - i\left(\frac{4}{3}\eta + \zeta\right)k^2\right]\delta Y - k^2\left(\frac{\partial P}{\partial \varrho}\right)_\Theta \delta\varrho(k) - k^2\left(\frac{\partial P}{\partial \Theta}\right)_\varrho \delta\Theta(k)$$
$$= \varrho k^2 \delta U(k) + E\varrho(\mathbf{k}\cdot\delta\mathbf{A}),$$

$$-\varrho S\,\delta Y + E\left[\varrho\left(\frac{\partial S}{\partial \varrho}\right)_\Theta + S\right]\delta\varrho(k)$$
$$+\left[E\varrho\left(\frac{\partial S}{\partial \Theta}\right)_\varrho - i\frac{\varkappa}{\Theta}k^2\right]\delta\Theta(k) = 0, \quad \text{(VI.48)}$$

$$m\varrho E\,\delta v_\perp^\alpha(k) = -i\eta k^2 \delta v_\perp^\alpha(k) - \varrho E\,\delta A^\alpha(k)_\perp.$$

If the viscosity is taken into account, it is important to remember that for the variations $\delta\langle\hat{A}\rangle$ we have from (V.32) the initial condition that $\delta\langle\hat{A}\rangle = 0$ at $t = -\infty$.

The variations $\delta\varrho(t, r)$, $\delta\Theta(t, r)$, $\delta v^\alpha(t, r)$, which are solutions of eqns. (VI.32), fulfill this initial condition if they are found from the equations (VI.32) written for negative times. The transition to negative times amounts to changing the signs of the kinetic coefficients in eqns. (VI.32) and (VI.48). Through the condition (V.32), the variations of the mean values can be expressed in terms of the retarded Green functions. We see that in the hydrodynamic region the retarded Green functions can be calculated from the hydrodynamic equations written for negative time.

Had we assumed the "initial" condition $\delta\langle\hat{A}\rangle = 0$ at $t = \infty$, the variations of the mean values would be expressed in terms of the advanced Green functions. On the other hand, in order to satisfy the initial condition these must be calculated from the hydrodynamic equations for positive time.

Having changed the sign of the kinetic coefficients we obtain

$$D(E) \approx m\varrho^2 \left(\frac{\partial S}{\partial \Theta}\right)_\varrho [E^3 + iA_2E^2 - c^2k^2E + iA_0]. \quad \text{(VI.49a)}$$

$$D_\varrho = \varrho^3 \left(\frac{\partial S}{\partial \Theta}\right)_\varrho \left[E - i\frac{\varkappa}{\varrho c_v}k^2\right](k^2\delta U(k)$$

$$-E(\boldsymbol{k}\cdot\delta \boldsymbol{A}(k))) = \frac{\varrho}{E}D_Y \quad \text{(VI.49b)}$$

where

$$A_2 = -\frac{k^2}{m\varrho}\left[\frac{4}{3}\eta + \zeta + \frac{m\varkappa}{c_v}\right],$$

$$A_0 = \frac{k^4}{m\varrho}\frac{1}{m}\left(\frac{\partial P}{\partial \varrho}\right)_\Theta \frac{m\varkappa}{c_v}, \quad c_v = \Theta\left(\frac{\partial S}{\partial \Theta}\right)_\varrho. \quad \text{(VI.50)}$$

In the expression for $D(E)$ we write an approximate equality since, in the coefficient of E, we have omitted terms proportional to the products of kinetic coefficients.

A solution is then required for the equation $D(E) = 0$, where $D(E)$ is given by (VI.49a). In first approximation we take as the solution to the equation, where $D(E)$ is given by (VI.45),

$$E \equiv \omega = \pm\omega_0 = \pm ck, \quad E = 0 \quad \text{(VI.51)}$$

and then the correction δ is calculated by means of Newton's method. Since we are interested in excitations for small k, the imaginary part of δ is most important. The real part involves higher powers of k. Hence, we have

$$\text{Im}\delta_\pm = -\text{Im}\frac{D(\pm ck)}{\left(\frac{\partial D}{\partial E}\right)_{\pm ck}} = -\text{Im}\frac{i(A_2c^2k^2 + A_0)}{2(c^2k^2 \pm ickA_2)}$$

$$= \frac{c^2k^2A_2 + A_0}{2(c^2k^2 + A_2^2)} \approx \frac{c^2k^2A_2 + A_0}{2c^2k^2} \equiv \varepsilon, \quad \text{(VI.52)}$$

$$\mathrm{Im}\,\delta_0 = -\mathrm{Im}\frac{D(0)}{\left(\frac{\mathrm{d}D}{\mathrm{d}E}\right)_0} = \frac{\varkappa}{\varrho c v}k^2 \equiv \varepsilon_0.$$

Using (VI.50) and

$$\frac{1}{m}\left(\frac{\partial P}{\partial \varrho}\right)_\Theta = \frac{1}{m}\frac{c_v}{c_p}\left(\frac{\partial P}{\partial \varrho}\right)_S = \frac{c_v}{c_p}c^2,$$

where c_v and c_p are the specific heats (per particle) at constant volume and pressure, we find

$$\varepsilon = \frac{k^2}{2m\varrho}\left\{\frac{4}{3}\eta+\zeta+\left(1-\frac{c_v}{c_p}\right)\frac{m\varkappa}{c_v}\right\} > 0. \tag{VI.53}$$

We now have for ω

$$\omega \approx \pm\omega_0+i\varepsilon, \qquad \omega = i\varepsilon_0. \tag{VI.54}$$

Thus, we can write (VI.49a) in the form

$$\begin{aligned} D(\omega) &\approx m\varrho^2\left(\frac{\partial S}{\partial \Theta}\right)_\varrho\left(\omega-\frac{i\varkappa}{\varrho c_v}k^2\right)(\omega-\omega_0-i\varepsilon)(\omega+\omega_0+i\varepsilon) \\ &\approx m\varrho^2\left(\frac{\partial S}{\partial \Theta}\right)_\varrho\left(\omega-\frac{i\varkappa}{\varrho c_v}k^2\right)(\omega^2-\omega_0^2-2i\omega\varepsilon). \end{aligned} \tag{VI.55}$$

Hence, by using (VI.49b) we obtain

$$\begin{aligned} &\frac{\delta\varrho(k)}{\delta U(k)} = \frac{\varrho k^2}{(\omega^2-\omega_0^2-2i\omega\varepsilon)}, \qquad \frac{\delta j_{k\parallel}}{\delta A(k)_\parallel} = -\frac{\omega^2\varrho}{\omega^2-\omega_0^2-2i\omega\varepsilon}, \\ &\frac{\delta\varrho_k}{\delta A(k)_\parallel} = -\frac{\varrho\omega k}{m(\omega^2-\omega_0^2-2i\omega\varepsilon)} = -\frac{1}{m}\frac{\delta j_{k\parallel}}{\delta U(k)}. \end{aligned} \tag{VI.56}$$

Thus, for suitable derivatives we have the expressions (VI.42) and the expressions (VI.47,56), obtained independently.

For an ideal fluid we have, for example,

$$\langle\langle\hat{\varrho}_k;\,\hat{\varrho}_{-k}\rangle\rangle_E = \frac{\varrho k^2}{2\pi m(\omega^2-\omega_0^2)}, \quad \begin{matrix}\varepsilon > 0,\\ \varepsilon \to 0.\end{matrix} \tag{VI.57}$$

For a viscous fluid (Galasiewicz, 1966b; Krasnikov, 1966; Petru, 1968):

$$\langle\langle\hat{\varrho}_k;\,\hat{\varrho}_{-k}\rangle\rangle_E = \frac{\varrho k^2}{2\pi m(\omega^2-\omega_0^2-2i\omega\varepsilon)},$$

$$\langle\langle\hat{\varrho}_k;(\hat{j}^0_{-k})_{||}\rangle\rangle_E = \langle\langle(\hat{j}^0_k)_{||};\varrho_{-k}\rangle\rangle_E = \frac{\varrho\omega k}{2\pi(\omega^2-\omega_0^2-2i\omega\varepsilon)}, \quad \text{(VI.58)}$$

$$\frac{\delta j_{k||}}{\delta A(k)_{||}} = \frac{\delta j^0_{k||}}{\delta A(k)_{||}} - \varrho = -\frac{2\pi}{m}\langle\langle(\hat{j}^0_k)_{||};(\hat{j}^0_{-k})_{||}\rangle\rangle_E - \varrho$$

$$= -\frac{\omega^2\varrho}{\omega^2-\omega_0^2-2i\omega\varepsilon},$$

$$\langle\langle j^\alpha_{k\perp};j^\beta_{-k\perp}\rangle\rangle_E = \frac{i\eta k^2}{2\pi\left(\omega - i\dfrac{\eta}{m\varrho}k^2\right)}\left(\delta_{\alpha\beta}-\frac{k^\alpha k^\beta}{k^2}\right)$$

where ε is given by (VI.53).

In formula (VI.57) we have the case of resonances; ε indicates the path of integration. In formula (VI.58) the poles of the Green functions have an imaginary part. Damping or, in other words, absorption of sound, is therefore present.

In deriving the hydrodynamic equations we have emphasized the smallness of the deviation of the system from thermodynamic equilibrium. This deviation leads to local equilibrium described by the local quantities $\varrho(t, r)$, $\Theta(t, r)$, $v^\alpha(t, r)$. The smallness of the deviations from local equilibrium was characterized by the parameter μ, introduced especially for this purpose.

Local equilibrium is a consequence of frequent collisions between particles in the system. Hence the hydrodynamic region is collision-dominated. In such a region the conventional methods of computation of Green functions cannot be successful. Conventional methods are based on expansions in terms of certain parameters describing the number of collisions in the system.

The advantage of the method presented here (Bogoliubov, 1963; Kadanoff and Martin, 1963; Hohenberg and Martin, 1965) lies in the fact that the Green functions are found from the hydrodynamic equations which properly describe a system in a collision-dominated region.

CHAPTER VII

The Bose superfluid

§ 1 Hydrodynamic equations without viscous terms†

For a Bose superfluid $\langle\psi(t,r)\rangle = \Phi(t,r) \neq 0$. Therefore the following equation for $\Phi(t,r)$:

$$i\frac{\partial\Phi(t,r)}{\partial t} = -\lambda\Phi(t,r) - \frac{\nabla^2}{2m}\Phi(t,r) + \int V(r-r')\langle\psi^+(t,r')$$

$$\times\psi(t,r')\psi(t,r)\rangle d^3\boldsymbol{r}' + U(t,r)\Phi(t,r) + \eta(t,r) + \frac{i}{m}(\boldsymbol{A}(t,r)\cdot\nabla\psi(t,r))$$

$$+\frac{i}{2m}\psi(t,r)\nabla\boldsymbol{A}(t,r) + \frac{1}{2m}\psi(t,r)A^2(t,r) \quad \text{(VII.1)}$$

must be considered together with the equations for ϱ, $\boldsymbol{j}$, $\varrho\varepsilon$ or S, and a similar equation for $\Phi^*(t,r)$. In place of these two equations it is convenient to have two equivalent equations, one for the amplitude a and another for the phase χ of $\Phi = ae^{i\chi}$ (we now write $\boldsymbol{A} = 0$):

$$\frac{\partial\chi}{\partial t} = \lambda + \frac{\nabla^2 a}{2ma} - \frac{1}{2m}\left(\frac{\partial\chi}{\partial \boldsymbol{r}}\right)^2 - U(t,r) - \frac{\zeta^*+\zeta}{2a}$$

$$-\frac{1}{2a^2}\int V(R)\{X_t(r,R)+X_t^*(r,R)\}d^3\boldsymbol{R}, \quad \text{(VII.2)}$$

and

$$\frac{\partial\varrho_c}{\partial t} + \operatorname{div}(\varrho_c\boldsymbol{v}_s) = i\int V(R)[X_t^*(r,R)-X_t(r,R)]d^3\boldsymbol{R} + i\sqrt{\varrho_c}(\zeta^*-\zeta) \quad \text{(VII.3)}$$

† The considerations of this section are based on work by Bogoliubov (1963, Section 3).

where

$$X_t(r, r'-r) = \langle \psi^+(t, r')\psi(t, r')\psi(t, r)\rangle \Phi^*(t, r),$$
$$\zeta(t, r) = \eta(t, r)e^{-i\chi(t, r)}, \qquad \text{(VII.4)}$$
$$\boldsymbol{v}_s = \frac{1}{m}\nabla\chi, \quad \varrho_c = a^2.$$

Here a new velocity, the velocity of the condensate $\boldsymbol{v}_s$ is introduced (for $\boldsymbol{A} \neq 0$, $\boldsymbol{v}_s = \frac{1}{m}(\nabla\chi - \boldsymbol{A})$, ϱ_c is the density of the condensate. For $\boldsymbol{v}_s$ we have the equation

$$m\frac{\partial \boldsymbol{v}_s^\alpha}{\partial t} = \frac{\partial}{\partial r_\alpha}\left\{\frac{\nabla^2 a}{2ma} - \frac{mv_s^2}{2} - U - \frac{\zeta^* + \zeta}{2a} - \frac{1}{2a^2}\int V(R)\,[X_t(r, R) + X_t^*(r, R)]\,\mathrm{d}^3\boldsymbol{R}\right\}. \qquad \text{(VII.5)}$$

Now consider a system in thermodynamic equilibrium with $U = 0$, $\eta = 0$. (These terms destroy translational invariance.) An ordinary fluid in thermodynamic equilibrium is described by ϱ, Θ and one velocity $\boldsymbol{v}$. Now we have two velocities: the velocity of the condensate $\boldsymbol{v}_s$ and the velocity of the "normal component" $\boldsymbol{v}_n$. Thus far, we have defined only $\boldsymbol{v}_s$ (VII.4). The expectation values for equilibrium with two velocities are denoted by

$$\langle \ldots \rangle_{v_S,\, v_n}.$$

These expectation values can be expressed conveniently in terms of expectation values in a coordinate system where the normal component is at rest

$$\langle \ldots \rangle_{v'_s,\,0}, \quad \boldsymbol{v}'_s = \boldsymbol{v}_s - \boldsymbol{v}_n, \quad \boldsymbol{v}'_n = 0.$$

To do this we must perform a Galilean transformation for the field operators

$$\psi \rightarrow \psi \exp[im(\boldsymbol{v}_n \cdot \boldsymbol{r})].$$

We start by considering a state characterized by the parameters $\varrho, \Theta, \boldsymbol{v}_s, \boldsymbol{v}_n = 0$. In this state

$$a = \text{const.} = a(\varrho, \Theta, u), \quad F = F(\varrho, \Theta, u),$$
$$\Lambda(\varrho, \Theta, u) = \frac{\partial(FN)}{\partial N} = F + \varrho\frac{\partial F}{\partial \varrho}, \quad u = \frac{\boldsymbol{v}_s^2}{2}.$$

Λ is the chemical potential per particle. Hohenberg and Martin (1965) used $\mu = \Lambda/m$.

We define a new current

$$j_\alpha = \varrho \frac{\partial F}{\partial v_s^{(\alpha)}} = \varrho \frac{\partial F}{\partial u} v_s^{(\alpha)}; \tag{VII.6}$$

if we write

$$\varrho \frac{\partial F}{\partial u} = \varrho_s m, \tag{VII.7}$$

we have

$$j_\alpha = m\varrho_s v_s^{(\alpha)}.$$

Equation (VII.7) is the definition of a new density ϱ_s, the density of the superfluid component ($\varrho_s \leqslant \varrho$).

We wish to demonstrate that j_α given by (VII.6) is equivalent to the normal definition of the current. For this reason, it is convenient to switch to a new coordinate system moving with velocity $\boldsymbol{v}_s$. Thus we now have $\boldsymbol{v}_s' = 0$, $\boldsymbol{v}_n' = -\boldsymbol{v}_s$ and the Hamiltonian of our system has the form

$$\hat{H} = \frac{1}{2m} \int \sum_a \left(\frac{\partial \psi^+}{\partial r_\alpha} - im\boldsymbol{v}_s^{(\alpha)} \psi^+ \right) \left(\frac{\partial \psi}{\partial r_\alpha} + im\boldsymbol{v}_s^{(\alpha)} \psi \right) \mathrm{d}^3\boldsymbol{r}$$

$$+ \int V(r-r')\psi^+(r)\psi^+(r')\psi(r')\psi(r)\,\mathrm{d}^3\boldsymbol{r}\,\mathrm{d}^3\boldsymbol{r}'.$$

We calculate $\varrho \dfrac{\partial F}{\partial v_s^{(\alpha)}}$ proceeding as in the derivation of (VI.16),

$$\varrho \frac{\partial F}{\partial v_s^{(\alpha)}} = \frac{1}{V} \left\langle \frac{\partial H}{\partial v_s^{(\alpha)}} \right\rangle_{0,-v_s} = \frac{1}{2V} \int \left\langle \left(\frac{\partial \psi^+}{\partial r_\alpha} - imv_s^{(\alpha)}\psi^+ \right) \psi \right.$$

$$\left. - \psi^+ \left(\frac{\partial \psi}{\partial r_\alpha} + imv_s^{(\alpha)} \psi \right) \right\rangle_{0,-v_s} \mathrm{d}^3\boldsymbol{r} = + \frac{i}{2V} \int \left\langle \frac{\partial \psi^+}{\partial r_\alpha} \psi - \psi^+ \frac{\partial \psi}{\partial r_\alpha} \right\rangle_{v_s,0} \mathrm{d}\boldsymbol{r}^3$$

$$= \frac{i}{2} \left\langle \frac{\partial \psi^+}{\partial r_\alpha} \psi - \psi^+ \frac{\partial \psi}{\partial r_\alpha} \right\rangle_{v_s,0} = j_\alpha = m\varrho_s v_s^{(\alpha)}.$$

At equilibrium the expectation value $\langle \dots \rangle$ does not depend on $\boldsymbol{r}$. Hence we have obtained the usual expression for j_α.

In thermodynamic equilibrium $X = X(r-r'|\varrho, \Theta, v_s)$. In this case (all quantities are time independent and constant in space), we have from (VII.2)

$$\frac{1}{2a^2} \int V(R) \{X(R|\varrho, \Theta, v_s) + X^*(R|\varrho, \Theta, v_s)\}\, \mathrm{d}^3\boldsymbol{R}$$

$$= \Lambda(\varrho, \Theta, u) - \frac{mv_s^2}{2}.$$

Equation (VII.3) gives

$$\frac{1}{a^2}\int V(R)X(R|\varrho,\Theta,v_s)\,d^3\boldsymbol{R} = \frac{1}{a^2}\int V(R)X^*(R|\varrho,\Theta,v_s)\,d^3\boldsymbol{R}$$

$$= \Lambda(\varrho,\Theta,u) - \frac{mv_s^2}{2}. \qquad \text{(VII.8)}$$

Now we would like to calculate the stress tensor $\mathcal{T}_{\alpha\beta}$ for thermodynamic equilibrium $(v_s, 0)$:

$$\mathcal{T}_{\alpha\beta}(\varrho,\Theta,v_s) = -\frac{1}{2m}\left\langle \frac{\partial\psi^+}{\partial r_\alpha}\frac{\partial\psi}{\partial r_\beta} + \frac{\partial\psi^+}{\partial r_\beta}\frac{\partial\psi}{\partial r_\alpha}\right\rangle_{v_s,0}$$

$$+\frac{1}{2}\int \frac{\partial V(R)}{\partial R_\alpha} R_\beta D(R|\varrho,\Theta,v_s)\,d^3\boldsymbol{R}.$$

Since the only preferred direction is that of $\boldsymbol{v}_s$, $\mathcal{T}_{\alpha\beta}$ must be of the form

$$\mathcal{T}_{\alpha\beta} = A(\varrho,\Theta,u)\,v_s^{(\alpha)}v_s^{(\beta)} + \delta_{\alpha\beta}B(\varrho,\Theta,u), \qquad \text{(VII.9)}$$

where A and B are scalar functions. In Chapter VI, § 1 it was shown that the diagonal part of the tensor $\mathcal{T}_{\alpha\alpha}$ is equal to $\frac{1}{V}\left(\frac{\partial NF}{\partial L}\right)_{L=1}$. If we now calculate this quantity we find

$$\left(\varrho\frac{\partial F}{\partial L}\right)_{L=1} = -\varrho^2\frac{\partial F}{\partial\varrho} - \varrho\frac{\partial F}{\partial u}\,(v_s^{(\alpha)})^2 = -P(\varrho,\Theta,u) - m\varrho_s(v_s^{(\alpha)}),$$

$$mv_s'^{(\alpha)} = \frac{1}{L}\frac{\partial\chi}{\partial r_\alpha},$$

$$\left[v_s'^{(\alpha)}\frac{\partial v_s'^{(\alpha)}}{\partial L}\right]_{L=1} = \left(-\frac{1}{L}\right)_{L=1}(v_s^{(\alpha)})^2 = -(v_s^{(\alpha)})^2.$$

From (VII.9) it follows that

$$\left(\varrho\frac{\partial F}{\partial L}\right)_{L=1} = \mathcal{T}_{\alpha\alpha} = A(v_s^{(\alpha)})^2 + B.$$

So we have

$$A = -m\varrho_s, \qquad B = -P(\varrho,\Theta,u),$$

and for $\mathcal{T}_{\alpha\beta}$

$$\mathcal{T}_{\alpha\beta} = -m\varrho_s\,v_s^{(\alpha)}v_s^{(\beta)} - \delta_{\alpha\beta}P(\varrho,\Theta,u).$$

Now we will consider a system with two velocities different from zero. According to the definitions we have

$$X(r-r'|\varrho, \Theta, v_s, v_n) = \langle\psi^+(r')\psi(r')\psi(r)\rangle_{v_s, v_n}\langle\psi^+(r)\rangle_{v_s, v_n},$$

$$D(r-r'|\varrho, \Theta, v_s, v_n) = \langle\psi^+(r)\psi^+(r')\psi(r')\psi(r)\rangle_{v_s, v_n}.$$

These quantities are invariant under the transformation

$$\psi \to \psi\exp[im(\boldsymbol{v}_n\cdot\boldsymbol{r})],$$

therefore

$$X(r-r'|\varrho, \Theta, v_s, v_n) = X(r-r'|\varrho, \Theta, v_s-v_n),$$

$$D(r-r'|\varrho, \Theta, v_s, v_n) = D(r-r'|\varrho, \Theta, v_s-v_n).$$

For the current we have

$$j_\alpha = \frac{i}{2}\left\langle\frac{\partial\psi^+}{\partial r_\alpha}\psi-\psi^+\frac{\partial\psi}{\partial r_\alpha}\right\rangle_{v_s, v_n} = \frac{i}{2}\left\langle\left(\frac{\partial\psi^+}{\partial r_\alpha}-imv_n^{(\alpha)}\psi^+\right)\psi\right.$$

$$\left.-\psi^+\left(\frac{\partial\psi}{\partial r_\alpha}+imv_n^{(\alpha)}\psi\right)\right\rangle_{v_s-v_n, 0} = mv_n^{(\alpha)}\varrho+m\varrho_s(v_s^{(\alpha)}-v_n^{(\alpha)})$$

$$= m\varrho_s v_s^{(\alpha)}+m\varrho_n v_n^{(\alpha)}. \qquad \text{(VII.10)}$$

We write

$$\varrho_n = \varrho-\varrho_s. \qquad \text{(VII.11)}$$

The local quantities j_α and ϱ are defined by (V.2), ϱ_s is defined by (VII.7), ϱ_n by (VII.11); hence (VII.10) becomes the definition of a new local quantity $\boldsymbol{v}_n(t, r)$, the velocity of the normal component.

We have introduced two densities connected with the velocity $\boldsymbol{v}_s$. The density of the condensate ϱ_c defined by (VII.4) and the density of the superfluid component defined by (VII.7). We are not allowed to identify ϱ_c with ϱ_s. The definition of ϱ_c is given independently of the velocity $\boldsymbol{v}_s$. On the other hand, the definition of ϱ_s is closely connected with $\boldsymbol{v}_s$. According to the estimates of Penrose and Onsager (1956; Penrose, 1958) for $\Theta = 0$, where $\varrho = \varrho_s$, the density ϱ_c is 8 per cent of ϱ.

We calculate the components of the stress tensor

$$\mathscr{T}_{\alpha\beta}(\varrho, \Theta, v_s, v_n) = \frac{1}{2}\int\frac{\partial V(R)}{\partial R_\alpha}R_\beta D(R|\varrho, \Theta, v_s, v_n)\,\mathrm{d}^3\boldsymbol{R}$$

$$-\frac{1}{2m}\left\langle\left(\frac{\partial\psi^+}{\partial r_\alpha}-imv_n^{(\alpha)}\psi^+\right)\left(\frac{\partial\psi}{\partial r_\beta}+imv_n^{(\beta)}\psi\right)+\left(\frac{\partial\psi^+}{\partial r_\beta}-imv_n^{(\beta)}\psi^+\right)\right.$$

$$\left.\left(\frac{\partial\psi}{\partial r_\alpha}+imv_n^{(\alpha)}\psi\right)\right\rangle_{v_s-v_n, 0} = -\frac{1}{2m}\left\langle\frac{\partial\psi^+}{\partial r_\alpha}\frac{\partial\psi}{\partial r_\beta}+\frac{\partial\psi^+}{\partial r_\beta}\frac{\partial\psi}{\partial r_\alpha}\right\rangle_{v_s-v_n, 0}$$

$$+\frac{1}{2}\int\frac{\partial V(R)}{\partial R_\alpha}R_\beta D(R|\varrho,\Theta,v_s-v_n)\,d^3\boldsymbol{R}-\frac{v_n^{(\alpha)}i}{2}\Big\langle\frac{\partial\psi^+}{\partial r_\beta}\psi$$

$$-\psi^+\frac{\partial\psi}{\partial r_\beta}\Big\rangle_{v_s-v_n,0}-\frac{v_n^{(\beta)}i}{2}\Big\langle\frac{\partial\psi^+}{\partial r_\alpha}\psi-\psi^+\frac{\partial\psi}{\partial r_\alpha}\Big\rangle_{v_s-v_n,0}-m\varrho v_n^{(\alpha)}v_n^{(\beta)}$$

$$=-\delta_{\alpha\beta}P(\varrho,\Theta,u)-m\varrho_s v_s^{(\alpha)}v_s^{(\beta)}-m\varrho_n v_n^{(\alpha)}v_n^{(\beta)} \qquad \text{(VII.12)}$$

where $u=(\boldsymbol{v}_s-\boldsymbol{v}_n)^2/2$.

For the mean energy we have

$$\varrho\varepsilon(\varrho,\Theta,v_s,v_n)=\tfrac{1}{2}\int V(R)D(R|\varrho,\Theta,v_s-v_n,0)\,d^3\boldsymbol{R}$$

$$-\frac{1}{4m}\Big\langle\Big[\Big(\frac{\partial}{\partial r}-im\boldsymbol{v}_n\Big)^2\psi^+\Big]\psi+\psi^+\Big[\Big(\frac{\partial}{\partial r}+im\boldsymbol{v}_n\Big)^2\psi\Big]\Big\rangle_{v_s-v_n,0}$$

$$=\varrho E(\varrho,\Theta,u)+\frac{m_\varrho v_n^2}{2}+m\varrho_s(\boldsymbol{v}_s-\boldsymbol{v}_n)\boldsymbol{v}_n \qquad \text{(VII.13)}$$

where $E(\varrho,\Theta,u)=F-\Theta\dfrac{\partial F}{\partial\Theta}$.

Now we wish to obtain the energy current $\mathscr{I}_\alpha$ for the case of thermodynamic equilibrium. Note that

$$G^{(\alpha)}(r'-r|\varrho,\Theta,v_s,v_n)=\frac{i}{4m}\Big\langle\psi^+(t,r')\Big\{\{\psi^+(t,r)\frac{\partial\psi(t,r)}{\partial r_\alpha}-\frac{\partial\psi^+(t,r)}{\partial r_\alpha}\psi(t,r)\Big\}\psi(t,r')\Big\rangle_{v_s,v_n}$$

$$=-\frac{1}{2m}\langle\psi^+(t,r')\hat{j}_\alpha(t,r)\psi(t,r')\rangle_{v_s-v_n,0}$$

$$=\frac{i}{4m}\Big\langle\psi^+(t,r')\Big\{\frac{\partial\psi(t,r)}{\partial r_\alpha}+imv_n^{(\alpha)}\psi(t,r)-\Big(\frac{\partial\psi^+(t,r)}{\partial r_\alpha}-imv_n^{(\alpha)}\psi^+(t,r)\Big)\psi(t,r)\Big\}\psi(t,r')\Big\rangle_{v_s-v_n,0}$$

$$=G^{(\alpha)}(r'-r|\varrho,\Theta,v_s-v_n)-\frac{v_n^{(\alpha)}}{2}D(r'-r|\varrho,\Theta,v_s-v_n).$$

For the vector $\mathscr{I}_\alpha$ we have

$$\mathscr{I}_\alpha(\varrho,\Theta,v_s,v_n)=-\frac{i}{4m^2}\Big\langle(\nabla^2\psi^+)\frac{\partial\psi}{\partial r_\alpha}-\frac{\partial\psi^+}{\partial r_\alpha}(\nabla^2\psi)\Big\rangle_{v_s,v_n}$$

$$+\int V(R)G^{(\alpha)}(R|\varrho,\Theta,v_s,v_n)\,d^3\boldsymbol{R}$$

$$-\sum_{\beta}\int\frac{\partial V(R)}{\partial R_\beta}R_\alpha G^{(\alpha)}(R|\varrho,\Theta,v_s,v_n)\,\mathrm{d}^3\boldsymbol{R}$$

$$=-\frac{i}{4m^2}\left\langle\left[\left(\frac{\partial}{\partial \boldsymbol{r}}-im\boldsymbol{v}_n\right)^2\psi^+\right]\left(\frac{\partial\psi}{\partial r_\alpha}+imv_n^{(\alpha)}\psi\right)\right.$$

$$\left.-\left(\frac{\partial\psi^+}{\partial r_\alpha}-imv_n^{(\alpha)}\psi^+\right)\left[\left(\frac{\partial}{\partial \boldsymbol{r}}+im\boldsymbol{v}_n\right)^2\psi\right]\right\rangle_{\boldsymbol{v_s}-\boldsymbol{v_n},\,0}$$

$$+\int V(R)G^{(\alpha)}(R|\varrho,\Theta,v_s-v_n)\,\mathrm{d}^3\boldsymbol{R}-\frac{v_n^{(\alpha)}}{2}\int V(R)D(R|\varrho,\Theta,v_s-v_n)\,\mathrm{d}^3\boldsymbol{R}$$

$$-\sum_{\beta}\frac{\partial V(R)}{\partial R_\beta}R_\alpha G^{(\beta)}(R|\varrho,\Theta,v_s-v_n)\,\mathrm{d}^3\boldsymbol{R}$$

$$+\frac{1}{2}\sum_{\beta}\int\frac{\partial V(R)}{\partial R_\beta}R_\alpha v_n^{(\beta)}D(R|\varrho,\Theta,v_s-v_n)\,\mathrm{d}^3\boldsymbol{R}$$

$$=-\frac{i}{4m^2}\left\langle(\nabla^2\psi^+)\frac{\partial\psi}{\partial r_\alpha}-\frac{\partial\psi^+}{\partial r_\alpha}(\nabla^2\psi)\right\rangle_{\boldsymbol{v_s}-\boldsymbol{v_n},\,0}$$

$$+\int V(R)G^{(\alpha)}(R|\varrho,\Theta,v_s-v_n)\,\mathrm{d}^3\boldsymbol{R}$$

$$-\sum_{\beta}\int\frac{\partial V(R)}{\partial R_\beta}R_\alpha G^{(\beta)}(R|\varrho,\Theta,v_s-v_n)\,\mathrm{d}^3\boldsymbol{R}$$

$$-v_n^{(\alpha)}\left\{-\frac{1}{4m}\left\langle\left[\left(\frac{\partial}{\partial \boldsymbol{r}}-im\boldsymbol{v}_n\right)^2\psi^+\right]\psi+\psi^+\left[\left(\frac{\partial}{\partial \boldsymbol{r}}+im\boldsymbol{v}_n\right)^2\psi\right]\right\rangle_{\boldsymbol{v_s}-\boldsymbol{v_n},\,0}\right.$$

$$\left.+\frac{1}{2}\int V(R)D(R|\varrho,\Theta,v_s-v_n)\,\mathrm{d}^3\boldsymbol{R}\right\}$$

$$+\sum_{\beta}v_n^{(\beta)}\left\{-\frac{1}{2m}\left\langle\frac{\partial\psi^+}{\partial r_\beta}\frac{\partial\psi^+}{\partial r_\alpha}+\frac{\partial\psi^+}{\partial r_\alpha}\frac{\partial\psi}{\partial r_\beta}\right\rangle_{\boldsymbol{v_s}-\boldsymbol{v_n},\,0}\right.$$

$$\left.+\frac{1}{2}\int\frac{\partial V(R)}{\partial R_\beta}R_\alpha D(R|\varrho,\Theta,v_s-v_n)\,\mathrm{d}^3\boldsymbol{R}\right\}$$

$$-\frac{v_n^2}{2}\frac{i}{2}\left\langle\frac{\partial\psi^+}{\partial r_\alpha}\psi-\psi^+\frac{\partial\psi}{\partial r_\alpha}\right\rangle_{\boldsymbol{v_s}-\boldsymbol{v_n},\,0}.$$

From the definitions of $\mathscr{I}_\alpha$ and j_α and the formulae (VII.12, 13) it follows that

$$\mathscr{I}_\alpha(\varrho, \Theta, v_s, v_n) = \mathscr{I}_\alpha(\varrho, \Theta, v_s - v_n, 0) - v_n^{(\alpha)}\varrho E(\varrho, \Theta, u)$$
$$- v_n^{(\alpha)}\frac{mv_n^2}{2}\varrho - v_n^{(\alpha)}\sum_\beta v_n^{(\beta)} m\varrho_s(v_s^{(\beta)} - v_n^{(\beta)})$$
$$+ \sum_\beta v_n^{(\beta)}\mathscr{T}_{\alpha\beta}(\varrho, \Theta, v_s - v_n, 0) - \frac{mv_n^2}{2}\varrho_s(v_s^{(\alpha)} - v_n^{(\alpha)})$$
$$= \mathscr{I}_\alpha(\varrho, \Theta, v_s - v_n, 0)$$
$$- v_n^{(\alpha)}\left[\varrho E + \frac{mv_n^2}{2}\varrho + P + m\varrho_s(\boldsymbol{v}_n \cdot (\boldsymbol{v}_s - \boldsymbol{v}_n))\right]$$
$$- m\varrho_s(v_s^{(\alpha)} - v_n^{(\alpha)})\left[(\boldsymbol{v}_s \cdot \boldsymbol{v}_n) - \frac{v_n^2}{2}\right]. \qquad \text{(VII.14)}$$

We write

$$\mathscr{I}_\alpha(\varrho, \Theta, v_s - v_n, 0) = [-\Lambda(\varrho, \Theta, u)\varrho_s + A(\varrho, \Theta, u)](v_s^{(\alpha)} - v_n^{(\alpha)}). \qquad \text{(VII.14')}$$

It will be shown later that $A(\varrho, \Theta, u) = A^0(\varrho, \Theta, u) \equiv 0$.

After obtaining expressions for some of the quantities in thermodynamic equilibrium we would like to derive the hydrodynamic equations for a superfluid. The basic idea of this procedure is similar to the one used to obtain equations for an ordinary fluid.

We again introduce a parameter μ characterizing the "slowness" of the process and the deviations from isotropy

$$\boldsymbol{\xi} = \mu \boldsymbol{r}, \qquad \tau = \mu t.$$

We put

$$\varrho(t, r) = \tilde{\varrho}(\tau, \xi), \quad \boldsymbol{j}(t, r) = \tilde{\boldsymbol{j}}(\tau, \xi), \quad \varepsilon(t, r) = \tilde{\varepsilon}(\tau, \xi),$$

$$\boldsymbol{v}_s(t, r) \approx \frac{\frac{\partial}{\partial \boldsymbol{r}}\langle\psi(t, r)\rangle}{m\langle\psi(t, r)\rangle} = \tilde{\boldsymbol{v}}_s(\tau, \xi),$$

$$a(t, r) = |\psi(t, r)| = \tilde{a}(\tau, \xi) \neq 0.$$

Since the gradients $\frac{\partial}{\partial t}, \frac{\partial}{\partial \boldsymbol{r}}$ of ϱ, $\boldsymbol{v}_s$ and $\boldsymbol{v}_n$ are of "first order of smallness" in μ, we conclude, from the continuity equation (V.4), that ζ, ζ^* are also of order μ. Hence

$$\zeta = \mu\zeta.$$

We use the identities for $\varrho(\tau, \xi), \boldsymbol{j}(\tau, \xi), \varepsilon(\tau, \xi), \boldsymbol{v}_s(\tau, \xi)$ to introduce the local variables $\Theta(\tau, \xi)$, $\boldsymbol{v}_n(\tau, \xi)$ which together with $\varrho(\tau, \xi)$, $\boldsymbol{v}_s(\tau, \xi)$ describe quasi-local equilibrium.

We take the functions

$$\varepsilon(\varrho, \Theta, v_s, v_n); \qquad \varrho_s(\varrho, \Theta, u) = \frac{1}{m}\varrho\frac{\partial F(\varrho, \Theta, u)}{\partial u};$$

$$\varrho_n = \varrho - \varrho_s(\varrho, \Theta, u), \qquad u = \tfrac{1}{2}(v_n - v_s)^2$$

which express ε, ϱ_s, ϱ_n for thermodynamic equilibrium and we define the functions $\Theta(\tau, \xi)$, $\boldsymbol{v}_n(\tau, \xi)$ by

$$\boldsymbol{\varepsilon}(\tau, \xi) = \varepsilon(\varrho, \Theta, v_s, v_n),$$

$$\boldsymbol{j}(\tau, \xi) = m\varrho_s(\varrho, \Theta, u)\boldsymbol{v}_s + m\varrho_n(\varrho, \Theta, u)\boldsymbol{v}_n.$$

After introducing $\Theta(\tau, \xi)$, $\boldsymbol{v}_n(\tau, \xi)$ we can also introduce the local quantities $\varrho_n(\tau, \xi)$, $\varrho_s(\tau, \xi)$ which are, by definition,

$$\varrho_s(\tau, \xi) = \varrho_s(\varrho, \Theta, u), \qquad \varrho_n(\tau, \xi) = \varrho_n(\varrho, \Theta, u).$$

Our assumption that the local state of non-equilibrium differs very little from the local quasi-equilibrium now has, in the "μ" formulation, the form

$$\begin{aligned} a(\tau, \xi) &= a(\varrho, \Theta, u) + \mu a^{(1)}(\tau, \xi) + \ldots, \\ X_r(\xi, R) &= X(R|\varrho, \Theta, v_s, v_n) + \mu X^{(1)}(\tau, \xi) + \ldots \\ &= X(R|\varrho, \Theta, v_s - v_n) + \mu X^{(1)}(\tau, \xi) + \ldots, \\ \mathscr{T}_{\alpha\beta}(\tau, \xi) &= \mathscr{T}_{\alpha\beta}(\varrho, \Theta, v_s, v_n) + \mu\mathscr{T}^{(1)}_{\alpha\beta}(\tau, \xi) + \ldots, \\ \mathscr{I}_\alpha(\tau, \xi) &= \mathscr{I}_\alpha(\varrho, \Theta, v_s, v_n) + \mu\mathscr{I}^{(1)}_\alpha(\tau, \xi) + \ldots \end{aligned} \tag{VII.15}$$

Equation (VII.5) and the other equations for local quantities can be written in the form

$$\begin{aligned} m\frac{\partial v_s^{(\alpha)}}{\partial\tau} &= \frac{\partial}{\partial\xi_\alpha}\left\{\mu^2\frac{\nabla_\xi^2 a}{2ma} - \frac{mv_s^2}{2} - U - \mu\frac{\zeta^* + \zeta}{2a}\right. \\ &\qquad\left. - \frac{1}{2a^2}\int V(R)\,[X_\tau(\xi, R) + X_\tau^*(\xi, R)]\,\mathrm{d}^3\boldsymbol{R}\right\}, \\ m\frac{\partial\varrho(\tau, \xi)}{\partial\tau} &+ \sum_\beta\frac{\partial j_\beta(\tau, \xi)}{\partial\xi_\beta} = iam(\zeta^* - \zeta), \\ \frac{\partial j_\alpha}{\partial\tau} &= \sum_\beta\frac{\partial\mathscr{T}_{\alpha\beta}}{\partial\xi_\beta} - \varrho\frac{\partial U}{\partial\xi_\alpha} + iam(\zeta^* - \zeta)v_s^{(\alpha)} \\ &\qquad + \frac{\mu}{2}(\zeta^* + \zeta)\frac{\partial a}{\partial\xi_\alpha} - \mu\frac{a}{2}\frac{\partial}{\partial\xi_\alpha}(\zeta^* + \zeta), \end{aligned} \tag{VII.16}$$

$$\frac{\partial(\varrho\varepsilon)}{\partial\tau}=\sum_{\beta}\frac{\partial\mathscr{I}_{\beta}}{\partial\xi_{\beta}}-\frac{1}{m}(\boldsymbol{j}\cdot\nabla_{\xi}U)+\frac{i}{a}\Big(\zeta^{*}\int V(R)X(\xi,R)\,\mathrm{d}^{3}\boldsymbol{R}$$
$$-\zeta\int V(R)X^{*}(\xi,R)\,\mathrm{d}^{3}\boldsymbol{R}\Big)+ia(\zeta^{*}-\zeta)\frac{mv_{s}^{2}}{2}$$
$$+\mu\frac{1}{2}(\zeta^{*}+\zeta)(\boldsymbol{v}_{s}\cdot\nabla_{\xi}a)-\mu\frac{a}{2}(\boldsymbol{v}_{s}\cdot\nabla_{\xi}(\zeta+\zeta^{*})).$$

Initially, we are concerned with an ideal fluid, therefore we omit terms proportional to μ (containing ζ, ζ^{*} and gradients of ϱ, Θ, $v_s^{(\alpha)}$, $v_n^{(\alpha)}$). In this approximation

$$m\frac{\partial v_{s}^{(\alpha)}}{\partial\tau}=-\frac{\partial}{\partial\xi_{\alpha}}\Big\{\frac{mv_{s}^{2}}{2}+U$$
$$+\frac{1}{2a^{2}(\varrho,\Theta,u)}\int V(R)\,[X(R|\varrho\,\Theta,v_{s}-v_{n})$$
$$+X^{*}(R|\varrho,\Theta,v_{s}-v_{n})]\,\mathrm{d}^{3}\boldsymbol{R}\Big\}.$$

Using (VII.8) we can write

$$m\frac{\partial v_{s}^{(\alpha)}}{\partial t}=-\frac{\partial}{\partial\xi_{\alpha}}\Big\{\frac{mv_{s}^{2}}{2}-\frac{m}{2}(\boldsymbol{v}_{s}-\boldsymbol{v}_{n})^{2}+\Lambda(\varrho,\Theta,u)+U\Big\},$$
$$u=\frac{(\boldsymbol{v}_{s}-\boldsymbol{v}_{n})^{2}}{2}. \qquad \text{(VII.17)}$$

Thus we find, from (VII.16), (VI.11),

$$\frac{\partial\varrho}{\partial\tau}+\sum_{\alpha}\frac{\partial(\varrho_{s}v_{s}^{(\alpha)}+\varrho_{n}v_{n}^{(\alpha)})}{\partial\xi_{\alpha}}=ia(\zeta^{*}-\zeta), \qquad \text{(VII.18)}$$

$$m\frac{\partial}{\partial\tau}(\varrho_{s}v_{s}^{(\alpha)}+\varrho_{n}v_{n}^{(\alpha)})=\sum_{\beta}\frac{\partial\mathscr{T}_{\alpha\beta}(\varrho,\Theta,v_{s},v_{n})}{\partial\xi_{\beta}}$$
$$+imv_{s}^{(\alpha)}(\zeta^{*}-\zeta)a-\varrho\frac{\partial U}{\partial\xi_{\alpha}}, \qquad \text{(VII.19)}$$

$$\frac{\partial(\varrho\varepsilon)}{\partial\tau}=\sum_{\beta}\frac{\partial\mathscr{I}_{\beta}(\varrho,\Theta,v_{s},v_{n})}{\partial\xi_{\beta}}-\frac{1}{m}(\boldsymbol{j}\cdot\nabla_{\xi}U)$$
$$+ia(\zeta^{*}-\zeta)\Big[\frac{mv_{s}^{2}}{2}+\Lambda(\varrho,\Theta,u)-\frac{(\boldsymbol{v}_{s}-\boldsymbol{v}_{n})^{2}}{2}m\Big].$$
$$\text{(VII.20)}$$

After substituting in (VII.19,20) the expressions for

$$\mathscr{T}_{\alpha\beta}(\varrho, \Theta, v_s, v_n), \quad \varepsilon(\varrho, \Theta, v_s, v_n) \text{ and } \mathscr{I}_{\beta}(\varrho, \Theta, v_s, v_n)$$

(formulae (VII.12–15)) we have

$$m\left\{\frac{\partial(\varrho_s v_s^{(\alpha)}+\varrho_n v_s^{(\alpha)})}{\partial\tau}+\sum_{\beta}\frac{\partial(\varrho_s v_s^{(\alpha)}v_s^{(\beta)}+\varrho_n v_n^{(\alpha)}v_n^{(\beta)})}{\partial\xi_\alpha}\right\}$$

$$= -\frac{\partial P}{\partial\xi_\alpha}-\varrho\frac{\partial U}{\partial\xi_\alpha}+imav_s^{(\alpha)}(\zeta^*-\zeta), \qquad \text{(VII.21)}$$

$$\frac{\partial}{\partial\tau}\left[\varrho E(\varrho, \Theta, u)+\frac{m\varrho v_n^2}{2}+m\varrho_s\big((\boldsymbol{v}_s-\boldsymbol{v}_n)\cdot\boldsymbol{v}_n\big)\right]$$

$$+\sum_{\beta}\frac{\partial}{\partial\xi_\beta}\left\{v_n^{(\beta)}\left[\frac{\varrho m v_n^2}{2}+\varrho_s m\big(\boldsymbol{v}_n\cdot(\boldsymbol{v}_s-\boldsymbol{v}_n)\big)+\varrho E+P\right]\right.$$

$$\left.+(v_s^{\prime(\beta)}-v_n^{(\beta)})\varrho_s\left[\Lambda+m\left((\boldsymbol{v}_s\cdot\boldsymbol{v}_n)-\frac{v_n^2}{2}\right)\right]\right\}$$

$$+\sum_{\beta}\frac{\partial U}{\partial\xi_\beta}(\varrho_s v_s^{(\beta)}+\varrho_n v_n^{(\beta)})-ia(\zeta^*-\zeta)\left[\Lambda+m\left((\boldsymbol{v}_s\cdot\boldsymbol{v}_n)-\frac{v_n^2}{2}\right)\right]$$

$$= -\sum_{\beta}\frac{\partial}{\partial\xi_\beta}(v_s^{(\beta)}-v_n^{(\beta)})A. \qquad \text{(VII.22)}$$

In eqns. (VII.17–20) we change from the auxiliary variables τ, $\boldsymbol{\xi}$ to the original variables t, $\boldsymbol{r}$. The system of hydrodynamic equations is

$$m\frac{\partial v_s^{(\alpha)}}{\partial t}+\frac{\partial}{\partial r_\alpha}\left\{m\left(\boldsymbol{v}_s\cdot\boldsymbol{v}_n-\frac{v_n^2}{2}\right)+\Lambda+U\right\}=0, \qquad \text{(VII.23)}$$

$$\frac{\partial\varrho}{\partial t}+\sum_{\beta}\frac{\partial(\varrho_s v_s^{(\beta)}+\varrho_n v_n^{(\beta)})}{\partial r_\beta}-ia(\zeta^*-\zeta)=0, \qquad \text{(VII.24)}$$

$$m\left\{\frac{\partial(\varrho_s v_s^{(\alpha)}+\varrho_n v_n^{(\alpha)})}{\partial t}+\sum_{\beta}\frac{\partial}{\partial r_\beta}(\varrho_s v_s^{(\alpha)}v_s^{(\beta)}+\varrho_n v_n^{(\alpha)}v_n^{(\beta)})\right\}$$

$$+\frac{\partial P}{\partial r_\alpha}+\varrho\frac{\partial U}{\partial r_\alpha}-ima(\zeta^*-\zeta)v_s^{(\alpha)}=0, \qquad \text{(VII.25)}$$

$$\frac{\partial}{\partial t}\left\{\varrho E+\frac{m\varrho v_n^2}{2}+m\varrho_s((\boldsymbol{v}_s-\boldsymbol{v}_n)\cdot\boldsymbol{v}_n)\right\}$$

$$+\sum_\beta \frac{\partial}{\partial r_\beta}\left\{v_n^{(\beta)}\left[\frac{m\varrho v_n^2}{2}+m\varrho_s((\boldsymbol{v}_s-\boldsymbol{v}_n)\cdot\boldsymbol{v}_n)+\varrho E+P\right]\right.$$

$$\left.+(v_s^{(\beta)}-v_n^{(\beta)})\varrho_s\left[\Lambda+m\left((\boldsymbol{v}_s\cdot\boldsymbol{v}_n)-\frac{v_n^2}{2}\right)\right]\right\}+\sum_\beta(\varrho_s v_s^{(\beta)}$$

$$+\varrho_n v_n^{(\beta)})\frac{\partial U}{\partial r_\beta}-ia(\zeta^*-\zeta)\left[\Lambda+m\left(\boldsymbol{v}_s\cdot\boldsymbol{v}_n-\frac{v_n^2}{2}\right)\right]$$

$$=-\sum\frac{\partial}{\partial r_\beta}(v_s^{(\beta)}-v_n^{(\beta)})A. \tag{VII.26}$$

Here we have

$$a=a(\varrho,\Theta,u),\quad A=A(\varrho,\Theta,u),\quad u=\frac{(\boldsymbol{v}_s-\boldsymbol{v}_n)^2}{2},$$

$$\varrho_s=\frac{1}{m}\varrho\frac{\partial F(\varrho,\Theta,u)}{\partial u},\quad \varrho_n=\varrho-\varrho_s,$$

$$E=F(\varrho,\Theta,u)-\Theta\frac{\partial F(\varrho,\Theta,u)}{\partial\Theta},\quad P=\varrho^2\frac{\partial F(\varrho,\Theta,u)}{\partial\varrho},$$

$$\Lambda=F(\varrho,\Theta,u)+\varrho\frac{\partial F(\varrho,\Theta,u)}{\partial\varrho},\quad \zeta(t,r)=\eta(t,r)e^{-i\chi(t,r)}.$$

Now, in order to obtain the equation for the entropy (per particle) we introduce $S=-\frac{\partial F}{\partial\Theta}$ in (VII.26). The first term in (VII.26) is

$$\frac{\partial}{\partial t}\left[\varrho E+\frac{m\varrho v_n^2}{2}+m\varrho_s((\boldsymbol{v}_n\cdot\boldsymbol{v}_s)-v_n^2)\right]=-\Theta\frac{\partial}{\partial t}\left(\varrho\frac{\partial F}{\partial\Theta}\right)$$

$$-\left(\Lambda+\frac{mv_n^2}{2}\right)\left[\frac{\partial}{\partial r_\beta}(\varrho_s v_s^{(\beta)}+\varrho_n v_n^{(\beta)})-ia(\zeta^*-\zeta)\right]$$

$$+m\varrho_s(v_s^{(\alpha)}-v_n^{(\alpha)})\frac{\partial v_s^{(\alpha)}}{\partial t}-m\varrho_n v_n^{(\alpha)}v_n^{(\beta)}\frac{\partial v_n^{(\alpha)}}{\partial r_\beta}$$

$$-m\varrho_s v_s^{(\beta)}v_n^{(\alpha)}\frac{\partial v_s^{(\alpha)}}{\partial r_\beta}-mv_n^{(\alpha)}(v_s^{(\alpha)}-v_n^{(\alpha)})\frac{\partial\varrho_s v_s^{(\beta)}}{\partial r_\beta}$$

$$+ima(\zeta^*-\zeta)(v_n^{(\alpha)}v_s^{(\alpha)}-v_n^2)-v_n^{(\alpha)}\frac{\partial P}{\partial r_\alpha}-(\varrho_s+\varrho_n)v_n^{(\alpha)}\frac{\partial U}{\partial r_\alpha}.$$

We also have

$$\frac{\partial}{\partial r_\beta}(v_n^{(\beta)}\varrho E) = \Lambda\frac{\partial \varrho v_n^{(\beta)}}{\partial r_\beta} - P\frac{\partial v_n^{(\beta)}}{\partial r_\alpha} - \Theta\frac{\partial}{\partial r_\beta}\left(\varrho v_n^{(\beta)}\frac{\partial F}{\partial \Theta}\right)$$
$$+m\varrho_s v_n^{(\beta)}(v_s^{(\alpha)}-v_n^{(\alpha)})\frac{\partial(v_s^{(\alpha)}-v_n^{(\alpha)})}{\partial r_\beta}.$$

The left-hand side of (VII.26) can be written as

$$\Theta\left[\frac{\partial(\varrho, S)}{\partial t} + \frac{\partial}{\partial r_\beta}(\varrho v_n^{(\beta)} S)\right].$$

So we have for the entropy

$$\Theta\left[\frac{\partial(\varrho, S)}{\partial t} + \sum_\beta \frac{\partial}{\partial r_\beta}(\varrho v_n^{(\beta)} S)\right] = \sum_\beta \frac{\partial}{\partial r_\beta}[(v_s^{(\beta)}-v_n^{(\beta)})A(\varrho, \Theta, u)]. \tag{VII.27}$$

We wish to demonstrate now that $A(\varrho, \Theta, u) \equiv 0$.

Consider the case of thermodynamic equilibrium where

$$U = 0, \quad \zeta = \zeta(r), \quad \Theta = \text{const.}, \quad \varrho = \varrho(r), \quad \boldsymbol{v}_n = 0,$$
$$\boldsymbol{v}_s = \boldsymbol{v}_s(r).$$

From (VII.27) we have

$$\sum_\beta \frac{\partial}{\partial r_\beta} v_r^{(\beta)} A\left(\varrho, \Theta, \frac{v_s^2}{2}\right) = 0 \tag{VII.28}$$

and from the continuity equation (VII.23) it follows that

$$\sum_\beta \frac{\partial}{\partial r_\beta}(\varrho_s v_s^{(\beta)}) = ia(\zeta^* - \zeta). \tag{VII.29}$$

Equation (VII.23) gives

$$\frac{\partial}{\partial r_\alpha}\Lambda\left(\varrho, \Theta, \frac{v_s^2}{2}\right) = 0. \tag{VII.30}$$

Hence we have three equations (VII.28–30) for two functions $\varrho(r)$, $v_s(r)$.

Consider the case where η, $\boldsymbol{v}_s$ and $\delta\varrho(r)$ are infinitesimal ($\varrho = \varrho_0 + \delta\varrho(r)$).

If we omit second-order terms, we have

$$\sum_\beta \frac{\partial v_s^{(\beta)}}{\partial r_\beta} = \frac{ia}{\varrho_s}(\zeta^* - \zeta); \quad A(\varrho, \Theta, 0)\sum_\beta \frac{\partial v_s^{(\beta)}}{\partial r_\beta} = 0$$

and it follows that

$$A(\varrho, \Theta, 0) = 0.$$

We can give a similar proof for the case when $\boldsymbol{v}_s = \boldsymbol{w} + \delta \boldsymbol{v}_s$, $\boldsymbol{w} \neq 0$ (see Bogoliubov, 1963).

Hence we have

$$\frac{\partial(\varrho S)}{\partial t} + \sum_{\beta} \frac{\partial(\varrho v_n^{(\beta)} S)}{\partial r_\beta} = 0. \tag{VII.31}$$

The system of hydrodynamic equations (VII.23–25,31) for a Bose superfluid, without viscous terms, was first derived by Landau (1941b, 1944), starting from phenomenological considerations. From eqn. (VII.31) we see that the transport of entropy is connected with the normal component only. It explains theoretically the existence of the thermomechanical effect.

§ 2. The hydrodynamic equations with viscous terms

In formulae (VI.11, 12), using the "μ" procedure, we have split off from $\mathcal{T}_{\alpha\beta}$ and $\mathcal{I}_\beta$ the terms $\mathcal{T}^{(1)}_{\alpha\beta}, \mathcal{I}^{(1)}_\beta = \overset{1}{\mathcal{I}}{}^{(1)}_\beta + \overset{2}{\mathcal{I}}{}^{(1)}_\beta$ which, in the case of superfluid, must be linear forms of $\eta, \eta^* \to \tilde{\eta}, \tilde{\eta}^*$ (which, as can be seen from (V.4), are of the same order of smallness as the gradients) and of the gradients of $\varrho, \Theta, \boldsymbol{v}_s, \boldsymbol{v}_n$ (Galasiewicz, 1963b). Owing to the covariance of the hydrodynamic equations the most general expression for this form is

$$\begin{aligned}\mathcal{T}^{(1)}_{\alpha\beta} = {} & \delta_{\alpha\beta}\{\zeta_1' \operatorname{div}_\xi(\boldsymbol{v}_s - \boldsymbol{v}_n) + \zeta_2' \operatorname{div}_\xi \boldsymbol{v}_n + aC_1\tilde{\eta} + aC_1^*\tilde{\eta}^*\} \\ & + C_2\left(\frac{\partial v_s^{(\alpha)}}{\partial \xi_\beta} + \frac{\partial v_s^{(\beta)}}{\partial \xi_\alpha}\right) \\ & + \eta\left(\frac{\partial v_n^{(\alpha)}}{\partial \xi_\beta} + \frac{\partial v_n^{(\beta)}}{\partial \xi_\alpha} - \frac{2}{3}\,\delta_{\alpha\beta} \operatorname{div}_\xi \boldsymbol{v}_n\right)\end{aligned} \tag{VII.32}$$

and

$$\begin{aligned}\overset{1}{\mathcal{I}}{}^{(1)}_\beta &= D\frac{\partial \varrho}{\partial \xi_\beta} + \varkappa \frac{\partial \Theta}{\partial \xi_\beta} + A^{(1)}(v_s^{(\beta)} - v_n^{(\beta)}), \\ A^{(1)} &= [\zeta_3 \operatorname{div} \varrho_s(\boldsymbol{v}_s - \boldsymbol{v}_n) + \zeta_4 \operatorname{div} \boldsymbol{v}_n + C_3\tilde{\eta}^* + C_3^*\tilde{\eta}]\varrho_s, \\ \overset{2}{\mathcal{I}}{}^{(1)}_\beta &= \sum_\alpha v_n^{(\alpha)} \mathcal{T}^{(1)}_{\alpha\beta}(\varrho, \Theta, v_s, v_n)\end{aligned} \tag{VII.33}$$

where ζ_i, C_i, D, $\varkappa$, η are coefficients characterizing the superfluid (D is the diffusion coefficient, $\varkappa$ is the thermal conductivity coefficient, η the viscosity coefficient and $\zeta_1 = \zeta_1' + 2C_2/\varrho_s$, $\zeta_2 = \zeta_2' + 2C_2$, ζ_3, ζ_4 are the second viscosity coefficients). The coefficients C_1, C_1^*, C_3, C_3^* and aB, aB^*, p_1, p_1^*, p_2, p_2^*, p_3, p_3^*, which will be introduced when we consider eqn. (VII.16), follow from the "μ" procedure and are introduced formally. We shall consider all these coefficients as given. In the case of weak interactions between the particles these coefficients could, in principle, be computed with the help of the Boltzmann equation. But in the general case they must be taken from experiment.

The form of $\overset{2}{\mathscr{I}}{}_\beta^{(1)}$ follows from the appearance of the factor $\sum_\beta v_n^{(\beta)} \mathscr{T}_{\alpha\beta}^{(0)}(\varrho, \Theta, v_s - v_n, 0)$ in (VII.14′). The elements of the stress tensor $\mathscr{T}_{\alpha\beta}^{(0)}(\varrho, \Theta, v_s - v_n, 0)$ are expressed in terms of the mean values $\langle\ \rangle_{v_s - v_n, 0}^{(0)}$. However, $\mathscr{I}_\beta^{(1)}$ is expressed in terms of the mean values $\langle\ \rangle_{v_s, v_n}^{(1)}$. Therefore in $\mathscr{I}_\beta$ the term $\sum_\beta v_n^{(\beta)} \mathscr{T}_{\alpha\beta}^{(1)}$ $(\varrho, \Theta, v_s, v_n)$ should appear.

In a paper by Galasiewicz (1963b), $\overset{1}{\mathscr{I}}{}_\alpha^{(1)}$ was expressed only by the first two terms. This is enough to obtain suitable linearized equations. It was, however, stated by Krasnikov (1967a) that in order to obtain a good agreement with the full equations derived by Khalatnikov (1952c) one must take $\overset{1}{\mathscr{I}}{}_\alpha^{(1)}$ in the form (VII.33).

In the case of neglecting viscous terms we demonstrated that in (VII.14′) $A(\varrho, \Theta, u) = A^{(0)} = 0$. Now $A = \mu A^{(1)}$, where the expression for $A^{(1)}$ is given by (VII.33).

In eqn. (VII.5) the following term appears:

$$-\frac{1}{2a^2}\int V(R)\,[X_t(r, R) + X_t^*(r, R)]\,\mathrm{d}^3\boldsymbol{R},$$

with the scalar function $X_t(r, R)$ which can be expanded in the parameter μ in the same way as was done for the vector and tensor functions $\mathscr{I}_\alpha$, $\mathscr{T}_{\alpha\beta}$ in (VII.15), (VII.32, 33),

$$X_\tau(\xi, R) = X(R|\varrho, \Theta, v_s - v_n) + \mu X^1(\tau, \xi) + \ldots$$

By considering thermodynamic equilibrium, we have obtained only the expression (VII.8) for the integral containing the term independent of μ. The most general form of terms of first order in μ is

$$\mu\frac{1}{2a^2}\int V(R) X_\tau^1(\xi, R)\,\mathrm{d}^3\boldsymbol{R} \equiv {}^1X$$
$$= \mu[p_1 \operatorname{div}_\xi \varrho_s(\boldsymbol{v}_s - \boldsymbol{v}_n) + p_2 \operatorname{div}_\xi \boldsymbol{v}_n - b\tilde{\eta}]. \qquad \text{(VII.34)}$$

Therefore, from (VII.34) we have

$$\frac{\mu}{2a^2}\int V(R)\,[X^1_\tau(\xi, R)+X^{(1)*}_\tau(\xi, R)]\,\mathrm{d}^3\boldsymbol{R} \tag{VII.35}$$
$$= [(p_1+p_1^*)\,\mathrm{div}_\xi\,\varrho_s(\boldsymbol{v}_s-\boldsymbol{v}_n)+(p_2+p_2^*)\,\mathrm{div}_\xi\,\boldsymbol{v}_n-b\tilde{\eta}-b^*\tilde{\eta}^*].$$

Now using (VII.17) and (VII.35) we obtain, from formula (VII.5), $(\zeta, \zeta^* \to \tilde{\zeta}, \tilde{\zeta}^*)$

$$m\frac{\partial v_s^{(\alpha)}}{\partial\tau} = -\frac{\partial}{\partial\xi_\alpha}\left[\Lambda(\varrho, \Theta, u)+\frac{mv_s^2}{2}-\frac{m}{2}(\boldsymbol{v}_s-\boldsymbol{v}_n)^2+U\right]$$
$$-\mu\left[(p_1+p_1^*)\frac{\partial}{\partial\xi_\alpha}\,\mathrm{div}_\xi\,\varrho_s(\boldsymbol{v}_s-\boldsymbol{v}_n)+(p_2+p_2^*)\frac{\partial}{\partial\xi_\alpha}\,\mathrm{div}\,\boldsymbol{v}_n\right.$$
$$\left.-b\frac{\partial\tilde{\eta}}{\partial\xi_\alpha}-b^*\frac{\partial\tilde{\eta}^*}{\partial\xi_\alpha}\right]-\mu\frac{\partial}{\partial\xi_\alpha}\frac{(\tilde{\zeta}+\tilde{\zeta}^*)}{2a}. \tag{VII.36}$$

After transformations analogous to those for an ordinary fluid given in Chapter VI, § 1 and by Bogoliubov (1963) for a superfluid (without considering the viscosity), we have, for the entropy,

$$\Theta\left[\frac{\partial(\varrho S)}{\partial\tau}+\frac{\partial(\varrho S v_n^{(\beta)})}{\partial r_\beta}\right] = \mu[R+\varkappa\nabla_\xi^2\Theta+D\nabla_\xi^2\varrho+f_\xi(\tilde{\eta}, \tilde{\eta}^*)] \tag{VII.37}$$

where

$$f_\xi(\tilde{\eta}, \tilde{\eta}^*) = \frac{\varrho_s}{2}\left(\frac{1}{a}-\frac{a}{\varrho_s}\right)((\boldsymbol{v}_s-\boldsymbol{v}_n)\cdot\nabla_\xi(\tilde{\zeta}+\tilde{\zeta}^*))$$
$$-\frac{1}{2}\left(\frac{1}{a^2}-\frac{1}{\varrho_s}\right)(\boldsymbol{v}_s\cdot\nabla_\xi a(\tilde{\zeta}+\tilde{\zeta}^*))+2ia(\tilde{\zeta}^{*1}X-\tilde{\zeta}^1X^*)$$
$$+\varrho_s((\boldsymbol{v}_s-\boldsymbol{v}_n)\cdot\nabla_\xi(b\tilde{\eta}+b^*\tilde{\eta}^*))+(\nabla_\xi\cdot[(\boldsymbol{v}_s-\boldsymbol{v}_n)\varrho_s(C_3\tilde{\eta}+C_3^*\tilde{\eta})]),$$

$$R = \frac{\partial v_n^{(\alpha)}}{\partial\xi_\beta}\mathscr{T}^{(1)}_{\alpha\beta}+\frac{\partial}{\partial\xi_\beta}[\varrho_s(v_s^{(\beta)}-v_n^{(\beta)})(\zeta_3\,\mathrm{div}\,\varrho_s(\boldsymbol{v}_s-\boldsymbol{v}_n)+\zeta_4\,\mathrm{div}\,\boldsymbol{v}_n]$$
$$+\varrho_s((\boldsymbol{v}_s-\boldsymbol{v}_n)\cdot\nabla_\xi[(^1X+{}^1X^*)+b\tilde{\eta}+b^*\tilde{\eta}^*]) \tag{VII.38}$$
$$= \eta\frac{\partial v_n^{(\alpha)}}{\partial\xi_\beta}\left(\frac{\partial v_n^{(\alpha)}}{\partial\xi_\alpha}+\frac{\partial v_n^{(\beta)}}{\partial\xi_\alpha}-\frac{2}{3}\delta_{\alpha\beta}\,\mathrm{div}\,\boldsymbol{v}_n\right)+\zeta_2(\mathrm{div}\,\boldsymbol{v}_n)$$
$$+\zeta_3[\mathrm{div}\,\varrho_s(\boldsymbol{v}_s-\boldsymbol{v}_n)]^2+(\zeta_1+\zeta_4)\,\mathrm{div}\,\boldsymbol{v}_n\,\mathrm{div}\,\varrho_s(\boldsymbol{v}_s-\boldsymbol{v}_n)$$
$$+\varrho_s((\boldsymbol{v}_s-\boldsymbol{v}_n)\cdot\{[(p_1+p_1^*)+\zeta_3]\nabla\varrho_s(\boldsymbol{v}_s-\boldsymbol{v}_n)+(p_2+p_2^*+\zeta_4)\nabla\boldsymbol{v}_n\}).$$

From the requirement that the dissipation function R must be a quadratic form of the divergences, it follows that

$$p_1+p_1^* = -\zeta_3, \quad p_2+p_2^* = -\zeta_4. \tag{VII.39}$$

Moreover, we have

$$\begin{aligned}&\frac{\partial v_n^{(\alpha)}}{\partial \xi_\beta}\left(\frac{\partial v_n^{(\alpha)}}{\partial \xi_\beta}+\frac{\partial v_n^{(\beta)}}{\partial \xi_\beta}-\frac{2}{3}\delta_{\alpha\beta}\nabla \boldsymbol{v}_n\right)\\ &=\frac{1}{2}\left(\frac{\partial v_n^{(\alpha)}}{\partial \xi_\beta}+\frac{\partial v_n^{(\beta)}}{\partial \xi_\alpha}-\frac{2}{3}\delta_{\alpha\beta}(\nabla\cdot\boldsymbol{v}_n)\right)^2\end{aligned} \tag{VII.40}$$

and we see that R is identical with the function given by Khalatnikov (1952c).

We transform, in eqns. (VII.16, 37), from the auxiliary variables τ, $\boldsymbol{\xi}$ to the original variables t, $\boldsymbol{r}$. As a result, we obtain the equations for a Bose superfluid with viscous terms

$$\frac{\partial \varrho}{\partial t}+\sum_\alpha \frac{\partial(\varrho_s v_s^{(\alpha)}+\varrho_n v_n^{(\alpha)})}{\partial r_\alpha} = ia(\tilde{\zeta}^*-\tilde{\zeta}), \tag{VII.41}$$

$$\begin{aligned}m\frac{\partial}{\partial t}(\varrho_s v_s^{(\alpha)}+\varrho_n v_n^{(\alpha)}) &= \sum_\beta \frac{\partial}{\partial r_\beta}\Big\{\delta_{\alpha\beta}[-P+\zeta_1'\operatorname{div}\varrho_s(\boldsymbol{v}_s-\boldsymbol{v}_n)\\ &+\zeta_2'\operatorname{div}\boldsymbol{v}_n+aC_1\tilde{\eta}+aC_1^*\tilde{\eta}^*]-m\varrho_s v_s^{(\alpha)}v_s^{(\beta)}-m\varrho_n v_n^{(\alpha)}v_n^{(\beta)}\\ &+C_2\left(\frac{\partial v_s^{(\alpha)}}{\partial r_\beta}+\frac{\partial v_t^{(\beta)}}{\partial r_\alpha}\right)+\eta\left(\frac{\partial v_n^{(\alpha)}}{\partial r_\beta}+\frac{\partial v_n^{(\beta)}}{\partial r_\alpha}-\frac{2}{3}\delta_{\alpha\beta}\operatorname{div}\boldsymbol{v}_n\right)\Big\}\\ &-\frac{1}{m}\varrho\frac{\partial U}{\partial r_\alpha}+iam(\tilde{\zeta}^*-\tilde{\zeta})v_s^{(\alpha)}+\frac{1}{2}(\tilde{\zeta}^*+\tilde{\zeta})\frac{\partial a}{\partial r_\alpha}-\frac{a}{2}\frac{\partial}{\partial r_\alpha}(\tilde{\zeta}^*+\tilde{\zeta}),\end{aligned} \tag{VII.42}$$

$$\begin{aligned}m\frac{\partial v_s^{(\alpha)}}{\partial t} &= -\frac{\partial}{\partial r_\alpha}\left[\Lambda(\varrho,\Theta,u)+\frac{mv_s^2}{2}-\frac{m}{2}(\boldsymbol{v}_s-\boldsymbol{v}_n)^2+U\right]\\ &+\left[\zeta_3\frac{\partial}{\partial r_\alpha}\operatorname{div}\varrho_s(\boldsymbol{v}_s-\boldsymbol{v}_n)+\zeta_4\frac{\partial}{\partial r_\alpha}\operatorname{div}\boldsymbol{v}_n+b\frac{\partial\tilde{\eta}}{\partial r_\alpha}+b^*\frac{\partial\tilde{\eta}^*}{\partial r_\alpha}\right]\\ &-\frac{\partial}{\partial r_\alpha}\left(\frac{\tilde{\xi}^*+\tilde{\xi}}{2a}\right).\end{aligned} \tag{VII.43}$$

$$\Theta\left[\frac{\partial(\varrho S)}{\partial t}+\sum_\beta\frac{\partial(\varrho S v_n^{(\beta)})}{\partial r_\beta}\right] = R+\varkappa\nabla^2\Theta+D\nabla^2\varrho+f(\tilde{\eta},\tilde{\eta}^*) \tag{VII.44}$$

where we now have

$$R = \frac{\eta}{2}\left(\frac{\partial v_n^{(\alpha)}}{\partial r_\beta} + \frac{\partial v_n^{(\beta)}}{\partial r_\alpha} - \frac{2}{3}\delta_{\alpha\beta}\operatorname{div}\boldsymbol{v}_n\right)^2 + \zeta_2(\operatorname{div}\boldsymbol{v}_n)^2$$

$$+\zeta_3[\operatorname{div}\varrho_s(\boldsymbol{v}_s-\boldsymbol{v}_n)]^2 + (\zeta_1+\zeta_4)\operatorname{div}\boldsymbol{v}_n \operatorname{div}\varrho_s(\boldsymbol{v}_s-\boldsymbol{v}_n). \quad \text{(VII.45)}$$

Equations (VII.41–44) (for $\tilde{\eta} = 0$) were obtained by Khalatnikov (1952c) starting from phenomenological considerations.

In eqns. (VII.41–44) there appear kinetic coefficients which are functions of ϱ, Θ, $v_s^{(\alpha)}$, $v_n^{(\alpha)}$. We are not interested in the form of these functions (we do not use the Boltzmann equation) and therefore it seems impossible to obtain relations among these coefficients, thereby decreasing their number. We must stress, however, that in order to obtain the Green functions we are interested only in the linearized, acoustic hydrodynamic equations. On the basis of these equations, and other considerations, we can obtain certain connexions between the kinetic coefficients.

§ 3 Hydrodynamic equations in the acoustic approximation

Consider now, with the help of eqns. (VII.41–44), an infinitesimal deviation from thermodynamic equilibrium (ϱ^0, Θ^0, $\boldsymbol{v}_s^0 = \boldsymbol{v}_n^0 = 0$) produced by infinitesimal sources $\delta\tilde{\eta}$, $\delta\tilde{\eta}^*$, an external scalar potential δU and, according to Krasnikov (1967b), a longitudina vector potential δA:

$$\begin{aligned} &\varrho^0 = \varrho+\delta\varrho(t,r), && S^0 = S+\delta S(t,r), && \Theta = \Theta^0+\delta\Theta(t,r),\\ &\boldsymbol{v}_s = \delta\boldsymbol{v}_s(t,r), && \boldsymbol{v}_n = \delta\boldsymbol{v}_n(t,r), && \tilde{\eta} = \delta\eta(t,r),\\ &\tilde{\eta}^* = \delta\eta^*(t,r), && U = \delta U, && A = \delta A. \end{aligned}$$

To second order of accuracy we have

$$\tilde{\eta} = \tilde{\xi} = \delta\eta.$$

So we pass now from the hydrodynamic equations to the linearized "acoustic equations". This means that in eqns. (VII.41–44) we neglect terms proportional to δv_s^2, δv_n^2, $\delta v_s\,\delta v_n$, $\delta\varrho\,\delta S$, ζ^2, ζ multiplied by gradients.

It must be emphasized that the "acoustic" approximation is completely different from an expansion in the small parameter μ which leads to the hydrodynamic equations for an ideal or viscous liquid. The acoustic approximation is a linearization of the obtained equations and can be applied to the equations both of an ideal and of a viscous liquid.

From the linearized equations we can find the variations $\delta\varrho$, $\delta\Theta$, $\delta v_s^{(\alpha)}$, $\delta v_n^{(\alpha)}$ describing deviations from thermodynamic equilibrium.

Using the relation $\frac{\partial v_s^{(\alpha)}}{\partial r_\beta} = \frac{\partial v_s^{(\beta)}}{\partial r_\alpha}$ we first write the linearized equation (VII.42) as

$$m\varrho_s \frac{\partial v_s^{(\alpha)}}{\partial t} + m\varrho_n \frac{\partial v_n^{(\alpha)}}{\partial t} = -\frac{\partial \delta P}{\partial r_\alpha} + \zeta_1' \varrho_s \frac{\partial}{\partial r_\alpha} \operatorname{div}(\boldsymbol{v}_s - \boldsymbol{v}_n)$$

$$+\zeta_2' \frac{\partial}{\partial r_\alpha} \operatorname{div} \boldsymbol{v}_n + 2C_2 \nabla^2 v_s^{(\alpha)} + \eta \sum_\beta \frac{\partial}{\partial r_\beta} \left(\frac{\partial v_n^{(\alpha)}}{\partial r_\beta} + \frac{\partial v_n^{(\beta)}}{\partial r_\alpha} \right.$$

$$\left. -\frac{2}{3} \delta_{\alpha\beta} \operatorname{div} \boldsymbol{v}_n \right) + \sqrt{\varrho_0}\, C \frac{\partial \tilde{\eta}}{\partial r_\alpha} + \sqrt{\varrho_0}\, C^* \frac{\partial \tilde{\eta}^*}{\partial r_\alpha} - \varrho \left(\frac{\partial U}{\partial r_\alpha} + \frac{\partial A^\alpha}{\partial t} \right),$$

$$\varrho_0 = (a^0)^2.$$

Since $\nabla^2 v_s^{(\alpha)} = \frac{\partial}{\partial r_\alpha} \operatorname{div} \boldsymbol{v}_s$ we see that the coefficient C_2 is superfluous and instead of ζ_1', ζ_2', C_2 we can introduce two coefficients $\zeta_1 = \zeta_1' + 2C_2/\varrho_s$ and $\zeta_2 = \zeta_2' + 2C_2$. ($C \equiv 0$; Galasiewicz, 1969c.)

The linearized equations for a viscous Bose liquid (the upper index of ϱ, Θ, S is omitted) also have the form

$$\frac{\partial \delta\varrho}{\partial t} + \varrho_s \operatorname{div} \boldsymbol{v}_s + \varrho_n \operatorname{div} \boldsymbol{v}_n = i \sqrt{\varrho_0} (\delta\eta^* - \delta\eta), \qquad \text{(VII.46)}$$

$$m\varrho_s \frac{\partial v_s^{(\alpha)}}{\partial t} + m\varrho_n \frac{\partial v_n^{(\alpha)}}{\partial t} = -\frac{\partial \delta P}{\partial r_\alpha} + \zeta_1 \varrho_s \frac{\partial}{\partial r_\alpha} \operatorname{div}(\boldsymbol{v}_s - \boldsymbol{v}_n)$$

$$+\zeta_2 \frac{\partial}{\partial r_\alpha} \operatorname{div} \boldsymbol{v}_n + \eta \sum_\beta \frac{\partial}{\partial r_\beta} \left(\frac{\partial v_n^{(\alpha)}}{\partial r_\beta} + \frac{\partial v_n^{(\beta)}}{\partial r_\alpha} - \frac{2}{3} \delta_{\alpha\beta} \operatorname{div} \boldsymbol{v}_n \right)$$

$$+\sqrt{\varrho_0}\, C \frac{\partial \tilde{\eta}}{\partial r_\alpha} + \sqrt{\varrho_0}\, C^* \frac{\partial \tilde{\eta}^*}{\partial r_\alpha} - \varrho \left(\frac{\partial U}{\partial r_\alpha} + \frac{\partial A^\alpha}{\partial t} \right), \qquad \text{(VII.47)}$$

where c is the sound velocity and ck the energy of the sound quanta;

$$m \frac{\partial v_s^{(\alpha)}}{\partial t} = -\frac{\partial \delta \Lambda}{\partial r_\alpha} + \zeta_3 \varrho_s \frac{\partial}{\partial r_\alpha} \operatorname{div}(\boldsymbol{v}_s - \boldsymbol{v}_n) + \zeta_4 \frac{\partial}{\partial r_\alpha} \operatorname{div} \boldsymbol{v}_n$$

$$+\sqrt{\varrho_0}\, B \frac{\partial \tilde{\eta}}{\partial r_\alpha} + \sqrt{\varrho_0}\, B^* \frac{\partial \tilde{\eta}^*}{\partial r_\alpha} - \left(\frac{\partial U}{\partial r_\alpha} + \frac{\partial A_\alpha}{\partial t} \right), \qquad \text{(VII.48)}$$

$$\varrho \frac{\partial \delta S}{\partial t} + S \frac{\partial \delta \varrho}{\partial t} + \varrho S \operatorname{div} \boldsymbol{v}_n = \frac{1}{\Theta} (D \nabla^2 \delta \varrho + \varkappa \nabla^2 \delta \Theta) \tag{VII.49}$$

where

$$\sqrt{\varrho_0} B = b + \frac{1}{2a}, \quad \sqrt{\varrho_0} B^* = b^* + \frac{1}{2a}, \quad \sqrt{\varrho_0} C = C_1 - \frac{a}{2},$$

$$\sqrt{\varrho_0} C^* = C_1^* - \frac{a}{2}, \quad \delta \Lambda = \frac{\partial \Lambda}{\partial \varrho} \delta \varrho + \frac{\partial \Lambda}{\partial \Theta} \delta \Theta = - S \delta \Theta + \frac{1}{\varrho} \delta P,$$

$$\sqrt{\varrho_0} = \langle \psi \rangle_{0,0} = \langle \psi^* \rangle_{0,0}, \quad P = \varrho^2 \frac{\partial F(\varrho, \Theta, u)}{\partial \varrho},$$

$$\Lambda = F + \varrho \frac{\partial F(\varrho, \Theta, u)}{\partial \varrho}, \quad S = - \frac{\partial F}{\partial \Theta}, \quad \left(\frac{\partial S}{\partial \Theta} \right)_\varrho = \frac{\Theta}{C_v},$$

$$E = F(\varrho, \Theta, u) - \Theta \frac{\partial F(\varrho, \Theta, u)}{\partial \Theta}. \tag{VII.50}$$

Nine coefficients appear in the equations in the acoustic approximation for a viscous liquid. But among them two are complex, therefore we have eleven real coefficients. They are only functions of Θ^0, ϱ^0, because we have $\boldsymbol{v}_s^0 = \boldsymbol{v}_n^0 = 0$. The question arises what is the maximal information about the relations among these eleven coefficients that can be obtained without the use of the Boltzmann equation. In order to obtain some relations, we make use of the general principle—the condition of possible thermodynamic equilibrium at $\delta\eta = \tilde{\eta}(r)$. In this case we have

$$\tilde{\eta} = \delta\eta(r), \quad \varrho = \varrho^0 + \delta(r), \quad \Theta = \Theta^0, \quad \boldsymbol{v}_s = \boldsymbol{v}_s(r),$$
$$\boldsymbol{v}_n = 0, \quad \delta U = 0, \quad \delta A = 0.$$

Applying (VII.50) we obtain eqns. (VII.46–49) in the form

$$\operatorname{div} \boldsymbol{v}_s = \frac{i\sqrt{\varrho_0}}{\varrho_s} (\tilde{\eta}^* - \tilde{\eta}), \tag{VII.51}$$

$$- \frac{\partial \delta P}{\partial r_\alpha} + \zeta_1 \varrho_s \frac{\partial}{\partial r_\alpha} \operatorname{div} \boldsymbol{v}_s + \sqrt{\varrho_0} C \frac{\partial \tilde{\eta}}{\partial r_\alpha} + \sqrt{\varrho_0} C^* \frac{\partial \tilde{\eta}^*}{\partial r_\alpha} = 0, \tag{VII.52}$$

$$- \frac{\partial \delta P}{\partial r_\alpha} + \varrho \zeta_3 \varrho_s \frac{\partial}{\partial r_\alpha} \operatorname{div} \boldsymbol{v}_s + \sqrt{\varrho_0} B \varrho \frac{\partial \tilde{\eta}}{\partial r_\alpha} + \sqrt{\varrho_0} B^* \varrho \frac{\partial \tilde{\eta}^*}{\partial r_\alpha} = 0, \tag{VII.53}$$

$$D\nabla^2 \varrho = 0. \tag{VII.54}$$

Introducing the expressions for $\operatorname{div} \boldsymbol{v}_s$ into (VII.52,53) we have

$$-\frac{\partial \delta P}{\partial r_\alpha} + \sqrt{\varrho_0}(C - i\zeta_1)\frac{\partial \tilde{\eta}}{\partial r_\alpha} + \sqrt{\varrho_0}(C^* + i\zeta_1)\frac{\partial \tilde{\eta}^*}{\partial r_\alpha} = 0,$$

$$-\frac{\partial \delta P}{\partial r_\alpha} + \varrho\sqrt{\varrho_0}(B - i\zeta_3)\frac{\partial \tilde{\eta}}{\partial r_\alpha} + \varrho\sqrt{\varrho_0}(B^* + i\zeta_3)\frac{\partial \tilde{\eta}^*}{\partial r_\alpha} = 0.$$

These equations give us

$$\begin{aligned} C &= \varrho B + i(\zeta_1 - \varrho\zeta_3), \\ C^* &= \varrho B^* - i(\zeta_1 - \varrho\zeta_3). \end{aligned} \tag{VII.55}$$

Now we take $\eta = \eta^* = 0$, $\delta U = \delta U(r)$. Equations (VII.46, 47,49) give us

$$-\frac{\partial \delta P}{\partial \varrho}\frac{\partial \delta \varrho}{\partial r_\alpha} + \varrho\frac{\partial \delta U(r)}{\partial r_\alpha} = 0, \quad D\nabla^2 \varrho = 0 \tag{VII.56}$$

and we see that

$$D = 0. \tag{VII.57}$$

From (VII.55,57) it follows that the number of coefficients introduced decreases from eleven to eight ($\zeta_1, \zeta_2, \zeta_3, \zeta_4, \varkappa, \eta, B, B^*$). Besides the connexions between the transport coefficients obtained from the general principle, i.e. the possibility of thermodynamic equilibrium (valid in our case for $\boldsymbol{v}_s^0 = \boldsymbol{v}_n^0 = 0$), there is the general connexion resulting from the Onsager relations. Such a connexion was obtained by Khalatnikov (1952c; 1965) and has the form

$$\zeta_1 = \zeta_4. \tag{VII.58}$$

Further considerations will allow us to obtain this relation in our scheme.

We now have the usual five independent coefficients: $\varkappa, \eta, \zeta_1, \zeta_2, \zeta_3$ and two new coefficients B, B^* which are introduced formally in order to account for all terms of order μ in the non-equilibrium averages. However, from the present considerations it does not follow that B, B^* are really new, independent coefficients. Later considerations show that they are linear functions of ζ_3.

After using (VII.55,57), eqns. (VII.46–49) finally have the form

$$\frac{\partial \delta \varrho}{\partial t} + \varrho_s \operatorname{div} \boldsymbol{v}_s + \varrho_n \operatorname{div} \boldsymbol{v}_n = i\sqrt{\varrho_0}(\tilde{\eta}^* - \tilde{\eta}), \tag{VII.59}$$

$$m\varrho_s \frac{\partial v_s^{(\alpha)}}{\partial t} + m\varrho_n \frac{\partial v_n^{(\alpha)}}{\partial t} = -\left(\frac{\partial P}{\partial \Theta}\right)_\varrho \frac{\partial \delta\Theta}{\partial r_\alpha} - \left(\frac{\partial P}{\partial \varrho}\right)_\Theta \frac{\partial \delta\varrho}{\partial r_\alpha}$$

$$+\zeta_1 \varrho_s \frac{\partial}{\partial r_\alpha} \operatorname{div} \boldsymbol{v}_s + \left(\zeta_2 + \frac{1}{3}\eta - \zeta_1 \varrho_s\right) \frac{\partial}{\partial r_\alpha} \operatorname{div} \boldsymbol{v}_n$$

$$+\eta \nabla^2 v_n^{(\alpha)} + \sqrt{\varrho_0}[\varrho B + i(\zeta_1 - \varrho\zeta_3)] \frac{\partial \tilde{\eta}}{\partial r_\alpha} + \sqrt{\varrho_0}[\varrho B^* - i(\zeta_1 - \varrho\zeta_3)] \frac{\partial \tilde{\eta}^*}{\partial r_\alpha}$$

$$-\varrho\left(\frac{\partial A^{(\alpha)}}{\partial t} + \frac{\partial U}{\partial r_\alpha}\right), \quad \text{(VII.60)}$$

$$m \frac{\partial v_s^{(\alpha)}}{\partial t} = \left[S - \frac{1}{\varrho}\left(\frac{\partial P}{\partial \Theta}\right)_\varrho\right] \frac{\partial \delta\Theta}{\partial r_\alpha} - \frac{1}{\varrho}\left(\frac{\partial P}{\partial \varrho}\right)_\Theta \frac{\partial \delta\varrho}{\partial r_\alpha}$$

$$+\zeta_3 \varrho_s \frac{\partial}{\partial r_\alpha} \operatorname{div} \boldsymbol{v}_s + (\zeta_4 - \zeta_3 \varrho_s) \frac{\partial}{\partial r_\alpha} \operatorname{div} \boldsymbol{v}_n$$

$$+\sqrt{\varrho_0} B \frac{\partial \tilde{\eta}}{\partial r_\alpha} + \sqrt{\varrho_0} B^* \frac{\partial \tilde{\eta}^*}{\partial r_\alpha} - \left(\frac{\partial A^{(\alpha)}}{\partial t} + \frac{\partial U}{\partial r_\alpha}\right), \quad \text{(VII.61)}$$

$$\varrho \frac{\partial \delta S}{\partial t} + S \frac{\partial \delta\varrho}{\partial t} + \varrho S \operatorname{div} \boldsymbol{v}_n = \frac{1}{\Theta} \varkappa \nabla^2 \delta\Theta. \quad \text{(VII.62)}$$

§ 4 The solution of the acoustic equations and the calculation of the Green functions

The aim of this section is to express $v_s^{(\alpha)}(t, r)$, $\delta\varrho(t, r)$, $\delta j(t, r)$ (on the basis of Chapter V, § 3) with the help of the Fourier components of the retarded Green functions and to calculate them independently from (VII.59–62).

For $\tilde{\eta} = \delta\eta(t, r)$, $\tilde{\eta}^* = \delta\eta^*(r, t)$, $U = \delta U(t, r)$, $A = \delta A(t, r)$ we have $\hat{H}_t^1[\tilde{\eta}, \tilde{\eta}^*, U, A] = \delta\hat{H}_t$. The small variation of the Hamiltonian (V.1) (Schrödinger representation)

$$\delta\hat{H}_\tau = \int \{\psi(0, r')\delta\eta^*(\tau, r') + \psi^+(0, r')\delta\eta(\tau, r')\} \mathrm{d}^3\boldsymbol{r}'$$

$$\text{(VII.63)}$$

$$+\int \delta U(\tau, r)\hat{\varrho}(0, r)\mathrm{d}^3\boldsymbol{r} - \frac{1}{m}\int (\hat{j}(0, r) \cdot \delta A(\tau, r))\mathrm{d}^3\boldsymbol{r}$$

is introduced adiabatically, and it is necessary to put $\delta\eta$, δU, $\delta A^{(\alpha)}$ in the form

$$\delta f(\tau, r') = e^{-i\omega\tau + \varepsilon\tau + i(k\cdot r')}\delta f(k) + e^{i\omega\tau + \varepsilon\tau - i(k\cdot r')}\delta f(-k), \quad \text{(VII.64)}$$

$$\delta f^*(\tau, r') = e^{-i\omega\tau+\varepsilon\tau+i(k\cdot r')}\delta f^*(-k)+e^{i\omega\tau+\varepsilon\tau-i(k\cdot r')}\delta f^*(k)$$

where $\varepsilon > 0$, $\varepsilon \to 0$. Therefore

$$\begin{aligned} \delta H_\tau &= e^{-i\omega\tau+\varepsilon\tau}V_{+\omega}+e^{i\omega\tau+\varepsilon\tau}V_{-\omega}, \\ V_\omega &= V_\omega^\eta+V_\omega^U+V_\omega^A, \end{aligned} \tag{VII.65}$$

where

$$\begin{aligned} V_{\pm\omega}^\eta &= \int e^{\pm i(k\cdot r')}\{\psi(0, r')\delta\eta^*(\mp k)+\psi^*(0, r')\delta\eta(\pm k)\}\,d^3\boldsymbol{r}', \\ V_{\pm\omega}^U &= \int e^{\pm i(k\cdot r')}\hat{\varrho}(0, r')\delta U(\mp k)\,d^3\boldsymbol{r}', \\ V_{\pm\omega}^A &= -\frac{1}{m}\int e^{\pm i(k\cdot r')}(\hat{\boldsymbol{j}}(0, r')\cdot\delta A(\pm k))\,d^3\boldsymbol{r}'. \end{aligned} \tag{VII.66}$$

We would like to consider as an example the variation $\delta\langle\psi(t, r)\rangle$ for $\delta U = 0$, $\delta A = 0$. Using (V.38) and (VII.66) we obtain

$$\begin{aligned} \delta\langle\psi(t, r)\rangle &= \delta\varphi(t, r) = \int e^{i(k\cdot r)}d^3\boldsymbol{r}'\left\{\int [\langle\langle\psi(t, r); \psi(\tau, r')\rangle\rangle^r\delta\eta^*_{-k}\right. \\ &\left.+\langle\langle\psi(t, r); \psi^+(\tau, r')\rangle\rangle^r\delta\eta_k]e^{-i\omega\tau+\varepsilon\tau}d\tau\right\} \\ &+\int e^{-i(k\cdot r')}d^3\boldsymbol{r}'\left\{\int [\langle\langle\psi(t, r); \psi(t, r')\rangle\rangle^r\delta\eta^*_k\right. \\ &\left.+\langle\langle\psi(t, r); \psi^+(\tau, r')\rangle\rangle^r\delta\eta_{-k}]e^{i\omega\tau+\varepsilon\tau}d\tau\right\} \\ &= e^{-i\omega t+\varepsilon t}2\pi\int e^{i(k\cdot r')}[\langle\langle\psi(r); \psi(r')\rangle\rangle^r_{\omega+i\varepsilon}\delta\eta^*_{-k} \\ &+\langle\langle\psi(r); \psi^+(r')\rangle\rangle^r_{\omega+i\varepsilon}\delta\eta_k]\,d^3\boldsymbol{r}' \\ &+e^{i\omega t+\varepsilon t}2\pi\int e^{-i(k\cdot r')}[\langle\langle\psi(r); \psi(r')\rangle\rangle^r_{-\omega+i\varepsilon}\delta\eta^*_k \\ &+\langle\langle\psi(r); \psi^+(r')\rangle\rangle^r_{-\omega+i\varepsilon}\delta\eta_{-k}]\,d^3\boldsymbol{r}'. \end{aligned} \tag{VII.67}$$

If we put

$$\begin{aligned} \delta\varphi(t, r) &= e^{-i\omega t+\varepsilon t+i(k\cdot r)}\,\delta\varphi(k)+e^{i\omega t+\varepsilon t-i(k\cdot r)}\delta\varphi(-k), \\ \delta\varphi(t, r)^* &= e^{-i\omega t+\varepsilon t+i(k\cdot r)}\,\delta\varphi(-k)^*+e^{i\omega t+\varepsilon t-i(k\cdot r)}\delta\varphi(k)^*, \end{aligned} \tag{VII.68}$$

then

$$\begin{aligned} \delta\varphi(k) &= 2\pi\{\langle\langle a_k; a_{-k}\rangle\rangle^r_E\delta\eta^*_{-k}+\langle\langle a_k; a_k^+\rangle\rangle^r_E\delta\eta_k\}, \\ \delta\varphi(-k)^* &= 2\pi\{\langle\langle a^+_{-k}; a_{-k}\rangle\rangle^r_E\delta\eta^*_{-k}+\langle\langle a^+_{-k}; a_k^+\rangle\rangle^r_E\delta\eta_k\}, \end{aligned} \tag{VII.69}$$

where $\langle\langle a_q; a_{-q}\rangle\rangle$, etc., denote the Fourier components of the retarded Green functions,

$$\langle\langle \psi(r); \psi(r')\rangle\rangle_E^r = \frac{1}{(2\pi)^3}\int \langle\langle a_q; a_{-q}\rangle\rangle_E^r e^{iq(r-r')} \mathrm{d}^3 q. \quad \text{(VII.70)}$$

Let us write the expression for the condensate current in terms of the wave function Φ of the condensate

$$j_c = \frac{i}{2}(\Phi\nabla\Phi^* - \Phi\nabla\Phi) - \Phi\Phi^* A = a^2(\nabla\chi - A) = \varrho_c m v_s \quad \text{(VII.71)}$$

where

$$\Phi = ae^{i\chi} = \sqrt{\varrho_c}\, e^{i\chi}, \quad v_s = \frac{1}{m}(\nabla\chi - A).$$

For thermodynamic equilibrium $a = \sqrt{\varrho_0}$, and therefore for small deviations from equilibrium we have

$$\Phi = \sqrt{\varrho_0} + \delta\Phi, \quad \Phi^* = \sqrt{\varrho_0} + \delta\Phi^*.$$

In this case

$$\delta j_c = \frac{i}{2}\sqrt{\varrho_0}(\nabla\delta\Phi^* - \nabla\delta\Phi) - \varrho_0 \delta A = \varrho_0 m \delta v_s$$

and

$$\delta v_s(t, r) = \frac{i}{2m\sqrt{\varrho_0}}\big(\nabla\delta\Phi^*(t, r) - \nabla\delta\Phi(t, r)\big) - \frac{1}{m}\delta A(t, r). \quad \text{(VII.72)}$$

Moreover,

$$\varrho_c = \Phi\Phi^* = \big(\sqrt{\varrho_0} + \delta\Phi\big)\big(\sqrt{\varrho_0} + \delta\Phi^*\big) = \varrho_0 + \sqrt{\varrho_0}(\delta\Phi + \delta\Phi^*).$$

Hence

$$\delta\varrho_c(t, r) = \sqrt{\varrho_0}\big(\delta\Phi(t, r) + \delta\Phi^*(t, r)\big). \quad \text{(VII.73)}$$

We now express the variations δv_s, $\delta\varrho_c$, $\delta\varrho$, δj, in the form (VII.64) and obtain, for the Fourier coefficients, the following general expressions ($\delta U \neq 0$, $\delta A \neq 0$):

$$\delta v_s^{(\alpha)}(k) = \frac{k_\alpha \pi}{m\sqrt{\varrho_0}}\Big\{[\langle\langle a_k; a_{-k}\rangle\rangle_E - \langle\langle a_{-k}^+; a_{-k}\rangle\rangle_E]\delta\eta^*(-k)$$
$$+ [\langle\langle a_k; a_k^+\rangle\rangle_E - \langle\langle a_{-k}^+; a_k^+\rangle\rangle_E]\delta\eta(k) + [\langle\langle a_k; \hat{\varrho}_{-k}\rangle\rangle_E - \langle\langle a_{-k}^+; \varrho_k^+\rangle\rangle]\delta U(k)$$
$$+ \frac{1}{m}[\langle\langle a_{-k}^+; \hat{j}_k^{(0)+}\rangle\rangle_E - \langle\langle a_k; \hat{j}_{-k}^{(0)}\rangle\rangle_E]\delta A\Big\} - \frac{1}{m}\delta A^{(\alpha)}, \quad \text{(VII.74)}$$

$$\delta\varrho_c(k) = 2\pi\sqrt{\varrho_0}\Big\{[\langle\langle a_k; a_{-k}\rangle\rangle_E + \langle\langle a^+_{-k}; a_{-k}\rangle\rangle_E]\delta\eta^*(-k)$$

$$+[\langle\langle a_k; a_k^+\rangle\rangle_E + \langle\langle a^+_{-k}; a_k^+\rangle\rangle_E]\delta\eta(k) + [\langle\langle a_k; \hat{\varrho}_{-k}\rangle\rangle_E + \langle\langle a^+_{-k}; \varrho_k^+\rangle\rangle_E]\delta U(k)$$

$$-\frac{1}{m}[\langle\langle a^+_{-k}; \hat{j}_k^{(0)+}\rangle\rangle_E + \langle\langle a_k; \hat{j}_{-k}^{(0)}\rangle\rangle_E]\delta A(k)\Big\}, \quad \text{(VII.75)}$$

$$\delta\varrho(k) = 2\pi\Big\{\langle\langle\hat{\varrho}_k; a_{-k}\rangle\rangle_E\delta\eta^*(-k) + \langle\langle\hat{\varrho}^+_{-k}; a_k^+\rangle\rangle_E\delta\eta(k)$$

$$+\langle\langle\hat{\varrho}_k; \hat{\varrho}_{-k}\rangle\rangle_E\delta U(k) - \frac{1}{m}\langle\langle\hat{\varrho}_k; \hat{j}_{-k}^{(0)}\rangle\rangle_E\delta A(k)\Big\}, \quad \text{(VII.76)}$$

$$\delta j_k = 2\pi\Big\{\langle\langle\hat{j}_k^{(0)}; a_{-k}\rangle\rangle_E\delta\eta^*(-k) + \langle\langle\hat{j}_\kappa^{(0)+}; a_k^+\rangle\rangle_E\delta\eta(k)$$

$$+\langle\langle\hat{j}_k^{(0)}, \hat{\varrho}_{-k}\rangle\rangle_E\delta U(k) - \frac{1}{m}\langle\langle j_k^{(0)}(j_{-k}^{(0)}\rangle\rangle_E\delta A(k))\Big\} - \varrho\delta A(k), \quad \text{(VII.77)}$$

$$\hat{\varrho}_k = \hat{\varrho}^+_{-k}, \quad \hat{j}_k = \hat{j}^+_{-k}. \quad \text{(VII.78)}$$

These formulae connect the hydrodynamic quantities which can be found from eqns. (VII.59–62) with the Green functions. We must stress, however, that since the hydrodynamic equations are valid only for "slow" changes of the hydrodynamic quantities ($\mu \ll 1$), this connexion has an asymptotic character.

From (VII.74–77) it is clear that if we find $\sum_\alpha k_\alpha\,\delta v_s^{(\alpha)}(k)$, $\delta\varrho(k)$, $\delta j(k)$, (VII.59–62) as functions of $\delta\eta_k$, $\delta\eta^*_{-k}$, $\delta U(k)$, $\delta A^{(\alpha)}(k)$, we can obtain the Fourier components of the retarded Green functions by comparing suitable derivatives with respect to $\delta\eta_k$, $\delta\eta^*_{-\kappa}$, $\delta U(k)$, $\delta A^{(\alpha)}(k)$.

Now, we write eqns. (VII.59–62) in terms of Fourier components (see (VII.64))

$$-E\delta\varrho(k) + \varrho_s\sum_\beta k^{(\beta)}\delta v_s^{(\beta)}(k) + \varrho_n\sum_\beta k^{(\beta)}\delta v_n^{(\beta)}(k)$$

$$= \sqrt{\varrho_0}(\delta\eta^*_{-k} - \delta\eta_k), \quad \text{(VII.79)}$$

$$Em[\varrho_s\delta v_s^{(\alpha)}(k) + \varrho_n\delta v_n^{(\alpha)}(k)] = k^{(\alpha)}\left[\left(\frac{\partial P}{\partial\varrho}\right)_\Theta\delta\varrho(k) + \left(\frac{\partial P}{\partial\Theta}\right)_\varrho\delta\Theta(k)\right]$$

$$-i\eta k^2\delta v_n^{(\alpha)}(k) - i\varrho_s\zeta_1 k^{(\alpha)}\sum_\beta k^{(\beta)}\delta v_s^{(\beta)}(k) - i\left(\zeta_2 + \frac{1}{3}\eta - \varrho_s\zeta_1\right)k^{(\alpha)}$$

$$\times \sum_\beta k^{(\beta)} \delta v_n^{(\beta)}(k) - k^{(\alpha)} \sqrt{\varrho_0} \{ [\varrho B + i(\zeta_1 - \varrho\zeta_3)] \delta\eta_k$$

$$+ [\varrho B^* - i(\zeta_1 - \varrho\zeta_3)\delta\eta^+_{-k}]\} + \varrho k^{(\alpha)} \delta U(k) - E\varrho \delta A^{(\alpha)}(k), \qquad \text{(VII.80)}$$

$$Em\delta v_s^{(\alpha)}(k) = k^{(\alpha)} \left\{ \left[\frac{1}{\varrho} \left(\frac{\partial P}{\partial \Theta} \right)_\varrho - S \right] \delta\Theta(k) + \frac{1}{\varrho} \left(\frac{\partial P}{\partial \varrho} \right)_\Theta \delta\varrho(k) \right\}$$

$$- i\varrho_s \zeta_3 k^{(\alpha)} \sum_\beta k^{(\beta)} \delta v_s^{(\beta)}(k) - i(\zeta_4 - \varrho_s \zeta_3) k^\alpha_\beta \sum_\beta k^{(\beta)} \delta v_n^{(\beta)}(k)$$

$$- k^{(\alpha)} \sqrt{\varrho_0} (B\delta\eta_k + B^* \delta\eta^*_{-k}) + k^{(\alpha)} \delta U(k) - E\delta A^{(\alpha)}(k), \qquad \text{(VII.81)}$$

$$E\left\{ \left[\varrho \left(\frac{\partial S}{\partial \varrho} \right)_\Theta + S \right] \delta\varrho(k) + \varrho \left(\frac{\partial S}{\partial \Theta} \right)_\varrho \delta\Theta(k) \right\}$$

$$= \varrho S \sum_\beta k^{(\beta)} \delta v_n^{(\beta)}(k) - i \frac{\varkappa}{\Theta} k^2 \delta\Theta(k). \qquad \text{(VII.82)}$$

We must note, however, that we have obtained (VII.74–77) for perturbations introduced and acting at negative times. Therefore we must obtain equations (VII.59–62) or (VII.79–82) for negative time. If the equations of hydrodynamics do not contain "dissipative terms" they are invariant under time reversal. If the equations contain these terms, they change their sign under time reversal. We must also change the signs of the kinetic coefficients in (VII.79–82).

We multiply (VII.80) and (VII.81) by $k^{(\alpha)}$ and sum over α. We denote

$$\sum_\beta k^{(\beta)} \delta v_s^{(\beta)} \equiv \delta X, \qquad \sum_\beta k^{(\beta)} \delta v_n^{(\beta)} \equiv \delta Y$$

and after changing the signs of the kinetic coefficients we write (VII.79–82) in the form

$$\varrho_s \delta X + \varrho_n \delta Y - E\delta\varrho + 0 \cdot \delta\Theta = \sqrt{\varrho_0} (\delta\eta^*_{-k} - \delta\eta_k), \quad \text{(VII.83)}$$

$$[Em\varrho_s - i\zeta_1 \varrho_s k^2] \delta X + [Em\varrho_n - i(4/3\eta + \zeta_2 - \zeta_1 \varrho_s) k^2] \delta Y$$

$$- k^2 \left(\frac{\partial P}{\partial \varrho} \right)_\Theta \delta\varrho - k^2 \left(\frac{\partial P}{\partial \Theta} \right)_\varrho \delta\Theta = k^2 \sqrt{\varrho_0} \{ [\varrho B + i(\zeta_1 - \varrho\zeta_3)] \delta\eta_k$$

$$+ [\varrho B^* - i(\zeta_1 - \varrho\zeta_3)] \delta\eta^*_{-k} \} = k^2 \varrho \sqrt{\varrho_0} (B\delta\eta_k$$

$$+ B^* \delta\eta^*_{-k}) - \sqrt{\varrho_0} i (\delta\eta^*_{-k} - \delta\eta_k)(\zeta_1 - \varrho\zeta_3) k^2 + \varrho k^2 \delta U(k)$$

$$- E\varrho(\boldsymbol{k} \cdot \delta \boldsymbol{A}(k)), \qquad \text{(VII.84)}$$

$$(Em\varrho - i\varrho\varrho_s\zeta_3 k^2)\,\delta X - i\varrho(\zeta_4 - \varrho_s\zeta_3)k^2\delta Y - k^2\left(\frac{\partial P}{\partial \varrho}\right)_\Theta \delta\varrho(k)$$

$$-k^2\left[\left(\frac{\partial P}{\partial \Theta}\right)_\varrho - \varrho S\right]\delta\Theta(k) = \sqrt{\varrho_0}\,k^2\varrho(B\delta\eta_k + B^*\delta\eta^*_{-k})$$

$$+\varrho k^2\delta U(k) - E\varrho(\boldsymbol{k}\cdot\delta\boldsymbol{A}(k)), \qquad \text{(VII.85)}$$

$$0\cdot\delta X - \varrho S\delta Y + E\left[\varrho\left(\frac{\partial S}{\partial \varrho}\right)_\Theta + S\right]\delta\varrho(k)$$

$$+\left[E\varrho\left(\frac{\partial S}{\partial \Theta}\right)_\varrho - i\frac{\varkappa}{\Theta}k^2\right]\delta\Theta(k) = 0. \qquad \text{(VII.86)}$$

We denote the determinant of the system of eqns. (VII.83–86) by $D(E)$ and the determinant obtained from $D(E)$, if we put in it the right-hand side of (VII.83–86) instead of the first (second, ...) column, by D_X, D_Y ..., respectively. Then we have

$$\delta X = \sum_\alpha k^{(\alpha)}\delta v_s^{(\alpha)}(k) = \frac{D_x}{D(E)},$$
$$\delta Y = \sum_\alpha k^{(\alpha)}\delta v_n^{(\alpha)}(k) = \frac{D_y}{D(E)}. \qquad \text{(VII.87)}$$

Consider now various special cases.

(i) Thermodynamic equilibrium ($E = 0$), ideal liquid approximation, $\delta U = 0$, $\delta A = 0$.

From (VII.83–86) (with the viscous terms equal to zero) we have

$$\delta X = \frac{\sqrt{\varrho_0}\,(\delta\eta^*_{-k} - \delta\eta_k)}{\varrho_s}, \qquad \delta\varrho = 0, \quad \delta\Theta = 0, \quad (\boldsymbol{k}\cdot\delta\boldsymbol{v}_n) = 0. \qquad \text{(VII.88)}$$

Since we have only a dependence on the difference $(\delta\eta^*_{-k} - \delta\eta_k)$, we must have (VII.74,75)

$$\delta\varrho_c = 2\pi\sqrt{\varrho_0}\,[\langle\langle a_k; a_{-k}\rangle\rangle_0^r + \langle\langle a^+_{-k}; a_{-k}\rangle\rangle_0^r]\,(\delta\eta^*_{-k} - \delta\eta_k), \qquad \text{(VII.89)}$$

$$\delta X = (\boldsymbol{k}\cdot\delta\boldsymbol{v}_s) = \frac{2\pi k^2}{m\sqrt{\varrho_0}}\,[\langle\langle a_k; a_{-k}\rangle\rangle_0^r - \langle\langle a^+_{-k}; a_{-k}\rangle\rangle_0^r]\,(\delta\eta^*_{-k} - \delta\eta_k),$$

$$\langle\langle a_k; a_{-k}\rangle\rangle_0^r + \langle\langle a^+_{-k}; a_{-k}\rangle\rangle_0^r = -\langle\langle a_k; a_k^+\rangle\rangle_0^r - \langle\langle a^+_{-k}; a_k^+\rangle\rangle_0^r,$$
$$\text{(VII.90)}$$
$$\langle\langle a_k; a_{-k}\rangle\rangle_0 - \langle\langle a^+_{-k}; a_{-k}\rangle\rangle_0^r = -\langle\langle a_k; a_k^+\rangle\rangle_0^r + \langle\langle a^+_{-k}; a_k^+\rangle\rangle_0^r.$$

Hence

$$\langle\langle a_k; a_{-k}\rangle\rangle_0^r = -\langle\langle a_k; a_k^+\rangle\rangle_0^r,$$

$$\langle\langle a_{-k}^+; a_{-k}\rangle\rangle_0^r = -\langle\langle a_{-k}^+; a_k^+\rangle\rangle_0^r. \qquad \text{(VII.91)}$$

Since for $E = 0$

$$\delta\varrho_c = \frac{\partial\varrho_c}{\partial\varrho}\,\delta\varrho + \frac{\partial\varrho_c}{\partial\Theta}\,\delta\Theta = 0, \qquad \text{(VII.92)}$$

we have from (VII.89)

$$\langle\langle a_k; a_{-k}\rangle\rangle_0^r = -\langle\langle a_{-k}^+; a_{-k}\rangle\rangle_0^r,$$

$$\langle\langle a_k; a_k^+\rangle\rangle_0^r = -\langle\langle a_{-k}^+; a_k^+\rangle\rangle_0^r. \qquad \text{(VII.93)}$$

Therefore from (VII.88)

$$\frac{\delta X}{\delta\eta_{-k}^*} = \frac{\sqrt{\varrho_0}}{\varrho_s}$$

and from (VII.89, 92)

$$\frac{\delta X}{\delta\eta_{-k}^*} = \frac{2\pi k^2}{m\sqrt{\varrho_0}}\langle\langle a_k; a_{-k}\rangle\rangle_0^r = -\frac{2\pi k^2}{m\sqrt{\varrho_0}}\langle\langle a_{-k}^+; a_{-k}\rangle\rangle_0^r$$

which gives

$$\langle\langle a_k; a_{-k}\rangle\rangle_0^r = \frac{m\varrho_0}{2\pi\varrho_s}\frac{1}{k^2} = -\langle\langle a_{-k}^+; a_{-k}\rangle\rangle_0^r. \qquad \text{(VII.94)}$$

We have obtained the "$1/k^2$ theorem" for single-particle Green functions (Bogoliubov, 1961). This theorem plays an essential role in examinations of long-range ordering in superfluid helium. On the basis of this theorem it was shown by Kadanoff and Kane (1967) that Yang's criterion for the existence of long-range order (Yang, 1962) is not satisfied in one and two dimensions.

(ii) $\Theta = 0$ (ideal liquid approximation). Consequently,

$$\varrho_s = \varrho, \quad \varrho_n = 0, \quad v_n = 0, \quad S = 0, \quad \frac{\partial S}{\partial\varrho} = 0,$$

$$\frac{\partial S}{\partial\Theta} = 0, \quad \frac{\partial P}{\partial\Theta} = 0.$$

(It is necessary to emphasize that the hydrodynamic equations have in this case only a formal meaning since the relaxation time becomes very large and the condition $ET \ll 1$ cannot be fulfilled.) We consider this example formally in order to determine the behaviour of our formulae as $\Theta \to 0$. In this case eqns. (VII.83–88) have the form

$$-E\delta\varrho(k)+\varrho\sum_{\beta} k^{(\beta)}v_s^{(\beta)}(k) = \sqrt{\varrho_0}\,(\delta\eta_{-k}^{*}-\delta\eta_k)$$

$$Em\varrho v_s^{(\alpha)}(k) = k^{(\alpha)}\left(\frac{\partial P}{\partial \varrho}\right)_{\Theta=0}\delta\varrho\,(k), \qquad \frac{1}{m}\left(\frac{\partial P}{\partial \varrho}\right)_{\Theta=0} = c^2.$$

From these equations we obtain

$$(\boldsymbol{k}\cdot\delta\boldsymbol{j}) = m\varrho_s\delta X = -\frac{mc^2k^2\sqrt{\varrho_0}}{(E^2-c^2k^2)}\,(\delta\eta_{-k}^{*}-\delta\eta_k),$$

$$\delta\varrho\,(k) = \frac{E\varrho}{k^2c^2}\,\delta X = -\frac{E}{E^2-c^2k^2}\sqrt{\varrho_0}\,(\delta\eta_{-k}^{*}-\delta\eta_k). \tag{VII.95}$$

Assuming that eqns. (VII.91,93) are valid also for small E, comparison of respective derivatives gives

$$\langle\langle a_k; a_{-k}\rangle\rangle_E \approx \frac{\varrho_0 c^2 m}{2\pi\varrho\,(E^2-c^2k^2)} = -\langle\langle a_{-k}^{+}; a_{-k}\rangle\rangle_E,$$

$$\langle\langle a_k, a_k^{+}\rangle\rangle_E \approx \frac{\varrho_0 c^2 m}{2\pi\varrho\,(E^2-c^2k^2)} = -\langle\langle a_{-k}^{+}; a_k^{+}\rangle\rangle_E. \tag{VII.96}$$

Since

$$\frac{\delta\varrho(k)}{\delta\eta_{-k}^{*}} = 2\pi\langle\langle\varrho_k; a_{-k}\rangle\rangle_E = \frac{E\varrho}{k^2c^2}\,\frac{\delta X}{\delta\eta_{-k}^{*}}, \tag{VII.97}$$

we see that (VII.95) gives the relation between the "three-leg" and "two-leg" Green functions

$$\langle\langle\varrho_k, a_{-k}\rangle\rangle_E = \frac{E\varrho}{mc^2\sqrt{\varrho_0}}\langle\langle a_k; a_{-k}\rangle\rangle_E. \tag{VII.98}$$

(iii) General case (ideal liquid approximation) $\delta U = 0$, $\delta A = 0$. We have

$$\delta X = \frac{D_x}{D(E)m} = \frac{k^2\Delta(k,E)\sqrt{\varrho_0}}{m\Omega\,(k,E)}\,(\delta\eta_{-k}^{*}-\delta\eta_k)$$

$$= \frac{\pi k^2}{m\sqrt{\varrho_0}}\{[\langle\langle a_k; a_{-k}\rangle\rangle_E^r-\langle\langle a_{-k}^{+}; a_{-k}\rangle\rangle_E]\delta\eta_{-k}^{*}$$

$$+[\langle\langle a_k; a_k^{+}\rangle\rangle_E^r-\langle\langle a_{-k}^{+}; a_k^{+}\rangle\rangle_E]\delta\eta_k\} \tag{VII.99}$$

where

$$D(E) = \tilde{D}\Omega(k, E), \qquad \tilde{D} = m^2\varrho^2\varrho_n\left(\frac{\partial S}{\partial \Theta}\right)_\varrho,$$

$$D_x = \tilde{D}k^2\Delta(k, E)\sqrt{\varrho_0}(\delta\eta^*_{-k} - \delta\eta_k),$$

$$\Omega(k, E) = (E^2 - c_1^2 k)(E^2 - c_2^2 k^2),$$

$$c_{1,2}^2 = \frac{1}{2m}\left(\frac{\partial P}{\partial \varrho}\right)_S + \frac{1}{2}\frac{S^2\varrho_s\Theta}{m\varrho_n c_v}$$

$$\pm\sqrt{\frac{1}{4}\left[\frac{1}{m}\left(\frac{\partial P}{\partial \varrho}\right)_S + \frac{S^2\varrho_s\Theta}{\varrho_n m c_v}\right]^2 - \frac{1}{m}\left(\frac{\partial P}{\partial \varrho}\right)_\Theta \frac{\varrho_s \Theta S^2}{m c_v}},$$

$$c_v = \Theta\left(\frac{\partial S}{\partial \Theta}\right)_\varrho,$$

$$\Delta(E, k) = \frac{1}{m}\left(\frac{\partial P}{\partial \varrho}\right)_\Theta k^2 \frac{S^2\Theta}{\varrho_n c_v} - E^2\left[\frac{1}{\varrho}\left(\frac{\partial P}{\partial \varrho}\right)_S - \left(\frac{\partial}{\partial \varrho}\varrho S^2\right)_\Theta \frac{\Theta}{\varrho c_v}\right]. \tag{VII.100}$$

c_1 (plus sign) is connected with the ordinary sound velocity. It tends to the sound velocity both if $\varrho \to 0$ and if $\Theta \to 0$. c_2 (minus sign) is the velocity of the second sound (thermal waves), unique to superfluids.

From (VII.99) we see that

$$\langle\langle a_k; a_{-k}\rangle\rangle_E - \langle\langle a^+_{-k}; a_{-k}\rangle\rangle_E = -\langle\langle a_k; a^+_k\rangle\rangle_E + \langle\langle a^+_{-k}; a^+_k\rangle\rangle_E.$$

Similarly as in case (ii) we assume that for small E, in which we are interested, formula (VII.93) is valid. This formula is equivalent to the assumption that the sources of particles do not affect the density of the condensate. Hence from (VII.59) and (VII.93) it follows that

$$\langle\langle a_k; a_{-k}\rangle\rangle_E = \frac{\Delta(E, k)\varrho_0}{2\pi\Omega(k, E)} = -\langle\langle a_k; a^+_{-k}\rangle\rangle_E. \tag{VII.101}$$

From (VII.100) we see that the single-particle Green functions have poles giving the spectrum of two kinds of elementary excitations, first and second sound quanta. For the other variations we have (Galasiewicz, 1967b)

$$\delta\varrho(k) = \frac{E\Delta\eta}{M_1 M_2}\left\{-E^2 + \frac{k^2 S\Theta}{m\varrho\varrho_n c_v}\left[\left(\frac{\partial P}{\partial \Theta}\right)_\varrho \varrho_n + \varrho_s \varrho S\right]\right\}, \tag{VII.102}$$

$$\delta Y = \frac{E^2k^2}{m\varrho_n M_1 M_2}\left\{-\frac{\varrho_n}{\varrho}\left(\frac{\partial P}{\partial \varrho}\right)_S\right.$$
$$\left.+\frac{S\Theta}{c_v}\left[\frac{S\varrho_s}{\varrho}-\left(\frac{\partial S}{\partial \varrho}\right)_\Theta(\varrho_n-\varrho_s)\right]\right\}\Delta\eta, \quad \text{(VII.103)}$$
$$M_i = (E^2-c_i^2k^2), \quad i = 1, 2; \quad \Delta\eta = \sqrt{\varrho_0}(\delta\eta^*_{-k}-\delta\eta_k).$$

Since, for He II, $c_p \approx c_v$ (where c_p, c_v are, respectively, the specific heats, per particle, at constant pressure and volume), we have

$$\left(\frac{\partial P}{\partial \Theta}\right)_\varrho \Big/ \left(\frac{\partial S}{\partial \Theta}\right)_\varrho = \left(\frac{\partial P}{\partial S}\right)_\varrho \sim \sqrt{c_p-c_v} \sim 0. \quad \text{(VII.104)}$$

Hence

$$\left(\frac{\partial P}{\partial \varrho}\right)_S = \left(\frac{\partial P}{\partial \varrho}\right)_\Theta - \left(\frac{\partial S}{\partial \varrho}\right)_\Theta \left(\frac{\partial P}{\partial \Theta}\right)_\varrho \Big/ \left(\frac{\partial S}{\partial \Theta}\right)_\varrho \approx \left(\frac{\partial P}{\partial \varrho}\right)_\Theta, \quad \text{(VII.105)}$$
$$c_1^2 \approx \frac{1}{m}\left(\frac{\partial P}{\partial \varrho}\right)_S, \quad c_2^2 \sim \frac{S^2\varrho_s\Theta}{m\varrho_n c_v}, \quad c_2^2 \ll c_1^2.$$

Moreover, we use the formulae

$$\varrho^2\left(\frac{\partial S}{\partial \varrho}\right)_\Theta = -\left(\frac{\partial P}{\partial \Theta}\right)_\varrho; \quad \left(\frac{\partial S}{\partial \Theta}\right)_\varrho = \frac{c_v}{\Theta}. \quad \text{(VII.106)}$$

Taking into account (VII.104–106) we obtain for $m\delta\varrho(k)$, δX, δY, $(\boldsymbol{k}\cdot\delta\boldsymbol{j}(k))$, $\delta\Theta$ the following expressions:

$$m\delta\varrho(k) = -\frac{mE}{M_1}\Delta\eta, \quad \text{(VII.107)}$$

$$\delta X = \frac{1}{m\varrho_s M_1 M_2} m(c_1k)^2\left[(c_2k)^2 - \frac{\varrho_s}{\varrho}E^2 - E^2\frac{\varrho_n}{\varrho}\left(\frac{c_2}{c_1}\right)^2\right]\Delta\eta, \quad \text{(VII.108)}$$

$$\delta Y = \frac{1}{m\varrho_n M_1 M_2} m(c_1k)^2\frac{\varrho_n}{\varrho}E^2\left[\left(\frac{c_2}{c_1}\right)^2 - 1\right]\Delta\eta, \quad \text{(VII.109)}$$

$$(\boldsymbol{k}\cdot\delta\boldsymbol{j}(k)) = m(\varrho_s X+\varrho_n Y) = -\frac{m(c_1k)^2}{M_1}\Delta\eta, \quad \text{(VII.110)}$$

$$\delta\Theta(k) = \frac{ES\Theta}{\varrho c_v M_2}\Delta\eta. \quad \text{(VII.111)}$$

From (VII.100) it follows that the equation $D(E) = 0$ has two roots. For the solution of this equation $E = c_1k$, corresponding to the wave of the first sound, we have $\delta X = \delta Y$ or $\delta \boldsymbol{v}_s = \delta \boldsymbol{v}_n$. We see that the normal and superfluid components move in the same direction. Moreover, in a first sound wave, the amplitudes of $\delta \varrho$ and $\delta \boldsymbol{j}$ have a sharp maximum.

From (VII.104), it follows that for He II, $\left(\frac{\partial P}{\partial \Theta}\right)_\varrho = 0$, therefore $\delta P(k) = \left(\frac{\partial P}{\partial \varrho}\right)_\Theta \delta \varrho(k) = mc_1^2 \, \delta \varrho(k)$ and the pressure amplitude has a sharp maximum, too.

The solution $E = c_2 k$, corresponding to a second sound wave, is not connected with a maximum of the variations $\delta \varrho$, $\delta \boldsymbol{j}$. In this wave, however, we have $\delta X = -\frac{\varrho_n}{\varrho_s} \delta Y$ or $\delta \boldsymbol{v}_s = -\frac{\varrho_n}{\varrho_s} \delta \boldsymbol{v}_n$. This means that the total current $\delta \boldsymbol{j}$ is equal to zero although both currents, of the normal and the superfluid components, are different from zero (Lifshitz, 1944a). We have two interpenetrating currents. As was stressed in the Introduction, the existence of such a current explains the thermal superconductivity observed in He II.

From (VII.76,77), (VII.107,110) we finally obtain

$$\frac{2\pi}{\sqrt{V}} \sum_{\boldsymbol{q}} (\boldsymbol{k} \cdot \boldsymbol{q}) \langle\langle a^+_{q-\frac{k}{2}} a_{q+\frac{k}{2}} ; a_- \rangle\rangle^r_E = -\frac{mc_1^2 k^2 \sqrt{\varrho_0}}{E^2 - c_1 k^2}$$

$$= 2\pi \boldsymbol{k} \langle\langle \boldsymbol{j}_k ; a_{-k} \rangle\rangle^r_E, \qquad \text{(VII.112)}$$

$$\frac{2\pi}{\sqrt{V}} \sum_{\boldsymbol{q}} \langle\langle a^+_{q-\frac{k}{2}} a_{q+\frac{k}{2}} ; a_{-k} \rangle\rangle^r_E = \frac{E\sqrt{\varrho_0}}{E^2 - c_1^2 k^2} = 2\pi \langle\langle \hat{\varrho}_k ; a_{-k} \rangle\rangle^r_E .$$

We see that Green functions are singular only for energies of the first sound quanta. The amplitudes of density and current variations can be expressed in terms of these functions. Due to the singularities of the Green functions, the amplitudes have a sharp maximum. Therefore the microscopic theory shows that only first sound contributes to the density oscillations amplitude. This suggests that second sound waves are not pressure waves ($\delta P \sim \delta \varrho$) and the nature of second sound must be quite different from the nature of first sound.

On the other hand, the expression for $\delta \Theta$ (VII.111) has poles for the energies of the second sound quanta, which indicates that second sound is connected with thermal waves. This was confirmed by Peshkov (1944) who produced and observed thermal waves in He II.

Formulae (VII.112, 113) are, in the Green functions language, the expressions of the results of Lifshitz (1944a).

(iv) Thermodynamic equilibrium ($E = 0$), $\delta U \neq 0$, $\delta A \neq 0$, viscous terms different from zero (Galasiewicz, 1968):

$$\delta X(k) = \frac{\sqrt{\varrho_0}}{\varrho_s}(\delta\eta^*_{-k} - \delta\eta_k), \tag{VII.114}$$

$$\delta\varrho(k) = -\frac{\varrho}{\left(\frac{\partial P}{\partial \varrho}\right)_\Theta}\{\delta U(k) + \sqrt{\varrho_0}[B_1(\delta\eta^*_{-k} + \delta\eta_k)$$
$$+ i(\zeta_3 - B_2)(\delta\eta^*_{-k} - \delta\eta_k)]\}, \qquad B = B_1 + iB_2.$$

Comparing with (VII.74,75) we see that

$$\frac{\delta X}{\delta U} = 0 = \langle\langle a_k; \varrho_{-k}\rangle\rangle_0 - \langle\langle a^+_{-k}; \varrho^+_k\rangle\rangle_0,$$
$$\frac{\delta\varrho}{\delta\eta^*_{-k}} = 2\pi\langle\langle\varrho_k; a_{-k}\rangle\rangle_0, \qquad \frac{\delta\varrho}{\partial\eta_k} = 2\pi\langle\langle\hat{\varrho}^+_{-k}; a^+_k\rangle\rangle_0. \tag{VII.115}$$

From (VII.115) and (V.59) we see that

$$\frac{\delta\varrho}{\delta\eta^*_{-k}} = \frac{\delta\varrho}{\delta\eta_k}. \tag{VII.116}$$

On the other hand, from (VII.114) we have

$$\frac{\delta\varrho}{\delta\eta^*_{-k}} = -\frac{\varrho\sqrt{\varrho_0}}{\left(\frac{\partial P}{\partial\varrho}\right)_\Theta}[B_1 + i(\zeta_3 - B_2)],$$

$$\frac{\delta\varrho}{\delta\eta_k} = -\frac{\varrho\sqrt{\varrho_0}}{\left(\frac{\partial P}{\partial\varrho}\right)_\Theta}[B_1 - i(\zeta_3 - B_2)]. \tag{VII.116a}$$

Comparing (VII.116) with (VII.116a) we obtain

$$B_2 = \zeta_3. \tag{VII.117}$$

Our considerations allow us to determine the imaginary part of the newly introduced kinetic coefficient. The real part cannot be determined from these considerations. We also obtain for the Green functions

$$\langle\langle\hat{\varrho}_k; a_{-k}\rangle\rangle_0 = -\frac{\varrho\sqrt{\varrho_0}}{2\pi\left(\frac{\partial P}{\partial\varrho}\right)_\Theta}B_1 = \langle\langle\varrho^+_{-k}; a^+_k\rangle\rangle_0. \tag{VII.118}$$

This type of function does not have a $1/k^2$ type singularity as $k \to 0$. The right-hand side is constant in the sense that it is independent of k. For an ideal Bose liquid this constant vanishes. For these reasons it seems that the right-hand side of (VII.118) must in general be equal to zero. Therefore we assume that

$$B_1 = 0. \tag{VII.118a}$$

(v) General case, $\delta U = 0$, $\delta A = 0$, viscous terms different from zero (Galasiewicz, 1963b).

Instead of $D(E)$ given by (VII.100) we have

$$D(E) \cong \tilde{D}[E^4 + iA_3 k^2 E^3 - (c_1 + c_2^2) k^2 E^2 + iA_1 k^4 E + c_1^2 c_2^2 k^4] \tag{VII.119}$$

where

$$A_1 = \frac{\varrho_s}{m^2 \varrho \varrho_n} \left(\frac{\partial P}{\partial \varrho}\right)_S \left[\frac{4}{3}\eta + \zeta_2 - \varrho(\zeta_1 + \zeta_4) + \zeta_3 \varrho^2\right]$$
$$+ \frac{S^2 \varrho_s}{m^2 \varrho_n \varrho} \frac{\Theta}{c_v} \left(\frac{4}{3}\eta + \zeta_2\right) + \frac{1}{m\varrho} \left(\frac{\partial P}{\partial \varrho}\right)_\Theta \frac{\varkappa}{\Theta} \frac{\Theta}{c_v}$$
$$- \frac{S \varrho_s}{m^2 \varrho^2 \varrho_n} \left[2\left(\frac{4}{3}\eta + \zeta_2\right) - \varrho(\zeta_1 + \zeta_4) + \zeta_3 \varrho^2\right] \left(\frac{\partial P}{\partial \Theta}\right)_\varrho \Bigg/ \left(\frac{\partial S}{\partial \Theta}\right)_\varrho, \tag{VII.120a}$$

$$A_3 = -\frac{1}{\varrho} \frac{\varkappa}{\Theta} \frac{\Theta}{c_v} + \frac{1}{m\varrho_n} \left[\frac{4}{3}\eta + \zeta_2 - \varrho_s(\zeta_1 + \zeta_4) + \varrho_s \zeta_3 \varrho\right]. \tag{VII.120b}$$

The expression for $D(E)$ is approximate in the sense that we do not consider changes of the coefficients of E^4, E^2, E^0. Those changes result from new terms appearing in (VII.83–86).

We can consider an ideal liquid as a problem of two oscillations forced by $\delta H_\tau(\omega)$ without introducing "friction" (and, consequently, damping phenomena). Hence a viscous liquid can be treated as a case of forced oscillations by taking into account damping phenomena. Therefore we are primarily interested in the damping coefficients and not in small changes in the oscillation frequencies which are given by the full coefficients of E^4, E^2, E^0.

From (VII.119) we see that the equation

$$D(E) = 0$$

has complex roots. The real parts of these roots will be very close to the roots of $D(E)$ given by (VII.100)

$$E_1^{\pm} = \omega_1^{\pm} = \pm c_1 k = \pm\omega_1, \quad E_2^{\pm} = \omega_2^{\pm} = \pm c_2 k = \pm\omega_2 \tag{VII.121}$$

and the imaginary parts will be small. Therefore we treat the roots of (VII.121) as a first approximation of the solution of (VII.119). The corrections obtained by Newton's method (we are interested in the imaginary part only) are

$$\operatorname{Im}\delta_\alpha^{\pm} = -\operatorname{Im}\frac{D(E_\alpha^{\pm})}{\left(\dfrac{\mathrm{d}D}{\mathrm{d}E}\right)_{E_\alpha^{\pm}}}, \quad \alpha = 1, 2. \tag{VII.122}$$

From (VII.119) we have

$$\begin{aligned} \operatorname{Im}\delta_1^{\pm} &= -\frac{A_3 c_1^2 + A_1}{2(c_1^2+c_2^2)}\,k^2 = \varepsilon_1, \\ \operatorname{Im}\delta_2^{\pm} &= +\frac{A_3 c_2^2 + A_1^2}{2(c_1^2-c_2^2)}\,k^2 = \varepsilon_2 \end{aligned} \tag{VII.123}$$

where A_1, A_3 are given by (VII.120a,b) and c_1^2, c_2^2 by (VII.100). From (VII.100) it follows that $c_1^2 - c_2^2 > 0$. The sign of δ_α depends on the sign of the numerator.

For D_X we have

$$D_x = \frac{1}{m}\,\tilde{D}k^2\{\Delta(k, E) - ik^2 E\Delta_0 - i\zeta_3\Delta_2(k, E)\}\Delta\eta \tag{VII.124}$$

where

$$\Delta(k, E) = \frac{k^2}{m}\left(\frac{\partial P}{\partial\varrho}\right)_\Theta \frac{S^2\Theta}{\varrho_n c_v} - E^2\left[\frac{1}{\varrho}\left(\frac{\partial P}{\partial\varrho}\right)_S - \left(\frac{\partial}{\partial\varrho}\varrho S^2\right)_\Theta \frac{\Theta}{\varrho c_v}\right], \tag{VII.125}$$

$$\begin{aligned} \Delta_0 = \Bigg\{ &-\left(\frac{\partial P}{\partial\varrho}\right)_\Theta \frac{\Theta\varkappa}{\varrho^2 c_v} - \frac{1}{m\varrho\varrho_n}\left(\frac{\partial P}{\partial\varrho}\right)_S\left[\frac{4}{3}\eta + \zeta_2 - \varrho(\zeta_1+\zeta_4)\right. \\ &\left.+\zeta_3\varrho^2\right] + \frac{S\left(\dfrac{\partial P}{\partial\Theta}\right)_\varrho}{m\varrho^2\varrho_n}\frac{\Theta}{c_v}\left[2\left(\frac{4}{3}\eta+\zeta_2\right) - \varrho(\zeta_1+\zeta_4) + \varrho\zeta_3\varrho_n\right] \\ &- \frac{S^2}{m\varrho\varrho_n}\frac{\Theta}{c_v}\left(\frac{4}{3}\eta + \zeta_2 - \varrho\zeta_3\varrho_s\right)\Bigg\}, \end{aligned} \tag{VII.126}$$

$$\Delta_2 = E^2 - k^2 \frac{\Theta}{c_v} \left[\frac{\varrho_s S^2}{m\varrho_n} + \frac{S\left(\dfrac{\partial P}{\partial \Theta}\right)_\varrho}{m\varrho} \right]. \tag{VII.127}$$

Formulae (VII.123) are obtained for a theoretical Bose fluid. In order to determine the sign of ε_α we introduce the properties of a real Bose fluid (helium II) (formulae (VI.104–106)). After simplifications which result from the properties of a real Bose fluid we have for A_1, A_3:

$$A_1 = \left\{ \frac{\varrho_s}{m\varrho\varrho_n} \left[\frac{4}{3}\eta + \zeta_2 - \varrho(\zeta_1+\zeta_4) + \varrho^2\zeta_3 \right] + \frac{\varkappa}{\varrho c_v} \right\} c_1^2,$$

$$A_3 = -\frac{1}{m\varrho_n} \left[\frac{4}{3}\eta + \zeta_2 - \varrho_s(\zeta_1+\zeta_4) + \varrho_s\varrho\zeta_3 \right] - \frac{\varkappa}{\varrho c_v}. \tag{VII.128}$$

Therefore we see that $\mathrm{Im}\,\delta_1^\pm$, $\mathrm{Im}\,\delta_2^\pm$ are, in fact, damping coefficients

$$\mathrm{Im}\,\delta_1^\pm = -\frac{1}{2}\left(A_3 + \frac{A_1}{c_1^2} \right) k^2 = \frac{1}{2m\varrho} \left(\frac{4}{3}\eta + \zeta_2 \right) \equiv \varepsilon_1 > 0,$$

$$\mathrm{Im}\,\delta_2^\pm = \frac{A_1}{2c_1^2} k^2 = \frac{1}{2m\varrho} \frac{\varrho_s}{\varrho_n} \left[\frac{4}{3}\eta + \zeta_2 - \varrho(\zeta_1+\zeta_4) + \varrho^2\zeta_3 + \frac{\varrho_n}{\varrho_s} \frac{m\varkappa}{c_v} \right] \equiv \varepsilon_2 > 0. \tag{VII.129}$$

c_v is the specific heat per particle; in the papers by Hohenberg and Martin (1965) and Khalatnikov (1952c, 1965) c_v is the specific heat per unit mass.

Instead of (VII.125–127)we now have

$$\Delta(k, E) = c_1^2 k^2 \frac{S^2\Theta}{\varrho_n c_v} - \frac{E^2}{\varrho} \left[mc_1^2 - \frac{S^2\Theta}{c_v} \right], \tag{VII.130}$$

$$k^2\Delta_0 + \zeta_3\Delta_2 \equiv \Delta_1(k, E) = -c_1^2 k^2 \left\{ \frac{\varkappa}{\varrho^2} \frac{\Theta}{c_v} + \frac{i}{\varrho\varrho_n} \left[\frac{4}{3}\eta + \zeta_2 - \varrho(\zeta_1+\zeta_4) + \zeta_3\varrho^2 \right] \right\} - c_2^2 k^2 \left(\frac{4}{3}\eta + \zeta_2 \right) \frac{1}{\varrho\varrho_s} + E^2\zeta_3. \tag{VII.131}$$

We can write $D(\omega)$ in the form

$$D(\omega) = \tilde{D}(\omega - i\varepsilon_1 - \omega_1)(\omega - i\varepsilon_1 + \omega_1)(\omega - i\varepsilon_2 - \omega_2)(\omega - i\varepsilon_2 + \omega_2)$$
$$\approx \tilde{D}(\omega^2 - \omega_1^2 - 2i\omega\varepsilon_1)(\omega^2 - \omega_2^2 - 2i\omega\varepsilon_2). \tag{VII.132}$$

We calculate δX from (VII.124), (VII.132) and (VII.118a,b) and obtain

$$\delta X = \frac{D_x}{D(\omega)} = \frac{k^2[\Delta(k,\omega)-i\omega\Delta_1(k,\omega)]}{m(\omega^2-\omega_1^2-2i\omega\varepsilon_1)(\omega^2-\omega_2^2-2i\omega\varepsilon_2)}\Delta\eta. \tag{VII.133}$$

Since for $E = 0$, $\delta\Theta$ and $\delta\varrho$ are also equal to zero, we assume that for small E formulae (VII.93) are valid. Therefore we obtain the Fourier components of the single-particle Green functions in the form

$$\langle\langle a_k; a_{-k}\rangle\rangle_E^r = \frac{\varrho_0[\Delta(k,E)-iE\Delta_1(k,E)]}{2\pi(E^2-\omega_1^2-2iE\varepsilon_1)(E^2-\omega_2^2-2iE\varepsilon_2)} = -\langle\langle a_k; a_k^+\rangle\rangle_E^r. \tag{VII.134}$$

From (VII.134) and (VII.129) it follows that for a superfluid with viscosity for the normal component, damping of the oscillations forced by δH_t appears (absorption of first and second sound). The imaginary part of the roots of the retarded Green functions give us the absorption coefficients of first and second sound. These coefficients depend on the kinetic coefficients.

(vi) General case, $\delta U \neq 0$, viscous terms different from zero. In order to obtain the Onsager–Khalatnikov relation we calculate the following derivatives (Petru, 1969b):

$$\left(\boldsymbol{k}\cdot\frac{\delta \boldsymbol{v}_s(k)}{\delta U(k)}\right) = \frac{\omega k^2}{m\Omega}\{\omega^2-k^2c_2^2-i\omega k^2k_1\}, \tag{VII.135}$$

$$\frac{\delta\varrho_k}{\delta\eta_{-k}^*} = \frac{\sqrt{\varrho_0}}{\Omega}\left\{-\omega(\omega^2-c_2^2k^2)+ik^2\omega^2K_2+B^*\frac{k^2\varrho}{m}(\omega^2-c_2^2k^2)\right.$$
$$\left.-i\zeta_3c_2^2k^4\frac{\varrho}{m}\right\}, \tag{VII.136}$$

$$\frac{\partial\varrho_k}{\partial\eta_k} = -\frac{\sqrt{\varrho_0}}{\Omega}\left\{-\omega(\omega^2-c_2^2k^2)+ik^2\omega^2K_2-B\frac{k^2\varrho}{m}(\omega^2-c_2^2k^2)\right.$$
$$\left.-i\zeta_3c_2^2k^4\frac{\varrho}{m}\right\}, \tag{VII.137}$$

where

$$\Omega = \omega^4-ik^2\omega^3K_3-k^2\omega^2[c_2^2+c_1^2]+ik^4\omega\left[c_1^2K_1+\frac{c_2^2}{m\varrho}\left(\frac{4}{3}\eta+\zeta_2\right)\right]$$
$$+k^4c_1^2c_2^2;$$

$$K_1 = \frac{1}{m\varrho_n}\left(\frac{4}{3}\eta - \varrho_s\zeta_1 + \zeta_2 + \varrho\varrho_s\zeta_3 - \varrho\zeta_4\right) + \frac{\varkappa}{\varrho c_v},$$

$$K_2 = \frac{1}{m\varrho_n}\left(\frac{4}{3}\eta - \varrho\zeta_1 + \zeta_2 + \varrho^2\zeta_3 - \varrho_s\zeta_4\right) + \frac{\varkappa}{c_v\varrho}, \qquad \text{(VII.138)}$$

$$K_3 = \frac{1}{m\varrho_n}\left(\frac{4}{3}\eta - \varrho_s\zeta_1 + \zeta_2 + \varrho\varrho_s\zeta_3 - \varrho_s\zeta_4\right) + \frac{\varkappa}{c_v\varrho}.$$

Considering the relation (V.59) between the Green functions $\langle\langle a_k; \varrho_{-k}\rangle\rangle_E$, $\langle\langle \varrho_k; a_{-k}\rangle\rangle_E$, we obtain

$$\left(\boldsymbol{k}\frac{\delta \boldsymbol{v}_s}{\delta U(k)}\right) = \frac{k^2}{2m\sqrt{\varrho_0}}\left[\frac{\delta\varrho_k}{\delta\eta_s} - \frac{\delta\varrho_s}{\delta\eta^*_{-k}}\right]. \qquad \text{(VII.139)}$$

From this it follows that

$$\omega^2\left(i\zeta_1 - i\zeta_4 - i\varrho\zeta_3 + \frac{B-B^*}{2}\varrho\right) + k^2c_2^2\varrho\left(i\zeta_3 - \frac{B-B^*}{2}\right) = 0, \qquad \text{(VII.140)}$$

and finally that

$$\zeta_1 = \zeta_4, \qquad \frac{B-B^*}{2} \equiv iB_2 = i\zeta_3. \qquad \text{(VII.141)}$$

In this way, we have obtained the Onsager–Khalatnikov relations and the relation (VII.117).

CHAPTER VIII

The Fermi superfluid

§ 1 Hydrodynamic equations

We draw attention to the fact that for the case of the coupled pair condensate (Bogoliubov, 1961)

$$\langle \psi(t, x_1)\psi(t_1, x_2)\rangle \neq 0.$$

In considering Bose systems, the velocity of the condensate is defined through the phase of the mean value $\langle \psi_B \rangle$, where the ψ_B are the Bose operators. Similarly, we introduce the velocity of the condensate of coupled pairs and obtain the equations for this velocity. Therefore we use (V.21) and introduce, instead of $\tilde{\Phi}$ and $\tilde{\Phi}^*$, two real functions, the phase χ and the modulus a. We write

$$\tilde{\Phi}(t|r_1, r_2) \equiv \sqrt{\varrho_0}\, a(t, r_1, r_2) e^{i\chi(t, r_1, r_2)} = \sqrt{\varrho_0}\, a(t, \boldsymbol{r}, \boldsymbol{R}) e^{i\chi(t, r, R)}, \tag{VIII.1}$$

where

$$\boldsymbol{r} = \tfrac{1}{2}(\boldsymbol{r}_1 + \boldsymbol{r}_2), \quad \boldsymbol{R} = \boldsymbol{r}_1 - \boldsymbol{r}_2,$$

and ϱ_0 is the density of the number of pairs in the condensate. We assume that for $R \to \infty$ $(R > R_0)$,

$$\chi(t, r_1, r_2) \to \chi_1(t, r_1) + \chi_2(t, r_2) \quad \text{and} \quad a(t, r, R) \to 0.$$

The external sources $\tilde{\eta}$ are taken in the form

$$\tilde{\eta}(t|r_1, r_2) = \tilde{\eta}(t|r, R) \equiv \vartheta(R)\eta_1(t, r), \tag{VIII.2}$$

where $\vartheta(R)$ is a function decreasing with distance (practically equal to zero for $R > R_0$). The radius of a pair in the condensate is R_0.

In the theory of superconductivity (thermodynamic equilibrium) (Bogoliubov, 1959)

$$a(R) = \frac{C}{(2\pi)^3} \int \frac{e^{i(p \cdot R)}}{2\sqrt{\xi^2(p)+C^2}}\, d^3 p,$$
$$\xi(p) = E'(p-p_F), \quad E' = \left(\frac{dE}{dp}\right)_{p_F}, \tag{VIII.3}$$

where C is the energy gap and p_F is the momentum for the Fermi surface. Recalling that we perform the integration in a thin layer, for $p \approx p_F$ (or in the thin layer $\varepsilon_F \pm \omega$, where ε_F is the energy of the Fermi surface, and ω the mean phonon energy), we have

$$a(R) \cong 8\pi p_F \frac{\sin p_F R}{R} \frac{C}{E'} \int_0^{\omega/c} \frac{\cos(R\xi C/E')}{\sqrt{\xi^2+1}}\, d\xi$$
$$= 8\pi p_F \frac{\sin p_F R}{R} \frac{C}{E'} K_0\left(\frac{C}{E'} R\right), \tag{VIII.4}$$

(the upper limit is taken to be $+\infty$) where K_0 is the Bessel function which (for $R \neq 0$) can be written in the form

$$K_0\left(\frac{C}{E'} R\right) = \sqrt{\frac{\pi}{2RC/E'}}\, e^{-RC/E'} \left(1 - \frac{1}{8RC/E'} + \frac{9}{128R^2(C/E')^2} - +\right). \tag{VIII.5}$$

Defining the radius of a pair as $R_0 = E'/C$, we see that for $R > R_0$

$$a(R) = \frac{8\pi p_F}{R} \frac{\sin(p_F E' R/CR_0)}{R/R_0} \sqrt{\frac{\pi}{2R/R_0}}\, e^{-R/R_0} \times \left(1 - \frac{1}{8R/R_0} + \frac{9}{128(R/R_0)^2} - \ldots + \ldots\right) \tag{VIII.6}$$

is a fast decreasing function of R/R_0. A similar expression for the radius of a pair was obtained by Abrikosov and Khalatnikov (1958), while considering the electrodynamics of the superconducting state.

The equations for χ and a are obtained from (V.21) by changing from $(\boldsymbol{r}_1, \boldsymbol{r}_2)$ to $(\boldsymbol{r}, \boldsymbol{R})$. We have:

$$\frac{\partial \chi(t, r, R)}{\partial t} = \frac{1}{4m}\left[\frac{1}{a(t, r, R)} (\nabla_R^2 + 4\nabla_R^2) a(t, r, R) - (\nabla_r \chi(t, r, R))^2 - (4\nabla_R \chi(t, r, R))^2\right] + 2\lambda - V(R)$$

$$-\frac{1}{2\varrho_0 a^2(t,r,R)}\int V(R')\{X_t(r,R,R')+X_t^*(r,R,R')\}\,d^3\boldsymbol{R}'$$
$$+S_\chi[\tilde{\eta},\tilde{\eta}^*], \quad \text{(VIII.7)}$$

$$i\frac{\partial a^2(t,r,R)}{\partial t}=-\frac{ia}{2m}[2(\nabla_r\chi\cdot\nabla_r a)+8(\nabla_R\chi\cdot\nabla_R a)+a(\nabla_r^2\chi$$
$$+4\nabla_R^2\chi)]+\frac{1}{\varrho_0}\int V(R')[X_t(r,R,R')-X_t^*(r,R,R')]\,d^3\boldsymbol{R}$$
$$+S_a[\tilde{\eta},\tilde{\eta}^*], \quad \text{(VIII.8)}$$

where

$$S_\chi=-\frac{1}{a\sqrt{\varrho_0}}\vartheta(R)[\tilde{\zeta}(t,r,R)-\tilde{\zeta}^*(t,r,R)]$$
$$-\frac{1}{a\sqrt{\varrho_0}}\int\vartheta(R)\left[\eta_1\left(t,r+\tfrac{1}{2}R-\tfrac{1}{2}\tilde{R}\right)e^{-i\chi(t,r,R)}\right.$$
$$\times\tilde{F}\left(t|r-\tfrac{1}{2}R,-R+\tilde{R}\right)+\eta_1\left(t,r-\tfrac{1}{2}R-\tfrac{1}{2}\tilde{R}\right)e^{-i\chi(t,r,R)}$$
$$\times\tilde{F}\left(t|r+\tfrac{1}{2}R,R+\tilde{R}\right)+\eta_1^*\left(t,r+\tfrac{1}{2}R-\tfrac{1}{2}\tilde{R}\right)e^{i\chi(t,r,R)}$$
$$\times\tilde{F}^*\left(t|r-\tfrac{1}{2}R,-R+\tilde{R}\right)+\eta_1^*\left(t,r-\tfrac{1}{2}R-\tfrac{1}{2}\tilde{R}\right)e^{i\chi(t,r,R)}$$
$$\left.\times\tilde{F}^*\left(t|r+\tfrac{1}{2}R,R+\tilde{R}\right)\right]d^3\boldsymbol{R}, \quad \text{(VIII.7a)}$$

$$S_a=\frac{2a}{\sqrt{\varrho_0}}\int\vartheta(\tilde{R})\left[\eta_1\left(t,r+\tfrac{1}{2}R-\tfrac{1}{2}\tilde{R}\right)e^{-i\chi(t,r,R)}\right.$$
$$\times\tilde{F}\left(t|r-\tfrac{1}{2}R,-R+\tilde{R}\right)+\eta_1\left(t,r-\tfrac{1}{2}R-\tfrac{1}{2}\tilde{R}\right)e^{-i\chi(t,r,R)}$$
$$\times\tilde{F}\left(t|r+\tfrac{1}{2}R,R+\tilde{R}\right)-\eta_1^*\left(t,r+\tfrac{1}{2}R-\tfrac{1}{2}\tilde{R}\right)e^{i\chi(t,r,R)}$$
$$\times\tilde{F}^*\left(t|r-\tfrac{1}{2}R,-R+\tilde{R}\right)-\eta_1^*\left(t,r-\tfrac{1}{2}R-\tfrac{1}{2}\tilde{R}\right)e^{i\chi(t,r,R)}$$
$$\left.\times\tilde{F}^*\left(t|r+\tfrac{1}{2}R,R+\tilde{R}\right)\right]d^3\boldsymbol{R}, \quad \text{(VIII.8a)}$$

$$\tilde{\zeta}(t,r,R)\equiv\eta_1(t,r)e^{-i\chi(t,r,R)},$$

$$\left[Y_t\left(r+\tfrac{1}{2}R,R',R\right)+Y_t\left(r-\tfrac{1}{2}R,R',-R\right)\right]$$
$$\times\sqrt{\varrho_0}\,a(t,r,R)e^{-i\chi(t,r,R)}\equiv X_t(r,R',R), \quad \text{(VIII.9)}$$

$$2Y_t(r,R',0)\sqrt{\varrho_0}\,a(t,r,0)e^{-i\chi(t,r,R)}=X_t(r,R,0)\equiv X_t^0(r,R').$$

We introduce the notation:

$$\chi(t, r, 0) \equiv \chi_0(t, r), \qquad a(t, r, 0) \equiv a_0(t, r),$$

$$(\nabla_R f(R))_{R=0} \equiv \mathring{\nabla}_R f(R), \qquad \tilde{\zeta}(t, r, 0) \equiv \tilde{\zeta}_0(t, r).$$

We define now the velocity of a condensate pair by means of the formula (Usui, 1964)

$$\boldsymbol{v}_s(t, r) \equiv \frac{1}{M} \nabla_r \chi_0, \quad M = 2m. \tag{VIII.10}$$

We also define the velocity

$$\boldsymbol{v}_r(t, r) \equiv \frac{1}{m_r} \mathring{\nabla}_R \chi, \quad m_r = \tfrac{1}{2} m. \tag{VIII.11}$$

$\boldsymbol{v}_r$ plays the role of the "relative" velocity of the particles of the pair.

According to (VIII.7) and (VIII.7a), the equation for $\boldsymbol{v}_s(t, r)$ has the form

$$M \frac{\partial \boldsymbol{v}_s(t, r)}{\partial t} = \nabla_r \Bigg\{ \frac{\mathring{\nabla}_r^2 a}{4 m a_0} - \frac{M v_s^2}{2} + \frac{\mathring{\nabla}_R^2 a}{m a_0} - \frac{m_r v_r^2}{2}$$

$$- \frac{1}{2 \varrho_0 a_0^2} \int V(R') \left[X_t(r, R', 0) + X_0^*(r, R', 0)\right] \mathrm{d}^3 \boldsymbol{R}'$$

$$- \frac{1}{a_0 \sqrt{\varrho_0}} \vartheta(0) \left[\tilde{\zeta}_0(t, r) + \tilde{\zeta}_0^*(t, r)\right] - \frac{2}{a_0 \sqrt{\varrho_0}} \int \vartheta(\tilde{R}) \Big[\eta_1\left(t, r - \tfrac{1}{2} \tilde{R}\right)$$

$$\times e^{-i \chi_0(t, r)} \tilde{F}(t|r, \tilde{R}) + \eta_1^*\left(t|r - \tfrac{1}{2} \tilde{R}\right) e^{i \chi_0(t, r)} \tilde{F}^*(t|r, \tilde{R})\Big] \mathrm{d}^3 \boldsymbol{R} \Bigg\}. \tag{VIII.12}$$

Consider the case of statistical equilibrium for a Fermi superfluid with $\tilde{\eta} = 0$. After introducing $\boldsymbol{v}_s$ we see that, as in the case of a Bose liquid, a Fermi superfluid is described by eight parameters. It is convenient to choose for these parameters the density of the number of particles ϱ, the temperature Θ, three components of the velocity of the condensate $\boldsymbol{v}_s$ and three components of the velocity of the "normal component" $\boldsymbol{v}_n$. If there are no sources, then these parameters are constant in thermodynamic equilibrium. If sources exist ($\tilde{\eta} = \tilde{\eta}(r)$), then $\varrho = \varrho(r)$, $\boldsymbol{v}_s = \boldsymbol{v}_s(r)$. For deviations from thermodynamic equilibrium we use these eight variables to describe the state of the system only if they vary slowly in time and space. We denote the mean values for thermodynamic equilibrium by $\langle \dots \rangle_{\varrho, \Theta, v_s, v_n}$. Later we shall deal with mean values of the type

$$\begin{aligned} \mathfrak{A} &= \langle (D_1 \gamma(t, r_1)), \dots (D_n \gamma(t, r_n)) \rangle_{\varrho, \Theta, v_s, v_n} \\ &= \mathfrak{A}(r_1 - r_2, \dots r_1 - r_n | \varrho, \Theta, v_s, v_n), \end{aligned} \tag{VIII.13}$$

where

$$\gamma(t, r_j) = \psi(t, r_j), \psi^+(t, r_j),$$

and where D_ν are linear forms of constants and spatial differential operators (see, for example, the expressions for D_t, $G_t^{(\alpha)}$, X_t).

It is convenient to express the mean values (VIII.13) in terms of the mean values

$$\langle \dots \rangle_{v_s', 0}, \quad v_s' = v_s - v_n, \quad v_n' = 0$$

(for the sake of simplicity we omit the indices ϱ, Θ). For this purpose we can use the Galilean transformation for the operators

$$\psi \to \psi \exp\big(im(\boldsymbol{v}_n \cdot \boldsymbol{r})\big).$$

Consider a state $(\varrho, \Theta, \boldsymbol{v}_s, 0)$ for which $\boldsymbol{v}_n = 0$. In this state

$$a = a(R/\varrho, \Theta, u), \qquad a_0 = a(\varrho, \Theta, u) = \text{const.},$$

$$u = \frac{1}{2} v_s^2,$$

$$\Lambda(\varrho, \Theta, u) = \frac{\partial(NF)}{\partial N} = F + \varrho \frac{\partial F}{\partial \varrho}, \qquad F = F(\varrho, \Theta, u),$$

where N is the number of particles, F is the free energy and Λ is the chemical potential per particle.

Since in thermodynamic equilibrium a is normalized to unity (Bogoliubov, 1961):

$$\int a^2(R) \mathrm{d}^3\boldsymbol{R} = 1,$$

the quantity

$$\sum_{\sigma_2} \int |\langle \psi(t, x_1) \psi(t, x_2) \rangle|^2 \mathrm{d}^3\boldsymbol{r}_2 = \varrho_0$$

gives the density of pairs in the condensate.

We introduce the current

$$j_\alpha = \varrho \frac{\partial F}{\partial v_s^{(\alpha)}} = \varrho \frac{\partial F}{\partial u} v_s^{(\alpha)}$$

and define

$$\frac{\varrho}{M} \frac{\partial F}{\partial u} = \varrho_s. \tag{VIII.14}$$

Therefore

$$j_\alpha = M\varrho_s v_s^{(\alpha)}, \quad M = 2m. \tag{VIII.15}$$

Similarly as in the papers by Bogoliubov (1963) (see Chapter VII), we can show that

$$j_\alpha = \frac{1}{2} i \sum_s \left\langle \frac{\partial \psi^+}{\partial r_\alpha} \psi - \psi^+ \frac{\partial \psi}{\partial r_\alpha} \right\rangle_{v_s, 0}. \tag{VIII.16}$$

In thermodynamic equilibrium $X^0 = X^0(R'|\varrho, \Theta, v_s)$. From eqn. (VIII.8) $(R = 0)$ we get

$$\frac{1}{\varrho_0 a_0^2} \int V(R') X^0(R'|\varrho, \Theta, v_s) \mathrm{d}^3 R' = -\frac{1}{2} M v_s^2 + 2\Lambda(\varrho, \Theta, u)$$

$$-V(0) + \frac{\mathring{\nabla}_R^2 a}{m a_0} - \frac{1}{2} m_r v_r^2 + \frac{i}{2a_0} (a_0 \mathring{\nabla}_R \boldsymbol{v}_r + 2\boldsymbol{v}_r \mathring{\nabla}_R a), \tag{VIII.17}$$

$$\frac{1}{\varrho_0 a_0^2} \int V(R') X^{0*}(R'|\varrho, \Theta, v_s) \mathrm{d}^3 \boldsymbol{R}' - \frac{1}{\varrho_0 a_0^2} \int V(R') X^0(R'|\varrho, \Theta, v_s) \mathrm{d}^3 \boldsymbol{R}$$

$$-\frac{i}{a_0} (a_0 \mathring{\nabla} \boldsymbol{v}_r + 2\boldsymbol{v}_r \mathring{\nabla}_r a). \tag{VIII.18}$$

Now we consider a state with two velocities $\boldsymbol{v}_s$, $\boldsymbol{v}_n$. We note that the expression

$$\langle \psi^+(t, x') \psi(t, x') \psi(t, x_1) \psi(t, x_2) \rangle_{v_s, v_n} \langle \psi^+(t, x_2) \psi^+(t, x_1) \rangle_{v_s, v_n}$$

is invariant under Galilean transformation and therefore

$$\psi \to \psi \exp\big(im(\boldsymbol{v}_n \cdot \boldsymbol{v}_r)\big), \quad X^0(R|\varrho, \Theta, v_s, v_n) = X^0(R|\varrho, \Theta, v_s - v_n).$$

Consequently, for a state with two velocities, in eqns. (VIII.17) we must write $(v_s - v_n)^2$ instead of v_s^2.

Consider now the expression for j_α,

$$j_\alpha = \frac{i}{2} \sum_s \left\langle \frac{\partial \psi^+}{\partial r_\alpha} \psi - \psi^+ \frac{\partial \psi}{\partial r_\alpha} \right\rangle_{v_s, v_n} = \frac{i}{2} \sum_s \left\langle \frac{\partial \psi^+}{\partial r_\alpha} \psi \right. \tag{VIII.19}$$

$$\left. - \psi^+ \frac{\partial \psi}{\partial r_\alpha} \right\rangle_{v_s - v_n, 0} + m\varrho v_n^{(\alpha)} = M\varrho_s v_s^{(\alpha)} + m\varrho v_n^{(\alpha)}, \quad M = 2m.$$

We have obtained (VIII.19) by using formulae (VIII.15,16) and introducing the density of the normal component ϱ_n by means of the definition

$$m\varrho_n = m(\varrho - 2\varrho_s). \tag{VIII.20}$$

Now let us consider hydrodynamic non-equilibrium processes

for which the non-equilibrium mean values of the type

$$\mathfrak{A}_n = \langle (D_1\gamma(t, r_1)) \dots (D_n\gamma(t, r_n))\rangle = \mathfrak{A}_n(t, r_1, r_1 - r_2, \dots r_1 - r_n) \tag{VIII.21}$$

vary extremely slowly under time and space translations. Indeed, we consider gradients $\frac{\partial}{\partial t}, \frac{\partial}{\partial \boldsymbol{r}}$ as of first order, and gradients like $\frac{\partial^2}{\partial r_\alpha \partial r_\beta}$ as of second order of smallness, etc. In order to formulate this assumption in a form suitable for calculations, a small parameter μ is introduced. With the help of this parameter we write (VIII.13) in the form

$$\mathfrak{A}_n(t, r_1, r_1 - r_2, \dots, r_1 - r_n) = \tilde{\mathfrak{A}}_n(\mu t, \mu r_1, r_1 - r_2, \dots, r_1 - r_n; \mu)$$

$$= \tilde{\mathfrak{A}}_n(\tau, \xi_1, R_{12}, \dots, R_{1n}; \mu), \quad \tau = \mu t, \quad \xi_1 = \mu r_1, \quad \boldsymbol{R}_{1j} = \boldsymbol{r}_1 - \boldsymbol{r}_j,$$

and further

$$\tilde{\mathfrak{A}}_n(\tau, \xi_1, R_{12}, \dots, R_{1n}; \mu) = \tilde{\mathfrak{A}}_n^{(0)}(\tau, \xi_1, R_{12}, \dots, R_{1n}) \tag{VIII.22}$$
$$+\mu\tilde{\mathfrak{A}}_n^{(1)}(\tau, \xi_1, R_{12}, \dots R_{1n}; \mu) + \mu^2\tilde{\mathfrak{A}}_n^{(2)}(\tau, \xi_1, R_{12}, \dots R_{1n}; \mu) + \dots$$

Terms of zero order in μ lead to the hydrodynamic equations for an ideal Fermi superfluid (see Chapter VII). We restrict ourselves to an ideal liquid, neglecting terms proportional to μ. We note that according to eqn. (V.14) we must consider $\tilde{\eta}$ or $\tilde{\zeta}$ as proportional to μ.

Consider, for example, eqn. (VIII.12). We have

$$M\frac{\partial v_s^{(\alpha)}}{\partial \tau} = \frac{\partial}{\partial \xi_\alpha}\left\{\mu^2 \frac{\mathring{\nabla}_\xi^2 a}{4ma_0(\tau, \xi)} - \frac{Mv_s^2}{2} + \frac{\mathring{\nabla}_R^2 a(\tau, \xi)}{ma_0(\tau, \xi)}\right.$$
$$-\frac{m_r v_r^2}{2} - \frac{1}{2\varrho_0 a_0^2}\int V(R')\,[X_\tau(\xi, R', 0) + X_\tau^*(\xi, R', 0)]\,\mathrm{d}^3\boldsymbol{R}$$
$$-\frac{\mu}{a_0\sqrt{\varrho_0}}\,\vartheta(0)\,[\tilde{\zeta}_0(\tau, \xi) + \tilde{\zeta}_0^*(\tau, \xi)]$$
$$\left.-\mu\frac{2\varrho}{a_0\sqrt{\varrho_0}}\int \vartheta(\tilde{R})\left[\tilde{\zeta}_0\left(\tau, \xi - \frac{1}{2}\mu\tilde{R}\right) - \tilde{\zeta}_0\left(\tau, \xi - \frac{1}{2}\mu\tilde{R}\right)\right]\right\}\mathrm{d}^3\tilde{\boldsymbol{R}}. \tag{VIII.23}$$

The last term in (VIII.23) is obtained by using the fact that the integrand contains the function $\vartheta(R)$ which results in integration in the neighbourhood of $|\tilde{R}| \approx 0$. Therefore $\tilde{F}(t, r, \tilde{R}) \cong \tilde{F}(t, r, 0) = \varrho(t, r)$, $\chi_0(t, r) \cong \chi_0(t, r-\tilde{R}/2)$.

For the ideal-liquid approximation we have

$$M\frac{\partial v_s^{(\alpha)}}{\partial \tau} = \frac{\partial}{\partial \xi_\alpha}\left\{-\frac{Mv_s^2}{2} + \frac{\mathring{\nabla}_R^2 a(R|\varrho, \Theta, u)}{ma_0(\varrho, \Theta, u)} - \frac{m_r v_r^2}{2}\right. \tag{VIII.24}$$

$$\left.-\frac{1}{2\varrho_0 a_0^2}\int V(R')\,[X^0(R'|\varrho, \Theta, v_s-v_n)+X^{0*}(R'|\varrho, \Theta, v_s-v_n)]\,d^3R'\right\}.$$

With the help of (VIII.17) we get

$$M\frac{\partial \boldsymbol{v}_s}{\partial \tau} = -\nabla_\xi\left\{\frac{1}{2}Mv_s^2 - \frac{1}{2}M(\boldsymbol{v}_s-\boldsymbol{v}_n)^2 + 2\Lambda(\varrho, \Theta, u)\right\}. \tag{VIII.25}$$

Let us consider now the remaining hydrodynamic equations. Comparison of eqns. (V.14–16) with corresponding equations for a Bose fluid shows that they differ by terms involving sources.

Therefore we shall consider the terms S_ϱ, $S_j^{(\alpha)}$, $S_{\varrho\varepsilon}$, in the approximation of an ideal liquid in which we are interested.

After using (V.19), (VIII.5,6) and (VIII.9) we obtain

$$S_\varrho = -\mu i4m\sqrt{\varrho_0}\,\mathscr{I}[\tilde{\zeta}-\tilde{\zeta}^*], \qquad \xi = \mu\tilde{\boldsymbol{r}}, \tag{VIII.26}$$

$$\mathscr{I}[\zeta] \equiv \int a(\tau, \tilde{\xi}, \tilde{R})\vartheta(\tilde{R})\tilde{\zeta}(\tau, \tilde{\xi}, \tilde{R})\,d^3\boldsymbol{R},$$

$$\boldsymbol{r}-\boldsymbol{r}' = \tilde{\boldsymbol{R}}, \qquad \tfrac{1}{2}(\boldsymbol{r}+\boldsymbol{r}') = \tilde{\boldsymbol{r}} = \boldsymbol{r}-\tfrac{1}{2}\tilde{\boldsymbol{R}}.$$

To find $S_j^{(\alpha)}$ we must first calculate

$$\sum_{S,S'}\int\left[\tilde{\eta}^*\frac{\partial \Phi}{\partial r_\alpha} - \Phi\frac{\partial \tilde{\eta}^*}{\partial r_\alpha}\right]d^3\boldsymbol{r}' = 4\sqrt{\varrho_0}\int a(t, \tilde{r}, \tilde{R})\tilde{\zeta}^*(t, \tilde{R}, \tilde{r})$$

$$\times\nabla_{\tilde{R}}^{(\alpha)}\vartheta(\tilde{R})\,d^3\tilde{\boldsymbol{R}} - 2\sqrt{\varrho_0}\int a(t, \tilde{r}, \tilde{R})\vartheta(\tilde{R})\tilde{\zeta}^*(t, \tilde{r}, \tilde{R})\nabla_r^{(\alpha)}\chi\,d^3\tilde{\boldsymbol{R}}$$

$$-2\sqrt{\varrho_0}\int\vartheta(\tilde{R})\tilde{\zeta}(t, \tilde{r}, \tilde{R})\,\nabla_r^{(\alpha)}a(t, \tilde{r}, \tilde{R})\,d^3\boldsymbol{R} \cong \overset{1}{S}{}_j^{(\alpha)}[\tilde{\zeta}^*]$$

$$-2\sqrt{\varrho_0}\,Mv_s^{(\alpha)}\mathscr{I}[\tilde{\zeta}^*] - 2\sqrt{\varrho_0}\int\vartheta(\tilde{R})\tilde{\zeta}(t, \tilde{r}, \tilde{R})\nabla_r^{(\alpha)}a(t, \tilde{r}, \tilde{R})\,d^3\tilde{\boldsymbol{R}},$$

$$\overset{1}{S}{}_j^{(\alpha)}[\tilde{\zeta}^*] \equiv 4\sqrt{\varrho_0}\int a(t, \tilde{r}, \tilde{R})\,\tilde{\zeta}^*(t, \tilde{r}, \tilde{R})\nabla_{\tilde{R}}^{(\alpha)}\vartheta(\tilde{R})\,d^3\tilde{\boldsymbol{R}}. \tag{VIII.27}$$

Note that the function $\vartheta(R)$ appears in the integrand and therefore we can take

$$\lim_{\tilde{R}\to 0} \nabla_{\tilde{r}}^{(\alpha)} \chi(t, \tilde{r}, \tilde{R}) = \nabla_r^{(\alpha)} \chi_0(t, r) = M v_s^{(\alpha)},$$

since for $\tilde{R} = 0$, $\tilde{r} = r$. We shall use this approximation again when calculating $S_{\varrho\varepsilon}$.

As the function ϑ, describing the external sources, is a given function of distance, we assume now that its gradients are small. In order to take this fact into account we shall use a small parameter μ and write

$$\nabla_R^n \vartheta(R) \to \mu^n \nabla_R \vartheta(\mu R).$$

From (VIII.27), using the smallness of $\nabla_{\tilde{R}}\vartheta$, we obtain

$$S_j^{(\alpha)} = \mu i 2\sqrt{\varrho_0}\, M v_s^{(\alpha)} \mathscr{I}[\tilde{\zeta} - \tilde{\zeta}^*] + \mu^2 \overset{1}{S}{}_j^{(\alpha)} [\tilde{\zeta} + \tilde{\zeta}^*]$$
$$-\mu^2 2\sqrt{\varrho_0} \int \vartheta(\mu R)\, [\tilde{\zeta}(\tau, \tilde{\xi}, \tilde{R}) + \tilde{\zeta}^*(\tau, \tilde{\xi}, \tilde{R})] \nabla_\xi a(\tau, \tilde{\xi}, R)\, \mathrm{d}^3\tilde{\boldsymbol{R}}. \tag{VIII.28}$$

We calculate $S_{\varrho\varepsilon}$ given by formula (V.18). The first term is

$$S_{\varrho\varepsilon}^{(1)} = \mu i 2\sqrt{\varrho_0} \int V(\tilde{R})\, a(\tau, \tilde{\xi}, \tilde{R})\, \vartheta(\mu R)$$
$$\times \left[\tilde{\zeta}\left(\tau, \xi - \tfrac{1}{2}\mu \tilde{R}, \tilde{R}\right) - \tilde{\zeta}^*\left(\tau, \xi - \tfrac{1}{2}\mu\tilde{R}, \tilde{R}\right)\right] \mathrm{d}^3\tilde{\boldsymbol{R}}$$
$$= \mu i 2\sqrt{\varrho_0}\, V(0) \mathscr{I}[\tilde{\zeta} - \tilde{\zeta}^*]. \tag{VIII.29}$$

Terms (V.18) containing the Laplace operator are calculated using the identity

$$(\nabla_r^2 + \nabla_{r'}^2)\, a(t, \tilde{r}, \tilde{R}) \exp\{i\chi(t, \tilde{r}, \tilde{R})\}$$
$$= e^{i\chi}\Big\{-\tfrac{1}{2} a (\nabla_{\tilde{r}}\chi)^2 - 2a(\nabla_{\tilde{R}}\chi)^2 + 2\nabla_{\tilde{R}}^2 a + \tfrac{1}{2}\nabla_{\tilde{r}}^2 a$$
$$+ i\left[4(\nabla_{\tilde{R}} a \cdot \nabla_{\tilde{R}}\chi) + \tfrac{1}{2} a \nabla_{\tilde{r}}^2 \chi + 2a\nabla_{\tilde{R}}^2\chi + \nabla_{\tilde{r}} a \nabla_{\tilde{r}}\chi\right]\Big\}, \tag{VIII.30}$$

$$\tilde{\boldsymbol{r}} = \tfrac{1}{2}(\boldsymbol{r} + \boldsymbol{r}'), \quad \tilde{\boldsymbol{R}} = \boldsymbol{r} - \boldsymbol{r}'.$$

Finally,

$$S_{\varrho\varepsilon}^{(2)} = \mu i \sqrt{\varrho_0}\left[M v_s^2 + m_r v_r^2 - \frac{2\mathring{\nabla}_{\tilde{R}}^2 a}{m a_0} - \mu^2 \frac{\mathring{\nabla}_\xi^2 a}{2 m a_0}\right] \mathscr{I}[\tilde{\zeta} - \tilde{\zeta}^*]$$
$$-\mu\sqrt{\varrho_0}\left[\frac{2}{a_0}(v_r \cdot \mathring{\nabla}_{\tilde{R}} a) + (\mathring{\nabla}_R \cdot v_r) + \mu \frac{2}{a_0}(v_s \cdot \mathring{\nabla}_\xi a)\right.$$
$$\left. + \mu(\nabla_\xi \cdot \boldsymbol{v}_s)\right] \mathscr{I}[\tilde{\zeta} + \tilde{\zeta}^*] + \mu^2 \overset{1}{S}_{\varrho\varepsilon}[\tilde{\zeta} - \tilde{\zeta}^*] + \mu^2 \overset{2}{S}_{\varrho\varepsilon}[\tilde{\zeta} + \tilde{\zeta}^*], \tag{VIII.31}$$

where

$$\overset{1}{S}_{\varrho\varepsilon}[\tilde{\zeta}] = -\frac{i4\sqrt{\varrho_0}}{m}\int[(\nabla_{\tilde{R}}a(\tau, \tilde{\xi}, \tilde{R})\cdot\nabla_{\tilde{R}}\vartheta(\mu R))]$$

$$+\mu a(\tau, \tilde{\xi}, \tilde{R})\nabla^2_{\tilde{R}}\vartheta(\mu\tilde{R})\tilde{\zeta}\left(\tau, \xi-\tfrac{1}{2}\mu\tilde{R}, \tilde{R}\right)\mathrm{d}^3\tilde{\boldsymbol{R}}, \tag{VIII.32}$$

$$\overset{2}{S}_{\varrho\varepsilon}[\tilde{\zeta}] = -2\sqrt{\varrho_0}\left([2\boldsymbol{v}_s-\boldsymbol{v}_r]\cdot\int[a(\tau, \tilde{\xi}, \tilde{R})\nabla_{\tilde{R}}\vartheta(\mu\tilde{R})]\right)$$

$$\times\tilde{\zeta}\left(\tau, \tilde{R}, \xi-\tfrac{1}{2}\mu\tilde{R}\right)\mathrm{d}^3\boldsymbol{R}.$$

We now calculate the last four terms of the formula (V.18). They can be written in the form (Galasiewicz, 1965, 1966a)

$$S^{(3)}_{\varrho\varepsilon} = \mu i\sqrt{\varrho_0}\left[-M(\boldsymbol{v}_s-\boldsymbol{v}_n)^2+4\Lambda-2V(0)+\frac{2\mathring{\nabla}^2_R a(R|\varrho, \Theta, u)}{ma_0(\varrho, \Theta, u)}\right.$$

$$\left.-m_r v_r^2\right]\mathscr{I}[\tilde{\zeta}-\tilde{\zeta}^*]+\mu\left[\frac{2}{a_0(\varrho, \Theta, u)}(\boldsymbol{v}_r\cdot\mathring{\nabla}_R a(R|\varrho, \Theta, u))\right.$$

$$\left.+(\mathring{\nabla}_R\cdot\boldsymbol{v}_r)\right]\sqrt{\varrho_0}\mathscr{I}[\tilde{\zeta}+\tilde{\zeta}^*]. \tag{VIII.33}$$

Finally, we obtain

$$S_{\varrho\varepsilon} = S^{(1)}_{\varrho\varepsilon}+S^{(2)}_{\varrho\varepsilon}+S^{(3)}_{\varrho\varepsilon} = \mu i\sqrt{\varrho_0}[-M(\boldsymbol{v}_s-\boldsymbol{v}_n)^2+M\boldsymbol{v}_s^2$$

$$+4\Lambda]\mathscr{I}[\tilde{\zeta}-\tilde{\zeta}^*]+\mu^2\sqrt{\varrho_0}(\nabla_\xi\cdot\boldsymbol{v}_s)\mathscr{I}[\tilde{\zeta}+\tilde{\zeta}^*]+\mu^2\overset{1}{S}_{\varrho\varepsilon}[\tilde{\zeta}-\tilde{\zeta}^*]$$

$$+\mu^2\overset{2}{S}_{\varrho\varepsilon}[\tilde{\zeta}-\tilde{\zeta}^*]+\mu^2\frac{2i}{\varrho_0 a_0^2}\left\{\int V(R')X^{(1)}_\tau(\xi, R', 0)^*\mathscr{I}[\zeta]\mathrm{d}^3\boldsymbol{R}'\right.$$

$$\left.-\int V(R')X^{(1)}_\tau(\xi, R', 0)\mathscr{I}[\zeta^*]\mathrm{d}^3\boldsymbol{R}'\right\}. \tag{VIII.34}$$

If we calculate the averages entering eqns. (V.14–16) as in Chapter VII, neglect terms proportional to μ, and change in all formulae from the auxiliary variables τ, $\boldsymbol{\xi}$ to the original variables t, $\boldsymbol{r}$, we obtain a system of hydrodynamic equations, without viscous terms, for a Fermi superfluid (Galasiewicz, 1966a, see also 1965):

$$m\frac{\partial\varrho}{\partial t}+\sum_\alpha\frac{\partial(M\varrho_s v_s^{(\alpha)}+m\varrho_n v_n^{(\alpha)})}{\partial r_\alpha} = -i4\sqrt{\varrho_0}\mathscr{I}[\tilde{\zeta}-\tilde{\zeta}^*], \tag{VIII.35}$$

$$\frac{\partial}{\partial t}(M\varrho_s v_s^{(\alpha)}+m\varrho_n v_n^{(\alpha)}) = -\sum \frac{\partial(M\varrho_s v_s^{(\alpha)} v_s^{(\beta)}+m\varrho_n v_n^{(\alpha)} v_n^{(\beta)})}{\partial r_\alpha}$$

$$-\frac{\partial P}{\partial r_\alpha}+i2M\sqrt{\varrho_0}\,\mathscr{I}[\tilde{\zeta}-\tilde{\zeta}^*]v_s^{(\alpha)}, \quad \text{(VIII.36)}$$

$$\frac{\partial}{\partial t}\left[\varrho E+\frac{1}{2}m\varrho v_n^2+M\varrho_s(\boldsymbol{v}_n\cdot(\boldsymbol{v}_s-\boldsymbol{v}_n))\right]+\sum_\beta \frac{\partial}{\partial r_\beta}\left\{v_n^{(\beta)}\left[\frac{1}{2}m\varrho v_n^2\right.\right.$$

$$\left.+M\varrho_s(\boldsymbol{v}_n\cdot(\boldsymbol{v}_s-\boldsymbol{v}_n))+\varrho E+P\right]+(v_s^{(\beta)}-v_n^{(\beta)})M\varrho_s\left[\frac{1}{M}\Lambda\right.$$

$$\left.\left.+\left((\boldsymbol{v}_s\cdot\boldsymbol{v}_n)-\frac{1}{2}v_n^2\right)\right]\right\}-i\sqrt{\varrho_0}\left[4\Lambda+2\left((\boldsymbol{v}_s\cdot\boldsymbol{v}_n)-\frac{1}{2}v_n^2\right)\right]$$

$$\times\mathscr{I}[\tilde{\zeta}-\tilde{\zeta}^*] = -\sum_\beta \frac{\partial}{\partial r_\beta}(v_s^{(\beta)}-v_n^{(\beta)})A, \quad \text{(VIII.37)}$$

$$M\frac{\partial v_s^{(\alpha)}}{\partial t} = -\frac{\partial}{\partial r_\alpha}\left\{\frac{1}{2}Mv_s^2-\frac{1}{2}M(\boldsymbol{v}_s-\boldsymbol{v}_n)^2-2\Lambda(\varrho,\Theta,u)\right\}. \quad \text{(VIII.38)}$$

In formula (VIII.37) ϱE is connected with $\varrho\varepsilon$ by the equation

$$\varrho\varepsilon(\varrho,\Theta,v_s,v_n) = \varrho E(\varrho,\Theta,u)+\frac{1}{2}m\varrho v_n^2+M\varrho_s((\boldsymbol{v}_s-\boldsymbol{v}_n)\cdot\boldsymbol{v}_n),$$

$$E(\varrho,\Theta,u) = F-\Theta\frac{\partial F}{\partial\Theta},$$

and the pressure P

$$P = \varrho^2\frac{\partial F(\varrho,\Theta,u)}{\partial\varrho}.$$

An unknown function $A(\varrho,\Theta,u)$ is introduced on the right-hand side of (VIII.37). Similarly, as in the case of a Bose liquid, we can prove that $A\equiv 0$.

Equation (VIII.37) is simplified if the entropy S (per particle) is introduced:

$$S = -\frac{\partial F}{\partial\Theta}.$$

After calculations analogous to those in Chapter VII we have

$$\frac{\partial(\varrho S)}{\partial t}+\sum_\beta\frac{\partial}{\partial r_\beta}(\varrho S v_n^{(\beta)}) = 0. \quad \text{(VIII.37a)}$$

The two-fluid hydrodynamic equations obtained for a Fermi superfluid (VIII.35–37a) are normally identical with those for a Bose superfluid. They differ in terms involving sources.

From eqn. (VIII.37a) results the mechanocaloric effect for a Fermi superfluid.

The possible application of the two-fluid model to electrons in superconducting metals was considered by Gorter (1955), Bardeen (1959), and Bardeen and Schrieffer (1961).

§ 2 The linearized hydrodynamic equations and the Green functions

We substitute in (VIII.35–37a):

$$\varrho = \varrho^0 + \delta\varrho, \quad S = S^0 + \delta S, \quad \boldsymbol{v}_s = \delta \boldsymbol{v}_s, \quad \boldsymbol{v}_n = \delta \boldsymbol{v}_n, \quad \tilde{\eta} = \delta \tilde{\eta}$$

and find

$$\eta_1 = \tilde{\xi}.$$

Thus, we change from the hydrodynamic equations to the linearized "acoustic equations". This means that in eqns. (VIII.30–33) we neglect terms proportional to $\boldsymbol{v}_s^2$, $\boldsymbol{v}_n^2$, $\boldsymbol{v}_s \cdot \boldsymbol{v}_n$, $\tilde{\zeta}^2$, $\tilde{\zeta} v_s^{(\alpha)}$, $\delta\varrho\,\delta\Theta$.

The hydrodynamic equations in the acoustic approximation, that is, the equations for the variations $\delta\varrho$, δS, $v_s^{(\alpha)}$, $v_n^{(\alpha)}$ of the parameters, have the form

$$\frac{\partial \delta\varrho}{\partial t} + 2\varrho_s \sum_\beta \frac{\partial v_s^{(\beta)}}{\partial r_\beta} + \varrho_n \sum_\beta \frac{\partial v_n^{(\beta)}}{\partial r_n} = -i4\sqrt{\varrho_0} \int a(\tilde{R})\,\vartheta(\tilde{R}) \times \left[\eta_1\left((t, r - \frac{1}{2}\tilde{R}\right) - \eta_1^*\left(t, r - \frac{1}{2}\tilde{R}\right)\right] \mathrm{d}^3\tilde{\boldsymbol{R}}, \quad \text{(VIII.39)}$$

$$M\varrho_s \frac{\partial v_s^{(\alpha)}}{\partial t} + m\varrho_n \frac{\partial v_n^{(\alpha)}}{\partial t} = -\frac{\partial \delta P}{\partial r_\alpha}, \quad \text{(VIII.40)}$$

$$M\frac{\partial v_s^{(\alpha)}}{\partial t} = -2\frac{\partial \delta\Lambda}{\partial r_\alpha}, \quad \delta\Lambda = -S\delta\Theta + \frac{1}{\varrho}\delta P, \quad \text{(VIII.41)}$$

$$\varrho \frac{\partial \delta S}{\partial t} + S\frac{\partial \varrho}{\partial t} + \varrho S \sum_\beta \frac{\partial v_n^{(\beta)}}{\partial r_\beta} = 0 \quad \text{(VIII.42)}$$

(the upper index ^{0} of ϱ, S is omitted).

For infinitesimal deviations from thermodynamic equilibrium we have

$$v_s(t, r) = \frac{i}{2Ma_0\sqrt{\varrho_0}}\left(\frac{\partial\delta\tilde{\Phi}^*(t, r, 0)}{\partial r} - \frac{\partial\delta\tilde{\Phi}(t, r, 0)}{\partial r}\right). \quad \text{(VIII.43)}$$

We assume that deviations from thermodynamic equilibrium are caused by adiabatically switching on infinitesimal sources. On the other hand, terms with sources in the Hamiltonian (V.10) $(\delta U = 0,\ \delta A = 0)$

$$\delta\hat{H}_\tau = \sum_{s,s'}\iint [\delta\tilde{\eta}(\tau|x_3, x_4)\psi^+(0, x_3)\psi^+(0, x_4) + \delta\tilde{\eta}^*(\tau|x_3, x_4)\psi(0, x_4)\psi(0, x_3)]\,d^3r_3\,d^3r_4$$

are switched on adiabatically if

$$\begin{aligned}\delta\eta_1(\tau|r) &= e^{-i\omega\tau+\varepsilon\tau+i(k\cdot r)}\delta\eta_k + e^{i\omega\tau+\varepsilon\tau-i(k\cdot r)}\delta\eta_{-k}\\ \delta\eta_1^*(\tau|r) &= e^{-i\omega\tau+\varepsilon\tau+i(k\cdot r)}\delta\eta_{-k}^* + e^{i\omega\tau+\varepsilon\tau-i(k\cdot r)}\delta\eta_k^*,\end{aligned} \quad \text{(VIII.44)}$$

where

$$\varepsilon > 0, \quad \varepsilon \to 0.$$

Now we can use formula (V.38) to obtain the mean value

$$\begin{aligned}\delta\langle\psi(t, x_1)\psi(t, x_2)\rangle &= \delta\Phi(t|x_1, x_2)\\ &= \varepsilon(s_1)\Delta(s_1+s_2)\delta\Phi(\tilde{t}|r, R)\end{aligned} \quad \text{(VIII.45)}$$

expressed in terms of the Fourier components of the retarded Green functions. Thus the velocity v_s can be computed from eqns. (VIII.39–42) and, independently, from (VIII.43) with the help of (V.38) and (VIII.45). This procedure leads to the calculation of the Fourier components of the retarded Green function.

From (VIII.6) and (VIII.44) it follows that

$$\delta\hat{H}_\tau = e^{-i\omega\tau+\varepsilon\tau}V_\omega + e^{i\omega\tau+\varepsilon\tau}V_{-\omega} \quad \text{(VIII.46)}$$

where

$$V_{\pm\omega} = \sum_{s,s'}\varepsilon(s)\Delta(s+s')\iint e^{\pm i(k\cdot(r+r'))}\vartheta(r-r') \times [\psi^+(0, x)\psi^+(0, x')\delta\eta_{\pm k} + \psi(0, x')\psi(0, x)\delta\eta_{\pm k}^*]\,d^3r\,dr^{3\prime}. \quad \text{(VIII.47)}$$

Using formula (V.38) we have

$$\varepsilon(s_1)\Delta(s_1+s_2)\,\delta\tilde{\Phi}(t|r, R) = e^{-i\omega t+\varepsilon t}2\pi\sum_{s_3, s_4}\varepsilon(s_3)\Delta(s_3+s_4)$$

$$\times\iint d^3\tilde{r}\,d^3\tilde{R}\,\vartheta(\tilde{R})e^{i(k\cdot r)}\Big[\langle\langle\psi\big(r+\tfrac{1}{2}R, s_1\big)\psi(r-\tfrac{1}{2}R, s_2);$$

$$\psi^+\left(\tilde{r}+\tfrac{1}{2}\tilde{R}, s_3\right)\psi^+\left(\tilde{r}-\tfrac{1}{2}\tilde{R}, s_4\right)\rangle\rangle^r_E\delta\eta_k+\langle\langle\psi\left(r+\tfrac{1}{2}R, s_1\right)$$
$$\times\psi\left(r-\tfrac{1}{2}R, s_2\right); \psi\left(\tilde{r}-\tfrac{1}{2}\tilde{R}, s_4\right)\psi\left(\tilde{r}+\tfrac{1}{2}\tilde{R}, s_3\right)\rangle\rangle^r_E\delta\eta^*_{-k}\Big]$$
$$+e^{i\omega t+\varepsilon t}2\pi\sum_{s_3,s_4}\varepsilon(S_2)\Delta(s_3+s_4)\int d^3\tilde{r}d^3\tilde{R}\vartheta(\tilde{R})e^{-i(k\cdot r)}$$
$$\times\Big[\langle\langle\psi\left(r+\tfrac{1}{2}R, s_1\right)\psi\left(r-\tfrac{1}{2}R, s_2\right); \psi^+\left(\tilde{r}+\tfrac{1}{2}\tilde{R}, s_3\right)$$
$$\times\psi^+\left(\tilde{r}-\tfrac{1}{2}\tilde{R}, s_4\right)\rangle\rangle^r_E\delta\eta_{-k}+\langle\langle\psi\left(r+\tfrac{1}{2}R, s_1\right)\psi\left(r-\tfrac{1}{2}R, s_2\right);$$
$$\psi\left(\tilde{r}-\tfrac{1}{2}\tilde{R}, s_4\right)\psi\left(\tilde{r}+\tfrac{1}{2}\tilde{R}, s_3\right)\rangle\rangle^r_E\delta\eta^*_k\Big] \qquad \text{(VIII.48)}$$

where

$$\boldsymbol{r}=\tfrac{1}{2}(\boldsymbol{r}_,+\boldsymbol{r}_2), \quad \boldsymbol{R}=\boldsymbol{r}_1-\boldsymbol{r}_2, \quad \tilde{\boldsymbol{r}}=\tfrac{1}{2}(\boldsymbol{r}_3+\boldsymbol{r}_4), \quad \tilde{\boldsymbol{R}}=\boldsymbol{r}_3-\boldsymbol{r}_4,$$

$$\langle\langle A; B\rangle\rangle^r_E=\frac{1}{2\pi}\int_{-\infty}^{+\infty}\langle\langle A(t); B(\tau)\rangle\rangle^r e^{iE(t-\tau)}d(t-\tau).$$

Now we transform to Fourier components and obtain

$$\delta\tilde{\Phi}(t, r, 0)=e^{-i\omega t+\varepsilon t+i(k\cdot r)}2\pi\sum_{p_1,p_3}\vartheta(p_3)$$
$$\times[\langle\langle a_{\frac{1}{2}k+p_1,+}a_{\frac{1}{2}k-p_1,-}; a^+_{\frac{1}{2}k+p_3,+}a^+_{\frac{1}{2}k-p_3,-}\rangle\rangle^r_E\delta\eta_k$$
$$+\langle\langle a_{\frac{1}{2}k+p_1,+}a_{\frac{1}{2}k-p_1,-}; a_{-\frac{1}{2}k+p_3,-}a_{-\frac{1}{2}k-p_3,+}\rangle\rangle^r_E\delta\eta^*_{-k}]$$
$$+e^{i\omega t+\varepsilon t-ikr}2\pi\sum_{p_1,p_3}\vartheta(p_3) \qquad \text{(VIII.49)}$$
$$\times[\langle\langle a_{-\frac{1}{2}k+p_1,+}a_{-\frac{1}{2}k-p_1,-}; a^+_{-\frac{1}{2}k+p_3,+}a^+_{-\frac{1}{2}k-p_3,-}\rangle\rangle^r_E\delta\eta_{-k}$$
$$+\langle\langle a_{-\frac{1}{2}k+p_1,+}a_{-\frac{1}{2}k+p_1,-}; a_{\frac{1}{2}k+p_3,-}a_{\frac{1}{2}k-p_3,+}\rangle\rangle^r_E\delta\eta^*_k]$$

where

$$\vartheta(p)=\frac{1}{V}\int\vartheta(R)e^{-i(p\cdot R)}d^3\boldsymbol{R}.$$

We introduce the Fourier components

$$v_s^{(\alpha)}(t, r)=e^{-i\omega t+\varepsilon t+i(k\cdot r)}v_s^{(\alpha)}(k)+e^{i\omega t+\varepsilon t-i(k\cdot r)}v_s^{(\alpha)}(-k), \qquad \text{(VIII.50)}$$

and use (VIII.43) and (VIII.49) to obtain

$$\boldsymbol{k}\cdot\boldsymbol{v}_s(k)=\frac{\pi k^2}{Ma_0\sqrt{\varrho_0}}\sum_{p_1,p_3}\vartheta(p_3) \qquad \text{(VIII.51)}$$

$$\times[\langle\langle a_{\frac{1}{2}k+p_1,+}\, a_{\frac{1}{2}k-p_1,-} - a^+_{-\frac{1}{2}k+p_1,+}\, a^+_{-\frac{1}{2}k-p_1,-};\, a_{-\frac{1}{2}k+p_3,-}$$

$$\times a_{-\frac{1}{2}k-p_3,+}\rangle\rangle^r_E \delta\eta^*_{-k} + \langle\langle a_{\frac{1}{2}k+p_1,+}\, a_{\frac{1}{2}k-p_1,-}$$

$$-a^+_{-\frac{1}{2}k+p_1,+}\, a^+_{-\frac{1}{2}k-p_1,-};\, a^+_{\frac{1}{2}k+p_3,+}\, a^+_{\frac{1}{2}k-p_3,-}\rangle\rangle^r_E \delta\eta_k].$$

According to the definition

$$\varrho_0 a_0^2 = \tilde{\Phi}^*(t, r, 0)\tilde{\Phi}(t, r, 0),$$

we obtain

$$\varrho_0 \delta a_0 = \frac{1}{2}[\delta\tilde{\Phi}^*(t, r, 0) + \delta\tilde{\Phi}(t, r, 0)]$$

$$= \varrho_0\left[\frac{\partial a_0}{\partial \varrho}\delta\varrho + \frac{\partial a_0}{\partial \Theta}\delta\Theta\right]. \qquad \text{(VIII.52)}$$

Again we transform to Fourier components and, using (VIII.49) and (VIII.52), now find

$$\delta a_0(t, r) = e^{-i\omega t+\varepsilon t+i(k\cdot r)}\delta a_k + e^{i\omega t+\varepsilon t-i(k\cdot r)}\delta a_{-k},$$

$$\varrho_0\delta a_k = 2\pi\sum_{p_1,p_3}\vartheta(p_3)[\langle\langle a_{\frac{1}{2}k+p_1,+}\, a_{\frac{1}{2}k-p_1,-}$$

$$+a^+_{-\frac{1}{2}k+p_1,+}\, a^+_{-\frac{1}{2}k-p_1,-};\, a^+_{\frac{1}{2}k+p_3,+}\, a^+_{\frac{1}{2}k-p_3,-}\rangle\rangle^r_E\delta\eta_k$$

$$+\langle\langle a_{\frac{1}{2}k+p_1,+}\, a_{\frac{1}{2}k-p_1,-} + a^+_{-\frac{1}{2}k+p_2,+}\, a^+_{-\frac{1}{2}k-p_1,-};$$

$$a_{-\frac{1}{2}k+p_3,-}\, a_{-\frac{1}{2}k-p_3,+}\rangle\rangle^r_E\delta\eta^*_{-k}]. \qquad \text{(VIII.53)}$$

We consider eqns. (VIII.39–42), in order to find the quantity $(\boldsymbol{k}\cdot\delta\boldsymbol{v}_s(k))$ as a function of $\delta\eta_k$, $\delta\eta^*_{-k}$. For this purpose we rewrite eqns. (VIII.39–42) in Fourier components. We introduce the notation

$$(\boldsymbol{k}\cdot\delta\boldsymbol{v}_s(k)) = \delta X, \quad (\boldsymbol{k}\cdot\delta\boldsymbol{v}_n(k)) = \delta Y,$$

$$\gamma(k) = \int a(R)\vartheta(R)e^{\frac{1}{2}i(k\cdot r)}\,d^3\boldsymbol{R}. \qquad \text{(VIII.54)}$$

The equations have the form

$$2\varrho_s\delta X + \varrho_n\delta Y - E\delta\varrho + 0\cdot\delta\Theta = 4\sqrt{\varrho_0}\,\gamma(k)(\delta\eta^*_{-k} - \delta\eta_k),$$

$$EM\varrho_s\delta X + Em\varrho_n\delta Y - k^2\left(\frac{\partial P}{\partial\varrho}\right)_\Theta\delta\varrho - k^2\left(\frac{\partial P}{\partial\Theta}\right)_\varrho\delta\Theta = 0,$$

$$EM\varrho\delta X+0\delta Y-k^2\left(\frac{\partial P}{\partial \varrho}\right)_{\Theta}\delta\varrho+k^2\left[\varrho S-\left(\frac{\partial P}{\partial \Theta}\right)_{\varrho}\right]\delta\Theta=0, \quad \text{(VIII.55)}$$

$$0\delta X-\varrho S\delta Y+E\left[S+\varrho\left(\frac{\partial S}{\partial\varrho}\right)_{\Theta}\right]\delta\varrho+E\varrho\left(\frac{\partial S}{\partial\Theta}\right)_{\varrho}\delta\Theta=0.$$

We have obtained a system of four equations with four unknowns. We denote the determinant of the system of eqns. (VIII.55) by $D(E)$ and the determinant derived from $D(E)$ by replacing the first column by the right-hand side of (VIII.55) by D_X. Thus we have

$$\delta X=(\boldsymbol{k}\cdot\delta\boldsymbol{v}_s(k))=\frac{D_X}{D(E)}. \quad \text{(VIII.56)}$$

From eqns. (VIII.56,51) it is clear that the roots of $D(E)$ give the poles of the Green functions and consequently the energies of the elementary excitations.

Considering first the special case of thermodynamic equilibrium ($E=0$) we have

$$\begin{aligned}(\boldsymbol{k}\cdot\delta\boldsymbol{v}_s(k))&=-\frac{2\sqrt{\varrho_0}}{\varrho_s}\gamma(k)(\delta\eta_k-\delta\eta^*_{-k}),\\ \delta\varrho(k)&=\delta\Theta(k)=0.\end{aligned} \quad \text{(VIII.57)}$$

From (VIII.52) it follows that $\delta a_k=0$. Therefore we obtain, from (VIII.53),

$$\begin{aligned}&\sum_{p_1}\langle\langle a_{\frac12 k+p_1,+}\,a_{\frac12 k-p_1,-};\,a^+_{\frac12 k+p,+}\,a^+_{\frac12 k-p,-}\rangle\rangle^r_{E=0}\\ &\quad=\sum_{p_1}\langle\langle a^+_{-\frac12 k+p_1,+}\,a^+_{-\frac12 k-p_1,-};\,a^+_{\frac12 k+p,+}\,a^+_{\frac12 k-p,-}\rangle\rangle^r_{E=0},\\ &\sum_{p_1}\langle\langle a_{\frac12 k+p_1,+}\,a_{\frac12 k-p_1,-};\,a_{-\frac12 k-p,+}\,a_{-\frac12 k+p,-}\rangle\rangle^r_{E=0}\\ &\quad=-\sum_{p_1}\langle\langle a^+_{-\frac12 k+p_1,+}\,a^+_{-\frac12 k-p_1,-};\,a_{-\frac12 k-p,+}\,a_{-\frac12 k+p,-}\rangle\rangle^r_{E=0}.\end{aligned} \quad \text{(VIII.58)}$$

If we take into account (VIII.58), formula (VIII.53) has the form

$$(\boldsymbol{k}\cdot\delta\boldsymbol{v}_s(k))=\frac{2\pi k^2}{Ma_0\sqrt{\varrho_0}}\sum_{p_1,p_3}\vartheta(p_1)[\langle\langle a_{\frac12 k+p_1,+}\,a_{\frac12 k-p_1,-};\;a_{-\frac12 k+p_3,-}\,a_{-\frac12 k-p_3,+}\rangle\rangle^r_{E=0}\delta\eta^*_{-k}$$

$$+\langle\langle a_{\frac{1}{2}k+p_1,+}\, a_{\frac{1}{2}k-p_1,+};\, a^+_{-\frac{1}{2}k+p_3,+}\, a^+_{\frac{1}{2}k-p_3,-}\rangle\rangle^r_{E=0}\,\delta\eta_k]$$

$$= -\frac{2\sqrt{\varrho_0}\gamma(k)}{\varrho_s}(\delta\eta_k - \delta\eta^*_{-k}). \quad \text{(VIII.59)}$$

By equating the coefficients of $\delta\eta^*_{-k}$ and $\delta\eta_k$ and taking into account the fact that the function $\gamma(k)$ contains the function ϑ (see (VIII.54)) we have

$$\sum_{p_1}\langle\langle a_{\frac{1}{2}k+p_1,+}\, a_{\frac{1}{2}k-p_1,-};\, a_{-\frac{1}{2}k+p,-}\, a_{-\frac{1}{2}k-p,+}\rangle\rangle^r_{E=0}$$

$$= \frac{M\varrho_0 a_0 \gamma_1\left(p+\dfrac{1}{2}k\right)}{\pi\varrho_s k^2}, \quad \text{(VIII.60)}$$

$$\sum_{p_1}\langle\langle a_{\frac{1}{2}k+p_1,+}\, a_{\frac{1}{2}k-p_1,-};\, a^+_{-\frac{1}{2}k+p_1,+}\, a^+_{\frac{1}{2}k-p,-}\rangle\rangle^r_{E=0}$$

$$= -\frac{M\varrho_0 a_0 \gamma_1\left(p+\dfrac{1}{2}k\right)}{\pi\varrho_s k^2}, \quad \gamma_1(p) = \int a(R)\exp\{i(\boldsymbol{p}\cdot\boldsymbol{R})\}\,\mathrm{d}^3\boldsymbol{R}.$$

Formula (VIII.60) gives the $1/k^2$ theorem (Bogoliubov, 1961) for Fermi systems (singularity of $1/k^2$ type) for $k \approx 0$. Formula (VIII.60) is valid for any p_3.

Consider now the general case. Equations (VIII.55) give

$$(\boldsymbol{k}\cdot\boldsymbol{v}_s(k)) = \frac{4\gamma(k)\sqrt{\varrho_0}\Delta(k,E)k^2}{m\Omega(k,E)}(\delta\eta^*_{-k} - \delta\eta_k) \quad \text{(VIII.61)}$$

where

$$\Omega(k,E) = (E^2 - c_1^2k^2)(E^2 - c_2^2k^2),$$

$$c_{1,2}^2 = \frac{1}{2m}\left(\frac{\partial P}{\partial\varrho}\right)_S + \frac{S^2\varrho_s\Theta}{m\varrho_n c_v}$$

$$\pm\sqrt{\frac{1}{4}\left[\frac{1}{m}\left(\frac{\partial P}{\partial\varrho}\right)_S + \frac{2S^2\varrho_s\Theta}{\varrho_n m c_v}\right]^2 - \frac{1}{m}\left(\frac{\partial P}{\partial\varrho}\right)_\Theta \frac{2\varrho_s\Theta S^2}{m\varrho_n c_v}},$$

$$c_v = \Theta\left(\frac{\partial S}{\partial\Theta}\right)_\varrho, \quad \text{(VIII.62)}$$

c_1 (plus sign) is the velocity of ordinary sound, c_2 (minus sign) is an analogue of the second sound velocity for a Bose superfluid. The quantity $\Delta(E, k)$ has the form

$$\Delta(E, k) = \frac{1}{m}\left(\frac{\partial P}{\partial \varrho}\right)_\Theta k^2 \frac{S^2\Theta}{\varrho_n c_v} - E^2\left[\frac{1}{\varrho}\left(\frac{\partial P}{\partial \varrho}\right)_S - \left(\frac{\partial}{\partial \varrho}\varrho S^2\right)_\Theta \frac{\Theta}{\varrho c_v}\right]. \tag{VIII.63}$$

After comparing the coefficients of $\delta\eta^*_{-k}$ and $\delta\eta_k$ in (VIII.51) and (VIII.61) and using the identities (VIII.58) (we assume that they are valid for small E) we have

$$\sum_{p_1} \langle\langle a_{\frac{1}{2}k+p_1,+} a_{\frac{1}{2}k-p_1,-} ; a_{-\frac{1}{2}k+p,-} a_{-\frac{1}{2}k-p,+} \rangle\rangle^r_E = \frac{4\varrho_0 a_0 \Delta(E, k)\gamma_1\left(p+\frac{1}{2}k\right)}{\Omega(k, E)}, \tag{VIII.64}$$

$$\sum_{p_1} \langle\langle a_{\frac{1}{2}k+p_1,+} a_{\frac{1}{2}k-p_1,-} ; a^+_{\frac{1}{2}k+p,+} a^+_{\frac{1}{2}k-p,-} \rangle\rangle^r_E = -\frac{4\varrho_0 a_0 \Delta(E, k)\gamma_1\left(p+\frac{1}{2}k\right)}{\Omega(k, E)}.$$

From (VIII.62) we see that the two-particle Green functions have poles which correspond to two kinds of elementary excitations:

$$E = c_1 k, \quad E = c_2 k. \tag{VIII.65}$$

Thus, on the basis of the microscopic theory, we have shown that in a system of identical Fermi particles with pairing forces (in this system there exists a condensate of coupled pairs) elementary excitations of two kinds appear, first and second sound quanta.

Bardeen and Schrieffer (1961) attempted to estimate the order of the velocity of the second sound in the superconductors on the basis of the two-fluid model. From the simplified formula (VII.105) they obtained $c_2 \approx 10$ cm/sec. For He II we have, for example, at 1.1°K, $c_2 \sim 2\times10^3$ cm/sec. In order to observe the second-sound frequency the mean free path of electrons must be of the order ~ 1 cm, which would be very difficult to realize in practice.

References

ABEL, W. R., ANDERSON, A. C., BLACK, W. and WHEATLEY, J. C. (1965) *Phys. Rev. Letters* **14**, 1291.

ABEL, W. R., ANDERSON, A. C. and WHEATLEY, J. C. (1966) *Proceedings of the 10th International Conference on Low Temperature Physics,* Moscow; enlarged version (1966) *Phys. Rev. Letters* **17**, 74.

ABRIKOSOV, A. A. (1952) *Dokl. Akad. Nauk* (*USSR*) **86**, 489.

(1957) *J. Exptl. Theoret. Phys.* (*USSR*) **32**, 1442; Soviet Phys.-JETP **5**, **1174).**

(1965) *Uspekhi Fiz. Nauk* (*USSR*) **87**, 125

ABRIKOSOV, A. A. and GOR'KOV, L. P. (1960) *J. Exptl. Theoret. Phys.* (*USSR*) **39**, 1781 (Soviet Phys.-JETP **12**, 1243).

ABRIKOSOV, A. A., GOR'KOV, L. P. and DZYALOSHINSKII, J. E. (1962) *Quantum Field Theoretical Methods in Statistical Mechanics,* Moscow; English translation published by Pergamon Press, 1965.

ABRIKOSOV, A. A. and KHALATNIKOV, I. M. (1958) *Uspekhi Fiz. Nauk* (*USSR*) **65**, 551.

AKHIEZER, A. I. and AKHIEZER, I. A. (1962) *J. Exptl. Theoret. Phys.* (*USSR*) **43**, 2208 (Soviet Phys.-JETP **16**, 1560).

AKHIEZER, A. I. and POMERANTSUK, I. Y. (1959) *J. Exptl. Theoret. Phys.* (*USSR*) **36**, 859 (Soviet Phys.-JETP **9**, 605).

ALLEN, J. F. and JONES, J. (1938) *Nature* **141**, 243.

ALLEN, J. F., PEIERLS, R. and UDDIN, M. ZAKI (1937) *Nature* **140**, 62.

AMBEGAOKAR, V. and KADANOFF, L. P. (1961) *Nuovo Cimento* **22**, 914.

ANDERSON, P. W. (1958) *Phys. Rev.* **110**, 827, 985.

(1959) *Phys. Rev. Letters* **3**, 325.

ANDERSON, P. W. and MOREL, P. (1961) *Phys. Rev.* **123**, 1911.

ANDERSON, P. W. and ROVELL, J. M. (1963) *Phys. Rev. Letters* **10**, 230.

ANDROES, G. M. and KNIGHT, D. W. (1959) *Phys. Rev. Letters* **2**, 386.

(1961) *Phys. Rev.* **121**, 779.

ANDRONIKASHVILI, E. L. (1946) *J. Phys.* (*USSR*) **10**, 201.

(1948) *J. Exptl. Theoret. Phys.* (*USSR*) **18**, 424.

ATKINS, K. R. and OSBORNE, D. V. (1950) *Phil. Mag.* **41**, 1078.

AUTLER, S. H. (1960) *Rev. Sci. Instrum.* **31**, 369.

BALIAN, R. and WERTHAMER, N. R. (1963) *Phys. Rev.* **131**, 1553.

BALTENSPERGER, W. and STRÄSLER, S. (1963) *Phys. Kondens. Materie* **1**, 20.

BARDEEN, J. (1955) *Phys. Rev.* **97**, 1724.

(1957) *Nuovo Cimento* **5,** 1764.
(1959) *Phys. Rev. Letters* **1,** 399.

BARDEEN, J. COOPER, L. and SCHRIEFFER, J. (1957a) *Phys. Rev.* **106,** 162.
(1957b) *Phys Rev.* **108,** 1175.

BARDEEN, J. and SCHRIEFFER, J. (1961) *Recent Developments in Superconductivity*, Progress in Low Temperature Physics, Vol. III, North Holland, Amsterdam.

BEAUMONT, C. F. and REEKIE, J. (1955) *Proc. Roy. Soc.* **A228,** 363.

BLATT, J. M. (1960) *Progr. Theor. Phys.* **24,** 851.
(1964) *Theory of Superconductivity*, Academic Press, New York.

BLOCH, C. and DE DOMINICIS, C. (1958) *Nuclear Phys.* **7,** 459.

BLOCH, C. and MESSIAH, A. (1962) *Nuclear Phys.* **39,** 95.

BLOCH, F. (1928) *Z. Phys.* **52,** 555.

BOGOLIUBOV, N. N. (1946) *The Problem of Dynamical Theory in Statistical Physics* (in Russian), Izdat. O.G.I.Z. Gostekhizdat, Moscow–Leningrad; English translation published in *Studies in Statistical Mechanics*, Vol. I, editors de Boer, J. and Uhlenbeck, G. E., North-Holland, 1962.
(1947a) *Lectures on Quantum Statistics* (in Ukrainian), Kiev; English translation published by Gordon & Breach, New York.
(1947b) *Izv. Akad. Nauk USSR,* **11,** 77; *J. Phys.* (*USSR*) **11,** 23.
(1958a) *Nuovo Cimento* **7,** 794.
(1958b) *J. Exptl. Theoret. Phys.* (*USSR*) **34,** 58, 73 (Soviet Phys.-JETP **7,** 41, 51).
(1958c) *Dokl. Akad. Nauk USSR* **119,** 52 (Soviet Phys.-Doklady **3,** 279).
(1959) *Uspekhi Fiz. Nauk* **67,** 549 (Soviet Phys.-Uspekhi **1,** 236).
(1960) *Physica, Suppl.* **26,** 1.
(1961) *The Quasi-averages in Problems of Statistical Physics*, Dubna preprint (in Russian), 2nd ed. 1963; English translation to be published by Gordon & Breach.
(1963) *On the Hydrodynamics of a Superfluid*, Lectures, Dubna preprint (in Russian); English translation to be published by Gordon & Breach.

BOGOLIUBOV, N. N. and SOLOVIEV, V. G. (1959) *Dokl. Akad. Nauk USSR* **124,** 1011 (Soviet Phys.-Doklady **4,** 143).

BOGOLIUBOV, N. N., TOLMACHEV, V. V. and SHIRKOV, D. V. (1958) *A New Method in the Theory of Superconductivity*, Suppl. II, Izdat. Acad. Nauk USSR, Moscow; English translation, Consultants Bureau, New York (1959); abbreviated version, *Fortschr. Phys.* **6,** 605 (1958).

BOGOLIUBOV, N. N., ZUBAREV, D. N. and TSERKOVNIKOV, YU. A. (1957) *Dokl. Akad. Nauk USSR,* **117,** 788 (Soviet Phys.-Doklady **2,** 535).
(1960) *J. Exptl. Theoret. Phys.* (*USSR*) **39,** 120 (Soviet Phys.-JETP **12,** 88).

BOHR, A., MOTTELSON, B. R. and PINES, D. (1958) *Phys. Rev.* **110,** 936.

BOHR, N. (1911) *Dissertation*, Copenhagen.

BROUT, R. (1957) *Phys. Rev.* **108,** 515.

BROWN, A., ZEMANSKY, M. W. and BOORSE, H. A. (1953) *Phys. Rev.* **92,** 52.

BRUECKNER, K. and GAMMEL, J. (1958) *Phys. Rev.* **109,** 1040.

BRUECKNER, K., SODA, T., ANDERSON, P. W. and MOREL, P. (1960) *Phys. Rev.* **118,** 42.

BUCKINGHAM, M. J. (1957) *Nuovo Cimento* **5,** 1763.

BYERS, N. and YANG, C. N. (1961) *Phys. Rev. Letters* **7,** 46.

CHAPMAN S. and COWLING, T. G. (1952) *The Mathematical Theory of Non-uniform Gases,* Cambridge University Press, London.
CHESTER, G. V. (1963) in *Liquid Helium*, (Ed.G. Carreri), Academic Press.
CLOGSTON, A. M., GOSSARD, A. C., JACCARINO, V. and YAFET, Y. (1962) *Phys. Rev. Letters* **9**, 262.
(1964) *Rev. Mod. Phys.* **36,** 170.
COHEN, M. L. (1964) *Phys. Rev.* **134,** 511.
COHEN, R. W. and ABRAHAMS, E. (1966) *Phys. Letters* **16,** 21.
COOPER, L. N. (1956) *Phys. Rev.* **104,** 1189.
(1960) *Am. J. Phys.* **2,** 91.
CORAK, W. S., GOODMAN, B. B., SATTERTHWAITE, C. B. and WEXLER, A. (1956) *Phys. Rev.* **102,** 656.
CORAK, W. S. and SATTERTHWAITE, C. B. (1954) *Phys. Rev.* **99,** 1660.
CZERWONKO, J. (1966a) *Physica* **32,** 1953.
(1966b) *Bull. Acad. Polon. Sci.* **14,** 463.
DAUNT, J. G. and MENDELSSOHN, K. (1939) *Nature* **143,** 719.
DEAVER, B. S., JR. and FAIRBANK, W. M. (1961) *Phys. Rev. Letters* **7,** 43.
DE GENNES, P. G. (1966) *Superconductivity of Metals and Alloys,* Benjamin, New York.
DE HAAS, W. J. and VOOGD, J. (1931) *Leiden Commun.* **214b.**
DOLL, R. and NÄBAUER, M. (1961) *Phys. Rev. Letters* **7,** 51.
EMERY, V. and SESSLER, A. (1960) *Phys. Rev.* **119,** 43.
FERRELL, R. A. (1959) *Phys. Rev. Letters* **3,** 262.
FEYNMAN, R. P. (1954) *Phys. Rev.* **94,** 262.
(1955) *Progress in Low Temperature Physics* **1,** Ch. 11, North Holland, Amsterdam.
(1963) *The Feynman Lectures on Physics*, Vol. 3, Ch. 19, Addison-Wesley, Reading, Massachusetts.
FISHER, J. C. (1960) *Austral. J. Phys.* **13,** 446.
FRÖHLICH, H. (1950) *Phys. Rev.* **79,** 845.
FULDE, P. and MAKI, K. (1965a) *Phys. Rev.* **139,** A788.
(1965b) *Phys. Rev.* **140,** A1586.
GALASIEWICZ, Z. (1960a) *Progr. Theor. Phys.* **23,** 197.
(1960b) *Acta Phys. Polon.* **19,** 467, 675.
(1963a) *Nuclear Phys.* **44,** 107.
(1963b) *Assymptotical Formulae for Green Functions in the Viscous Liquid Approximation for the Superfluid Systems of Bose Particles*, Dubna preprint (in Russian); revised and enlarged version in *Acta Univ. Vratislaviensis* **80,** 75, Materials of the III Winter School of Theoretical Physics, Karpacz 1966.
(1965) *Phys. Letters* **15,** 39.
(1966a) *Nuclear Phys.* **76,** 145.
(1966b) *Acta Phys. Polon.* **30,** 567.
(1967a) *Bull. Acad. Polon. Sci.* **15,** 185 and ibid. **15,** "Erratum" to No. 10.
(1967b) *Bull. Acad. Polon. Sci.* **15,** 191.
(1968) *Bull. Acad. Polon. Sci.* **16,** 751.
(1969a) "Energy Gap and the Stability of the Solution Describing the Superconducting State", *Bull. Acad. Polon. Sci.,* in print.
(1969b), *Helium 4,* Selected Readings in Physics, Pergamon Press, Oxford, in print.

(1969c)" Remarks about a Kinetic Coefficient for the Bose Superfluid", *Bull. Acad. Polon. Sci.*, in print.
GAUDIN, M. (1960) *Nuclear Phys.* **15**, 89.
GEBALLE, T. H., MATTHIAS, B. T., HULL, G. W. and CORENZWIT, E. (1961) *Phys. Rev. Letters* **6**, 275.
GELL-MANN, M. and BRUECKNER, K. A. (1957) *Phys. Rev.* **106**, 364.
GIAEVER, I. (1960) *Phys. Rev. Letters* **5**, 147, 464.
GINZBURG, V. L. (1953) *Fortschr. Phys.* **1**, 101.
GINZBURG, V. L. and KIRZHNITS, D. A. (1964) *J. Exptl. Theoret. Phys.* (*USSR*) **46**, 397 (Soviet. Phys.-JETP **19**, 269).
GINZBURG, V. L. and LANDAU, L. D. (1950) *J. Exptl. Theoret. Phys.* (*USSR*) **20**, 1064 English translation in D. ter Haar, Men of Physics, *L. D. Landau* Vol. I, Pergamon Press, 1965.
GOODMAN, B. B. (1953) *Proc. Phys. Soc.* **A56**, 217.
GOR'KOV, L. P. (1958) *J. Exptl. Theoret. Phys.* (*USSR*) **34**, 735 (Soviet Phys.-JETP **7**, 505).
(1959) *J. Exptl. Theoret. Phys.* (*USSR*) **36**, 1918 (Soviet Phys.-JETP **9**,1364)
GOR'KOV, L. P. and PITAEVSKII, L. P. (1960) *J. Exptl. Theoret. Phys.* (*USSR*) **42**, 600 (Soviet Phys.-JETP **15**,417.
GORTER, C. J. (1955) *Progress in Low Temperature Physics*, North Holland, Vol. 1, Amsterdam.
GORTER, C. J. and CASIMIR, H. B. G. (1934a) *Physica* **1**, 306.
(1934b) *Phys. Z.* **35**, 963.
(1934c) *Z. techn. Phys.* **15**, 539.
GRADSTEIN, I. S. and RYZHIK, I. M. (1963) *Tables of Integrals, Sums, Series and Derivatives*, Gos. Izdat. Fiz.-Mat. Literat., Moscow; English translation published by Academic Press, 1965.
GROSS, E. P. (1961) *Nuovo Cimento* **20**, 454.
GUREVICH, V. L., LARKIN, A. I. and FIRSOV, Y. A. (1962) *Fiz. Tverdogo Tela* **4**, 185 (Soviet Phys.-Solid State **4**, 131).
HAMMOND, R. H. and KELLY, G. M. (1964) *Rev. Mod. Phys.* **36**, 185.
(1967) *Phys. Rev. Letters* **18**, 156 (1967).
HEIN, R. A., GIBSON, J. W., MAZELSKY, R., MILLER, R. C. and HULM, J. K. (1964) *Phys. Rev. Letters* **12**, 320.
HENSHAW, D. and WOODS, A. (1961) *Phys. Rev.* **121**, 1266.
HOHENBERG, P. C. and MARTIN, P. C. (1965) "Microscopic Theory of Superfluid Helium", *Ann. Phys.* (*N.Y.*) **34**, 291–359.
JOSEPHSON, B. D. (1962) *Phys. Letters* **1**, 251.
KADANOFF, L. P. and KANE, J. W. (1967) *Phys. Rev.* **155**, 80.
KADANOFF, L. P. and MARTIN, P. C. (1961) *Phys. Rev.* **124**, 670.
(1963) "Hydrodynamic Equations and Correlation Function", *Ann. Phys-* (*N.Y.*) **24**, 419–469.
KAMERLINGH ONNES, H. (1911a) *Leiden Commun.* **122b**, **124c**.
(1911b) *Proc. Roy. Acad. Amsterdam* **13**, 1903.
KAMERLINGH ONNES, H. and BOKS, J. D. A. (1924) *Leiden Commun.* **170a**.
KAPITZA, P. L. (1938) *Nature* **141**, 74; *Collected Papers*, Pergamon 1967, Vol. II.
(1941a) *J. Phys.* (*USSR*) **4**, 181; *Collected Papers*, Pergamon 1967, Vol. II; *J. Exptl. Theoret. Phys.* (*USSR*) **11**, 1; *Collected Papers*, Pergamon 1967, Vol. II.

(1941b) *J. Phys.* (*USSR*) **5**, 59; *Collected Papers*, Pergamon 1967, Vol. II; *J. Exptl. Theoret. Phys.* (*USSR*) **11**, 581; *Collected Papers*, Pergamon 1967, Vol. II.

*KEESOM, W. H. (1942) *Helium*, Elsevier Press, Amsterdam.

KEESOM, W. H. and CLUSIUS K. (1932a) *Proc. Roy. Acad. Amsterdam* **35**, 307.
(1932b) *Leiden Commun.* **219c**.

KEESOM, W. H. and KEESOM, A. P. (1936) *Physica* **3**, 359.

KEESOM, W. H. and MAC WOOD, J. E. (1938) *Physica* **5**, 737.

KEESOM, W. H. and VAN DEN ENDE, J. N. (1932) *Leiden Commun.* **219b**.

KEESOM, W. H. and WOLFKE, M. (1928a) *Leiden Commun.* **190b**.
(1928b) *Proc. Roy. Acad. Amsterdam* **31**, 90.

KHALATNIKOV, I. M. (1950) *J. Exptl. Theoret. Phys.* (*USSR*) **20**, 243.
(1952a) *J. Exptl. Theoret. Phys.* (*USSR*) **23**, 8.
(1952b) *J. Exptl. Theoret. Phys.* (*USSR*) **23**, 21.
(1952c) *J. Exptl. Theoret. Phys.* (*USSR*) **23**, 169.
(1965) *Introduction to the Theory of Superfluidity*, Izdat. Nauka, Moscow (in Russian); English translation published by Benjamin, New York 1966.

KHALATNIKOV, I. M. and ABRIKOSOV, A. A., (1958) *Uspekhi Fiz. Nauk* **65**, 551; English translation in *Adv. Phys.* **8**, 45 (1959).

KOPPE, H. (1950) *Ergeb. exakt. Naturw.* **23**, 283.

KRASNIKOV, W. A. (1966) *Dokl. Akad. Nauk USSR* **170**, 807 (Soviet Phys.-Doklady **11**, 850).
(1967a) *Dissertation* (private communication).
(1967b) *Dokl. Akad. Nauk USSR* **174**, 1037 (Soviet Phys.-Doklady **12**).

KUBO, R. and OBATA, Y. (1956) *J. Phys. Soc. Japan* **11**, 547.

KUNZLER, J. W., BUEHLER, E., HSU, F. S. L. and WERNICK, J. H. (1961) *Phys. Rev. Letters* **6**, 89.

*KUPER, CH. G. (1968) *An Introduction to the Theory of Superconductivity*, Clarendon Press, Oxford.

LANDAU, L. D. (1937) *Phys. Z. SSSR* **11**, 129; *Collected Papers*, Pergamon 1965, p. 217.
(1938) *Nature* **141**, 688; *Collected Papers*, Pergamon 1965, p. 226.
(1941) *J. Phys.* (*USSR*) **5**, 71; *J. Exptl. Theoret. Phys.* (*USSR*) **11**, 592; *Collected Papers*, Pergamon 1965. p. 301.
(1944) *J. Exptl. Theoret. Phys.* (*USSR*) **14**, 112; *Collected Papers*, Pergamon 1965, p. 446.
(1947) *J. Phys.* (*USSR*) **11**, 91.
(1956) *J. Exptl. Theoret. Phys.* (*USSR*) **30**, 1058. *Collected Papers*, Pergamon 1965, p. 723.
(1957) *J. Exptl. Theoret. Phys.* (*USSR*) **32**, 59. *Collected Papers*, Pergamon 1965. p. 731.

LANDAU, L. D. and KHALATNIKOV, I. M. (1949) *J. Exptl. Theoret. Phys.* (*USSR*) **19**, 709. Landau's *Collected Papers*, Pergamon 1965, p. 511.

LANDAU, L. D. and LIFSHITZ, E. M. (1957) *Electrodynamics of Continuous Media*, Izdat. Techn.-Teoret. Literat., Moskow; English translation, Pergamon 1960.

*LANE, C. T. (1962) *Superfluid Physics*, McGraw-Hill, New York.,

LARKIN, A. I. and MIGDAL, A. B. (1963) *J. Exptl. Theoret. Phys.* (*USSR*) **45**, 1036 (Soviet Phys.-JETP **18**, 717).

References

LEE, T. D. and MOHLING, F. (1959) *Phys. Rev. Letters* **2**, 284.
LIFSHITZ, E. M. (1944a) *J. Phys. (USSR)* **8**, 110.
(1944b) *J. Exptl. Theoret. Phys. (USSR)* **14**, 116.
LITTLE, W. A. (1965) *Scientific American* **212**, 21.
LONDON, F. (1938) *Nature* **141**, 643.
(1938) *Phys. Rev.* **54**, 947.
(1950) *Superfluids*, Vol. I, Wiley, New York.
(1954) *Superfluids*, Vol. II, Ch. 13, Wiley, New York.
LONDON, H. and LONDON, F. (1935) *Proc. Roy. Soc. London* **A149**, 71.
(1935) *Physica* **2**, 341.
LUTTINGER, J. M. and WARD, J. C. (1960) *Phys. Rev.* **118**, 1417.
*LYNTON, E. A. (1963) *Superconductivity*, Methuen, London.
MARTIN, D. L., BENZ, M. G., BRUCH, C. A. and ROSNER, C. H. (1963) *Cryogenics* **3**, 114.
MARTIN, P. and SCHWINGER, J. (1959) *Phys. Rev.* **115**, 1342.
MATHUR, V. S., PANCHAPAKESAN, N. and SAXENA, R. P. (1966) *Nuovo Cimento* **42**, 1.
MATSUBARA, T. (1955) *Progr. Theor. Phys.* **14**, 351.
MATTHIAS, B. T., SUHL, H. and CORENZWIT, E. (1958) *Phys. Rev. Letters* **1**, 449.
MATTHIAS, B. T. and SUHL, H. (1960) *Phys. Rev. Letters* **4**, 51.
MAXWELL, E. (1950) *Phys. Rev.* **78**, 477.
MEISSNER, W. and OCHSENFELD, R. (1933) *Naturwiss.* **21**, 787.
MESHKOVSKY, A. G. and SHALNIKOV, A. I. (1947) *J. Phys. (USSR)* **11**, 1.
MIGDAL, A. B. (1958) *J. Exptl. Theoret. Phys. (USSR)* **34**, 1438 (Soviet Phys.-JETP, **7** (**34**), 996 (1958)).
(1965) *Theory of the Finite Fermi Systems and the Properties of the Atomic Nuclei* (in Russian), Izdat. Nauka, Moscow; English translation published by Benjamin, New York.
MILLER, A., PINES, D. and NOZIÈRES, P. (1962) *Phys. Rev.* **127**, 1452.
MOREL, P. and NOZIÈRES, P. (1962) *Phys. Rev.* **126**, 1909.
MOTTELSON, B. R. (1959) *Nuclear Structure*, Ch. 2, The Many-body Problem, Dunod, Paris.
NAMBU, Y. (1960) *Phys. Rev.* **117**, 648.
NOER, R. J. and KNIGHT, W. D. (1961) *Bull. Am. Phys. Soc.* **2**, 122.
(1964) *Rev. Mod. Phys* **36**, 177.
OSBORNE, D.V. (1950) *Proc. Phys. Soc.* **A 63**, 909.
PALEVSKY, H., OTNES, K. and LARSSON, K. E. (1959) *Phys. Rev.* **112**, 11.
PENROSE, O. (1958) *Nuovo Cimento* **9**, 256.
PENROSE, O. and ONSAGER, L. (1956) *Phys. Rev.* **104**, 576.
PESHKOV, V. P. (1944) *J. Phys. (USSR)* **8**, 131.
(1946a) *J. Phys. (USSR)* **10**, 389.
(1946b) *J. Phys. (USSR)* **16**, 1000.
(1948) *J. Exptl. Theoret. Phys. (USSR)* **18**, 857, 951.
(1949) *J. Exptl. Theoret. Phys. (USSR)* **19**, 270.
(1960) *J. Exptl. Theoret. Phys. (USSR)* **38**, 799 (Soviet Phys.-JETP **11**, 508).
(1964) *J. Exptl. Theoret. Phys. (USSR)* **46**, 1510 (Soviet Phys.-JETP **19**, 1023).
PETRU, Z. (1968) *Bull. Acad. Polon. Sci.* **16**, 67.
(1969a) *Acta Phys. Polon.* **35**, 57.
(1969b) *Acta Phys. Polon.* **35**, 49.

*PINES, D. (1958) *Phys. Rev.* **109**, 280.

PINES, D. and NOZIÈRES, P. (1965) *Theory of Quantum Liquids*, Vol. I, Benjamin, New York.

PIPPARD, A. B. (1953) *Proc. Roy. Soc. London* **A216**, 547.

PITAEVSKII, L. P. (1954) *J. Exptl. Theoret. Phys. (USSR)* **31**, 262 (Soviet Phys.-JETP **4**).

(1959) *J. Exptl. Theoret. Phys. (USSR)* **37**, 1794 (Soviet Phys.-JETP **10**, 1267).

(1961) *J. Exptl. Theoret. Phys. (USSR)* **40**, 646 (Soviet Phys.-JETP **13**, 451).

PRIVOROTSKY, I. A. (1962) *J. Exptl. Theoret. Phys. (USSR)* **43**, 2255 (Soviet Phys.-JETP **16**, 1593).

REIF, F. (1957) *Phys. Rev.* **106**, 208.

REIF, F. and WOOLF, M. A. (1962) *Phys. Rev. Letters* **9**, 315.

REYNOLDS, C. A., SERIN, B., WRIGHT, W. H. and NESBITT, L. B. (1950) *Phys. Rev.* **78**, 487.

RICE, T. M. (1965) *Phys. Rev.* **140**, A1889.

RICKAYZEN, G. (1959) *Phys. Rev.* **115**, 795.

(1965) *Theory of Superconductivity*, Interscience Publishers, Wiley, New York.

*SAINT-JAMES, D., SARMA, G. and THOMAS E. J. (1969) *Type II Superconductivity*, Pergamon, Oxford.

SALZANO, F. and STRONGIN, M. (1967) *Phys. Rev.* **153**, 533.

SAWADA, K. (1957) *Phys. Rev.* **106**, 372.

SAWADA, K., BRUECKNER, K. A., FUKUDA, N. and BROUT, R. (1957) *Phys. Rev.* **108**, 507.

SCHOOLEY, J. F., HOSLER, W. R. and COHEN, M. L. (1964) *Phys. Rev. Letters* **12**, 474.

SCHRIEFFER, J. R. (1964) *Theory of Superconductivity*, Benjamin, New York.

SHAPIRO, S., JAMES, A. R. and HOLLY, S. (1964) *Rev. Mod. Phys.* **36**, 223.

*SHOENBERG, D. (1965) *Superconductivity*, Cambridge University Press.

SOLOVIEV, V. G., (1958) *Dokl. Akad. Nauk USSR* **123**, 437.

(1958) *Nuovo Cimento* **10**, 1022.

THOULESS, M. (1957) *Phys. Rev.* **107**, 1162.

*TINKHAM, M. (1965) *Superconductivity*, New York.

TISZA, L. (1938a) *Nature* **141**, 913.

(1938b) *C. R.* **207**, 1035, 1186.

(1940) *J. physique radium* (8) **1**, 164, 350.

(1947) *Phys. Rev.* **72**, 838.

TYABLIKOV, S. V. (1958) *Dokl. Akad. Nauk USSR* **121**, 250 (Soviet Phys.-Doklady **3**, 722).

(1965) *Methods of the Quantum Theory of Magnetism* (in Russian), Izdat. Nauka, Moscow; English translation published by Plenum Press.

USUI, T. (1964) *Progr. Theor. Phys.* **32**, 190.

VALATIN, J. G. (1958) *Nuovo Cimento* **7**, 843.

(1961) *Phys. Rev.* **122**, 1012.

VAN LEEUWEN, J. H. (1931) *J. phys.* (6) **2**, 361.

VONSOVSKY, S. W. and SVIRSKY, M. S. (1964a) *Izv. Akad. Nauk USSR* **38**, 418.

(1964b) *J. Exptl. Theoret. Phys. (USSR)* **46**, 1619 (Soviet Phys.-JETP **19**, 1095).

References

WELLER, W., (1964) *Z. Naturforschg*. **19a**, 410.

WIGNER, E. (1934) *Phys. Rev*. **46**, 1002.
(1938) *Trans. Faraday Soc*. **34**, 678.

*WILKS J. (1967) *The Properties of Liquid and Solid Helium,* Clarendon Press, Oxford.

YANG, C. N. (1962) *Rev. Mod. Phys*. **34**, 694.

YARNELL, J. L., ARNOLD, G. P., BENDT, P. J. and KERR, E. C. (1959) *Phys. Rev*. **113**, 1379.

YOSIDA, K. (1958) *Phys. Rev*. **110**, 769.
(1959) *Progr. Theor. Phys*. **21**, 731.

ZAVARITSKY, N. W. (1952) *Dokl. Akad. Nauk USSR* **86**, 501.

ZUBAREV, D. N. (1961) *Uspekhi Fiz. Nauk* (*USSR*) **71**, 71 (Soviet Phys.-Uspekhi **3**, 320, 1962).

* Asterisk denotes monographs which are not quoted in the text.

Index

Abrikosov's model for type II superconductors 20
Approximate second quantization 85, 89, 103

Bloch states 4, 22, 54
Bogoliubov transformation 33, 34, 62, 71, 73, 74
  generalized 34, 44
Bose–Einstein condensation 19, 24, 119
Bose fluid, ordinary 153
Bose superfluid 171
Bound state 24, 59
Buckingham's sum rule 107
Bulk viscosity coefficient 162, 191

Chemical potential 34, 38, 160, 172, 173, 213
Coherence length 17, 21
Collective oscillations 85
  of uncharged particles 93
  of charged particles 95
Condensate 24
  of coupled pairs 209, 226
  wave function of 138, 194
Condensation energy 25
Cooper pair 17, 21, 29
Critical current 6
  magnetic field 5, 11, 13, 14, 19, 20, 22, 24, 81
  temperature 5, 7, 14, 22, 23, 29, 31, 69
  velocity 52, 53, 120, 125, 126
Current
  carrying state 52, 53
  diamagnetic 16
  of the condensate 194

Damping coefficients 206
  effects 170, 204
Debye frequency 70
Debye model of phonons 123
Demagnetizing factor 9
Density
  of a condensate 172, 175, 200
  of a normal component 175, 214
  of pairs in the condensate 213
  of states of electrons in superconductors 51
  of a superfluid component 173, 175
Diamagnetic picture of a superconductor 8, 12
Diamagnetism, perfect 9, 15, 16
Diffusion coefficient 162, 163, 184
Dissipation function 163, 186

Effective mass 18, 129
Effective wave function 18
Electrodynamics of superconductors 103
Electron–phonon coupling constant 46
  interaction 46
Energy current 159, 162, 176, 178, 184
Energy gap 23, 24, 30, 52, 57, 59, 93, 123, 128;
  dependence on temperature 66, 68
Energy of a quasi-particle 45, 65, 85
Entropy 14, 80, 160, 182;
  of a superconductor 78, 82, 219
  waves 122

Fermi fluid, ordinary 153
Fermi superfluid 209
Fermi momentum 25, 49, 50, 52, 128, 129, 210

Index

Fermi sea 24
Feynmann's vortices 126
First sound 122, 200, 225
Flux quantization 30
Fountain (thermomechanical) effect 119, 120, 121
Free energy 13, 158, 213
Fröhlich's Hamiltonian 46

Galilean transformation of operators 153, 157, 172, 213, 214
Gap equation 50, 58, 94
Gapless superconductivity 29
Gases, permanent 3
Gauge invariance of a current 108, 112
Gauge transformation
  time independent 147
  time dependent 150
Ginsburg–Landau theory 17, 19
Green functions
  advanced 148, 168
  retarded 142, 163, 192, 207, 220

Hartree–Fock variational method 33, 41, 54, 58, 61
  generalized 34, 38, 57
Heisenberg representation 143, 146, 147
Helium I, II 117
Helium three 127

Impurities 4, 20, 29
Infinite conductivity 6, 7
Interaction representation 145, 147
Intermediate state 11, 17
Isotope effect 22, 23

Josephson's effect 30

Knight shift 27

Lagrange multipliers 40, 56
Latent heat 13, 117
Lattice vibrations 4
Linearized hydrodynamic equations 163, 188, 220
London's equations 15

Magnetic induction 6, 7, 10
Magnetization 9
Mean energy of a superconductor 64
Mean free path 17, 130, 154, 166
Meissner–Ochsenfeld effect 8, 12, 16, 19, 108, 113
Mottelson's model 60

Non-local theory 16, 17, 23
Normal current 15
  product 35, 62, 63
Nuclear pairing correlations 29

Onsager's relations 112, 191
Onsager–Khalatnikov relation 207, 208
Orbital paramagnetism 28
Order, long range 32, 198
Ordinary Bose and Fermi fluids 153

Pairing correlations, parallel spins 91, 99
Penetration depth 15, 17
Phase transitions
  of the first order 12, 13, 20, 21, 117
  of the second order 12, 14, 18, 20, 21, 81, 113
Phonons (sound quanta) 4, 22, 46, 47, 120, 123, 167, 202
Pippard's theory 16, 17, 23
Plasma frequency 97

Quantization of magnetic flux 30

Relaxation time 154, 166, 198
Rotons 120, 123, 124

Schafroth's condensation 25
Schrödinger representation 143, 146

Second sound 122, 200, 225, 226
viscosity coefficients 185, 191
Sound
first 122, 207, 226,
velocity of 122, 125, 200, 202
ordinary 122,
velocity of 129
second 122, 207,
velocity of 122, 125, 200, 202, 225, 226
zero 130
Sources
of pairs 139, 209
of particles 133, 147, 162, 200
Specific heat 14, 21, 23, 24, 78, 80, 117, 122, 129, 169, 201, 206
jump of 14, 24, 30, 81, 117
Spin–orbit coupling 27
Spin waves 31
Stability of the superconducting state 55
Statistical operator (density matrix) 62, 143
Stress tensor 157, 162, 174, 175
Structure factor 125
Superconducting magnets 20, 31
Superconductivity
in long organic molecules 31, 32
in polar conductors 32
surface 31, 32
Superconductors
type I 19
type II 19
Supercurrent 6, 15, 16, 17, 26
Superfluid
Bose 171
Fermi 209
Surface energy 17, 26
superconductivity 31, 32
Susceptibility 9, 27, 29, 99, 104

Thermal conductivity 118, 121,
Thermal coefficient 162, 184, 202
Thermal superconductivity 118, 121, 202
Thermodynamical potential 13, 76, 78
Thermomechanical (fountain) effect 119, 121, 220
Time ordered product 146
Tunelling of electrons 30
Two-fluid model 120, 220, 226

Velocity
of a condensate 172, 212
of a normal component 172
Viscosity
coefficient 162
of helium 118, 120

Ward's identities 151
Wave function for condensate 138, 194

Zero point oscillations 127
Zero sound 130

ERRATA

Page, line	For	Read
50^{12}	$\frac{g^2}{2}$	g^2
64_9	$[T(f,f')+\zeta(f,f')]$	$[T(f,f')+\zeta(f,f')]_{F(f,f')}$
$104^{10,11}$	$\boldsymbol{H}$	$\mathscr{H}$
155_{12}	$r = \mu t$	$\tau = \mu t$
155_{12}	$= \mu \boldsymbol{r}$	$\xi = \mu r$
160^8	$\sum_{\alpha,\beta} \frac{\partial}{\partial r_\beta} (\tilde{v}_\alpha \mathscr{T}_{\alpha\beta}(\tilde{\varrho},\tilde{\Theta}, 0))$	$\sum_{\alpha,\beta} \frac{\partial}{\partial \xi_\beta} (v_\alpha \mathscr{T}_{\alpha\beta}(\tilde{\varrho},\tilde{\Theta}, 0)) - \frac{1}{m} \sum_\alpha \tilde{j}_\alpha \frac{\partial U}{\partial \xi_\alpha}$
164^5	$m \frac{\partial v_\alpha}{\partial t}$	$m\varrho \frac{\partial v_\alpha}{\partial t}$
176_6	$\frac{\partial \psi(t,r)}{\partial r_\alpha} + imv_n^{(\alpha)} \psi(t,r)$	$\psi^+(t,r) \left(\frac{\partial \psi^+(t,r)}{\partial r_\alpha} + imv_n^{(\alpha)} \psi(t,r) \right)$
178_7	$\frac{\frac{\partial}{\partial \boldsymbol{r}} \langle \psi(t,r) \rangle}{m \langle \psi(t,r) \rangle}$	$-i \frac{\frac{\partial}{\partial \boldsymbol{r}} \langle \psi(t,r) \rangle}{m \langle \psi(t,r) \rangle}$
186_3	$\zeta_2(\mathrm{div}\, \boldsymbol{v}_n)$	$\zeta_2(\mathrm{div}\, \boldsymbol{v}_n)^2$
$189_{4,3}$	where ... quanta;	expunge
207_8	k_1	K_1

Galasiewicz: *Superconductivity and Quantum Fluids*

OTHER TITLES IN THE SERIES IN NATURAL PHILOSOPHY

Vol. 1. DAVYDOV—Quantum Mechanics
Vol. 2. FOKKER—Time and Space, Weight and Inertia
Vol. 3. KAPLAN—Interstellar Gas Dynamics
Vol. 4. ABRIKOSOV, GOR'KOV and DZYALOSHINSKII—Quantum Field Theoretical Methods in Statistical Physics
Vol. 5. OKUN'—Weak Interaction of Elementary Particles
Vol. 6. SHKLOVSKII—Physics of the Solar Corona
Vol. 7. AKHIEZER *et al*—Collective Oscillations in a Plasma
Vol. 8. KIRZHNITS—Field Theoretical Methods in Many-body Systems
Vol. 9. KLIMONTOVICH—The Statistical of Non-equilibrium Processes in a Plasma
Vol. 10. KURTH—Introduction to Stellar Statistics
Vol. 11. CHALMERS—Atmospheric Electricity (2nd Edition)
Vol. 12. RENNER—Current Algebras and their Applications
Vol. 13. FAIN and KHANIN—Quantum Electronics, Volume 1—Basic Theory
Vol. 14. FAIN and KHANIN—Quantum Electronics, Volume 2—Maser Amplifiers and Oscillators
Vol. 15. MARCH—Liquid Metals
Vol. 16. HORI—Spectral Properties of Disordered Chains and Lattices
Vol. 17. SAINT JAMES, THOMAS and SARMA—Type II Superconductivity
Vol. 18. MARGENAU and KESTNER—Theory of Intermolecular Forces
Vol. 19. JANCEL—Foundations of Classical and Quantum Statistical Mechanics
Vol. 20. TAKAHASHI—An Introduction to Field Quantization
Vol. 21. YVON—Correlations and Entropy in Classical Statistical Mechanics
Von. 22. PENROSE—Foundations of Statistical Mechanics
Vol. 23. VISCONTI—Quantum Field Theory, Volume 1
Vol. 24. FURTH—Fundamental Principles of Theoretical Physics
Vol. 25. ZHELESNYAKOV—Radioemission of the Sun and Planets
Vol. 26. GRINDLAY—An Introduction to the Phenomenological Theory of Ferroelectricity
Vol. 27. UNGER—Introduction to Quantum Electronics
Vol. 28. KOGA—Introduction to Kinetic Theory Stochastic Processes in Gaseous Systems